CAMBRIDGE TRACTS IN MATHEMATICS

General Editors

204 Group Cohomology and Algebraic Cycles

CAMBRIDGE TRACTS IN MATHEMATICS

A complete list of books in the series can be found at www.cambridge.org/mathematics.
Recent titles include the following:

170. Polynomials and Vanishing Cycles. By M. Tibăr
171. Orbifolds and Stringy Topology. By A. Adem, J. Leida, and Y. Ruan
172. Rigid Cohomology. By B. Le Stum
173. Enumeration of Finite Groups. By S. R. Blackburn, P. M. Neumann, and G. Venkataraman
174. Forcing Idealized. By J. Zapletal
175. The Large Sieve and Its Applications. By E. Kowalski
176. The Monster Group and Majorana Involutions. By A. A. Ivanov
177. A Higher-Dimensional Sieve Method. By H. G. Diamond, H. Halberstam, and W. F. Galway
178. Analysis in Positive Characteristic. By A. N. Kochubei
179. Dynamics of Linear Operators. By F. Bayart and É. Matheron
180. Synthetic Geometry of Manifolds. By A. Kock
181. Totally Positive Matrices. By A. Pinkus
182. Nonlinear Markov Processes and Kinetic Equations. By V. N. Kolokoltsov
183. Period Domains over Finite and p-adic Fields. By J.-F. Dat, S. Orlik, and M. Rapoport
184. Algebraic Theories. By J. Adámek, J. Rosický, and E. M. Vitale
185. Rigidity in Higher Rank Abelian Group Actions I: Introduction and Cocycle Problem. By A. Katok and V. Niţică
186. Dimensions, Embeddings, and Attractors. By J. C. Robinson
187. Convexity: An Analytic Viewpoint. By B. Simon
188. Modern Approaches to the Invariant Subspace Problem. By I. Chalendar and J. R. Partington
189. Nonlinear Perron–Frobenius Theory. By B. Lemmens and R. Nussbaum
190. Jordan Structures in Geometry and Analysis. By C.-H. Chu
191. Malliavin Calculus for Lévy Processes and Infinite-Dimensional Brownian Motion. By H. Osswald
192. Normal Approximations with Malliavin Calculus. By I. Nourdin and G. Peccati
193. Distribution Modulo One and Diophantine Approximation. By Y. Bugeaud
194. Mathematics of Two-Dimensional Turbulence. By S. Kuksin and A. Shirikyan
195. A Universal Construction for R-Free Groups. By I. Chiswell and T. Müller
196. The Theory of Hardy's Z-Function. By A. Ivić
197. Induced Representations of Locally Compact Groups. By E. Kaniuth and K. F. Taylor
198. Topics in Critical Point Theory. By K. Perera and M. Schechter
199. Combinatorics of Minuscule Representations. By R. M. Green
200. Singularities of the Minimal Model Program. By J. Kollár
201. Coherence in Three-Dimensional Category Theory. By Nick Gurski
202. Canonical Ramsey Theory on Polish Spaces. By V. Kanovei, M. Sabok, and J. Zapletal
203. A Primer on the Dirichlet Space. By O. El-Fallah, K. Kellay, J. Mashreghi, and T. Ransford

Group Cohomology and Algebraic Cycles

BURT TOTARO
University of California, Los Angeles

CAMBRIDGE
UNIVERSITY PRESS

CAMBRIDGE
UNIVERSITY PRESS

32 Avenue of the Americas, New York, NY 10013-2473, USA

Cambridge University Press is part of the University of Cambridge.

It furthers the University's mission by disseminating knowledge in the pursuit of education, learning, and research at the highest international levels of excellence.

www.cambridge.org
Information on this title: www.cambridge.org/9781107015777

First published 2014

A catalog record for this publication is available from the British Library.

Library of Congress Cataloging in Publication Data
Totaro, Burt.
Group cohomology and algebraic cycles / Burt Totaro, University of California, Los Angeles.
pages cm. – (Cambridge tracts in mathematics)
Includes bibliographical references and index.
ISBN 978-1-107-01577-7 (hardback : alk. paper)
1. Homology theory. 2. Algebra. I. Title.
QA612.3.T68 2014
512′.64–dc23 2014002457

ISBN 978-1-107-01577-7 Hardback

for Susie

Contents

Preface

Group cohomology reveals a deep relation between algebra and topology. A group determines a topological space in a natural way, its classifying space. The cohomology ring of a group is defined to be the cohomology ring of its classifying space. The challenges are to understand how the algebraic properties of a group are related to its cohomology ring, and to compute the cohomology rings of particular groups.

A fundamental fact is that the cohomology ring of any finite group is finitely generated. So there is some finite description of the whole cohomology ring of a finite group, but it is not clear how to find it. A central problem in group cohomology is to find an upper bound for the degrees of generators and relations for the cohomology ring. If we can do that, then there are algorithms to compute the cohomology in low degrees and therefore compute the whole cohomology ring.

Peter Symonds made a spectacular advance in 2010: for any finite group G with a faithful complex representation of dimension n at least 2 and any prime number p, the mod p cohomology ring of G is generated by elements of degree at most n^2 [130]. Not only is this the first known bound for generators of the cohomology ring; it is also nearly an optimal bound among arbitrary finite groups, as we will see.

This book proves Symonds's theorem (Corollary 4.3) and several new variants and improvements of it. Some involve algebro-geometric analogs of the cohomology ring. Namely, Morel-Voevodsky and I independently showed how to view the classifying space of an algebraic group G (e.g., a finite group) as a limit of algebraic varieties in a natural way. That allows the definition of the Chow ring of algebraic cycles on the classifying space BG [107, proposition 2.6]; [138]. A major goal of algebraic geometry is to compute the Chow ring for varieties of interest, since that says something meaningful about all subvarieties of the variety.

The fact that not all the cohomology of BG is represented by algebraic cycles (even for abelian groups G) is the source of Atiyah-Hirzebruch's counterexamples to the integral Hodge conjecture [8, 137, 138]. It is a natural problem of "motivic homotopy theory" to understand the Chow ring and more generally the motivic cohomology of classifying spaces BG. Concretely, computing the Chow ring of BG amounts to computing the Chow groups of the quotients by G of all representations of G. Such quotients are extremely special among all varieties, but they have been fundamental examples in algebraic geometry for more than 150 years. Computing their Chow groups is a fascinating problem. (Rationally, the calculations are easy; the interest is in integral or mod p calculations.)

Bloch generalized Chow groups to a bigraded family of groups, now called motivic cohomology. A great achievement of motivic homotopy theory is the proof by Voevodsky and Rost of the Bloch-Kato conjecture [145, theorem 6.16]. A corollary, the Beilinson-Lichtenbaum conjecture (Theorem 6.9), says that for any smooth variety over a field, a large range of motivic cohomology groups with finite coefficients map isomorphically to etale cohomology. Etale cohomology is a more computable theory, which coincides with ordinary cohomology in the case of complex varieties. Thus the Beilinson-Lichtenbaum conjecture is a powerful link between algebraic geometry and topology.

Chow groups are the motivic cohomology groups of most geometric interest, but they are also farthest from the motivic cohomology groups that are computed by the Beilinson-Lichtenbaum conjecture. A fundamental difficulty in computing Chow groups is "etale descent": for a finite Galois etale morphism $X \to Y$ of schemes, how are the Chow groups of X and Y related? This is easy after tensoring with the rationals; the hard case of etale descent is to compute Chow groups integrally, or with finite coefficients. Etale descent is well understood for etale cohomology, and hence for many motivic cohomology groups with finite coefficients.

The problem of etale descent provides some motivation for trying to compute the Chow ring of classifying spaces of finite groups G. Computing the Chow ring of BG means computing the Chow ring of certain varieties Y which have a covering map $X \to Y$ with Galois group G (an approximation to $EG \to BG$) such that X has trivial Chow groups. Thus the Chow ring of BG is a model case in seeking to understand etale descent for Chow groups.

Chow rings can be generalized in various ways, for example, to algebraic cobordism and motivic cohomology. Another direction of generalization leads to unramified cohomology, cohomological invariants of algebraic groups [47], and obstructions to rationality for quotient varieties [17, 76]. All of these invariants are worth computing for classifying spaces, but we largely focus on the most classical case of Chow rings. Some of our methods will certainly be useful for these more general invariants. For example, finding generators for the

Chow ring of a smooth variety automatically gives generators of its algebraic cobordism, by Levine and Morel [96, theorem 1.2.19].

This book mixes algebraic geometry and algebraic topology, and few readers will have all the relevant background. With that in mind, I include brief introductions to several of the theories we use. Chapter 1 introduces group cohomology. Chapter 2 summarizes the basic properties of the Chow ring of a smooth variety without proof, and then introduces equivariant Chow rings in more detail, including some calculations. I hope this allows topologists who have seen a little algebraic geometry to get some feeling for Chow rings. However, large parts of the book are devoted to group cohomology, including many new results, and topologists may prefer to concentrate on those parts.

An explicit bound for the degrees of generators of the Chow ring of BG, of the same form as Symonds's bound for cohomology, was given in 1999 [138, theorem 14.1]. The first new result of this book is to improve the earlier bound for the Chow ring by about a factor of two: for any finite group G with a faithful complex representation of dimension n at least 3, the Chow ring of BG is generated by elements of degree at most $n(n-1)/2$. Moreover, this improved bound is optimal, for all n (Chapter 5).

For a p-group, Chapter 7 gives a stronger bound for the degrees of generators of the cohomology ring and the Chow ring. For the cohomology ring of a p-group, this result goes well beyond Symonds's general bound. The case of p-groups is central in the cohomology theory of finite groups, with many questions reducing to that case. It may be that these bounds for p-groups can be improved further.

Chapter 8 proves some of the fundamental theorems on the cohomology and Chow ring of a finite group. First, there is Quillen's theorem that, up to F-isomorphism (loosely, "up to pth powers"), the cohomology ring of a finite group is determined by the inclusions among its elementary abelian subgroups. We prove Yagita's theorem that the Chow ring of a finite group, up to F-isomorphism, has the same description in terms of the elementary abelian subgroups. It follows that the cycle map from the Chow ring of a finite group to the cohomology ring is an F-isomorphism.

Next, we give a strong bound for the degrees of generators of the Chow ring of a finite group modulo transfers from proper subgroups. In particular, for a group with a faithful representation of dimension n and any prime number p, the mod p Chow ring is generated by elements of degree less than n modulo transfers from proper subgroups (Corollary 10.5). (In fact, we only need to consider transfers from a particular class of subgroups, centralizers of elementary abelian p-subgroups.) This result reduces the problem of finding generators for the Chow ring of a given group to the problem of finding generators for the Chow groups of certain low-dimensional quotient varieties. Symonds proved the analogous very strong bound for the cohomology ring of a finite group modulo transfers from proper subgroups, and we give a version of his argument (Corollary 10.3).

In examples, the Chow ring of a finite group G always turns out to be simpler than the cohomology ring, and it seems to be closely related to the complex representation theory of G. In that direction, I conjectured that the Chow ring of any finite group was generated by transfers of Euler classes (top Chern classes) of complex representations [138]. That was disproved by Guillot for a certain group of order 2^7, the extraspecial 2-group 2_+^{1+6} [62]. It would be good to find similar examples at odd primes. Nonetheless, the theorem on the Chow ring modulo transfers gives a class of p-groups for which the question has a positive answer. Namely, the Chow ring of a p-group with a faithful complex representation of dimension at most $p + 2$ consists of transferred Euler classes (Theorem 11.1). This includes all 2-groups of order at most 32, and all p-groups of order at most p^4 with p odd.

We extend Symonds's theorem on the Castelnuovo-Mumford regularity of the cohomology ring to the Chow ring of the classifying space of a finite group (Theorem 6.5). We also bound the regularity of motivic cohomology (Theorem 6.10). It follows, for example, that all our bounds on generators for the Chow ring also lead to bounds on the relations. In each case, our upper bound for the degree of the relations is twice the bound for the degree of the generators. Another application is an identification of the motivic cohomology of a classifying space BG in high weights with the ordinary (or etale) cohomology. This statement goes beyond the range where motivic cohomology and etale cohomology are the same for arbitrary varieties, as described by the Beilinson-Lichtenbaum conjecture.

Let G be a finite group with a faithful complex representation of dimension n. Chapter 12 shows that the cohomology of G is determined by the cohomology of certain subgroups (centralizers of elementary abelian subgroups) in degrees less than $2n$. This was conjectured by Kuhn, who was continuing a powerful approach to group cohomology developed by Henn, Lannes, and Schwartz [86, 69]. We also prove an analogous result for the Chow ring: the Chow ring of a finite group is determined by the cohomology of centralizers of elementary abelian subgroups in degrees less than n. This is a strong computational tool, in a slightly different direction from the bounds for degrees of generators. The proof is inspired by Kuhn's ideas on group cohomology.

For a finite group G, Henn, Lannes, and Schwartz found that much of the complexity of the cohomology ring of G is described by one number, the "topological nilpotence degree" d_0 of the cohomology ring. This number is defined in terms of the cohomology ring as a module over the Steenrod algebra, but it is also equal to the optimal bound for determining the cohomology of G in terms of the low-degree cohomology of centralizers of elementary abelian subgroups.

Section 13.5 gives the first calculations of the topological nilpotence degree d_0 for some small p-groups, such as the groups of order p^3. In these examples, d_0 turns out to be much smaller than known results would predict. Improved bounds for d_0 would be a powerful computational tool in group cohomology.

To understand the cohomology of finite groups, it is important to compute the cohomology of large classes of p-groups. The cohomology of particular finite groups such as the symmetric groups and the general linear groups over finite fields F (with coefficients in $\mathbf{F}_p$ for p invertible in F) were computed many years ago by Nakaoka and Quillen. The calculations were possible because the Sylow p-subgroups of these groups are very special (iterated wreath products). To test conjectures in group cohomology, it has been essential to make more systematic calculations for p-groups, such as Carlson's calculation of the cohomology of all 267 groups of order 2^6 [26, appendix]. More recently, Green and King computed the cohomology of all 2328 groups of order 2^7 and all 15 groups of order 3^4 or 5^4 [51, 52]. In that spirit, we begin the systematic calculation of Chow rings of p-groups. Chapter 13 computes the Chow rings of all 5 groups of order p^3 and all 14 groups of order 16. Chapter 14 computes the Chow ring for all 15 groups of order $3^4 = 81$, and for 13 of the 15 groups of order p^4 with $p \geq 5$. Most of the proofs use only Chow rings, but the hardest cases also use calculations of group cohomology by Leary and Yagita.

One tantalizing example for which the Chow ring is not yet known is the group G of strictly upper triangular matrices in $GL(4, \mathbf{F}_p)$, which has order p^6. The machinery in this book should at least make that calculation easier. For p odd, Kriz and Lee showed that the Morava K-theory $K(2)^*BG$ is not concentrated in even degrees, disproving a conjecture of Hopkins, Kuhn, and Ravenel [83, 84]. It seems to be unknown whether the complex cobordism of BG is concentrated in even degrees in this example. Until this is resolved, it remains a possibility that the Chow ring of BG may map isomorphically to the quotient $MU^*(BG) \otimes_{MU^*} \mathbf{Z}$ of complex cobordism for every complex algebraic group G (including finite groups), as conjectured in [138]. Yagita strengthened this conjecture to say that algebraic cobordism Ω^*BG should map isomorphically to the topologically defined MU^*BG for every complex algebraic group G [154, conjecture 12.2].

Chapter 15 gives examples of p-groups for any prime number p such that the geometric and topological filtrations on the complex representation ring are different. When $p = 2$, Yagita has also given such examples [156, corollary 5.7]. A representation of G determines a vector bundle on BG, and these two filtrations describe the "codimension of support" of a virtual representation in the algebro-geometric or the topological sense. Atiyah conjectured that the (algebraically defined) γ-filtration of the representation ring was equal to the topological filtration [6], but that was disproved by Weiss, Thomas, and (for p-groups) Leary and Yagita [93]. Since the geometric filtration lies between the

γ and topological filtrations, the statement that the geometric and topological filtrations can be different is stronger. The examples use Vistoli's calculation of the Chow ring of the classifying space of $PGL(p)$ for prime numbers p [143].

Chapter 16 constructs an Eilenberg-Moore spectral sequence in motivic cohomology for schemes with an action of a split reductive group. The spectral sequence was defined by Krishna with rational coefficients [82, theorem 1.1]. We give an integral result, as far as possible. The Eilenberg-Moore spectral sequence in ordinary cohomology is a basic tool in homotopy theory. Given the cohomology of the base and total space of a fibration, the spectral sequence converges to the cohomology of a fiber. The reason for including the motivic Eilenberg-Moore spectral sequence in this book is to clarify the relation between the classifying space of an algebraic group and its finite-dimensional approximations.

Finally, Chapter 17 considers the Chow Künneth conjecture: for a finite group G and a field k containing enough roots of unity, the natural map $CH^*BG_k \otimes_{\mathbf{Z}} CH^*X \to CH^*(BG_k \times X)$ should be an isomorphism for all smooth schemes X over k. This would in particular imply that the Chow ring of BG_K is the same for all field extensions K of k. Although there is no clear reason to believe the conjecture, we prove some partial results for arbitrary groups, and prove the second version of the conjecture completely for p-groups with a faithful representation of dimension at most $p+2$. Chapter 18 is a short list of open problems. The Appendix tabulates several invariants of the Chow rings of p-groups of order at most p^4.

I thank Ben Antieau and Peter Symonds for many valuable suggestions.

1

Group Cohomology

This chapter gives the topological and algebraic definitions of group cohomology. We also define equivariant cohomology.

Although we give the basic definitions, a beginner may have to refer to other sources. Brown [24] is an excellent introduction to group cohomology. Group cohomology is also treated in general texts on homological algebra such as Weibel [149]. Some of the main advanced books on the cohomology of finite groups are Adem-Milgram [1], Benson [12], and Carlson [26].

Group cohomology unified many earlier ideas in algebra and topology. It was defined in 1943–1945 by Eilenberg and MacLane, Hopf and Eckmann, and Freudenthal.

1.1 Definition of group cohomology

Group cohomology arises from the fact that any group determines a topological space, as follows. Let G be a topological group. The special case where G is a discrete group is a rich subject in itself. Say that G acts *freely* on a space X if the map $G \times X \to X \times X$, $(g, x) \mapsto (x, gx)$, is a homeomorphism from $G \times X$ onto its image. By Serre, if a Lie group G acts freely on a metrizable topological space X, then the map $X \to X/G$ is a principal G-bundle, meaning that it is locally a product $U \times G \to U$ [109, section 4.1].

There is always a contractible space EG on which G acts freely. The *classifying space* of G is the quotient space of EG by the action of G, $BG = EG/G$. Any two classifying spaces for G that are paracompact are homotopy equivalent [72, definition 4.10.5, exercise 4.9]. If G is a discrete group, a classifying space of G can also be described as a connected space with fundamental group G whose universal cover is contractible, or as an Eilenberg-MacLane space $K(G, 1)$.

The cohomology of the classifying space of a topological group G is well-defined, because the classifying space is unique up to homotopy equivalence. In particular, for any commutative ring R, the cohomology $H^*(BG, R)$ is a graded-commutative R-algebra that depends only on G. For a discrete group G, we call $H^*(BG, R)$ the *cohomology of* G with coefficients in R; confusion should not arise with the cohomology of G as a topological space, which is uninteresting for G discrete. A fundamental challenge is to understand the relation between algebraic properties of a group and algebraic properties of its cohomology ring.

The cohomology of a group G manifestly says something about the cohomology of certain quotient spaces. More generally, for any space X on which G acts freely, there is a fibration

$$X \to (X \times EG)/G \to BG,$$

where the total space is homotopy equivalent to X/G. The resulting spectral sequence $H^*(BG, H^*X) \Rightarrow H^*(X/G)$, defined by Hochschild and Serre, shows that the cohomology of G gives information about the cohomology of any quotient space by G.

Another role of the classifying space of a group G is that it classifies principal G-bundles. By definition, a principal G-bundle over a space X is a space E with a free G-action such that $X = E/G$. The classifying space BG classifies principal G-bundles in the sense that for any CW-complex X, there is a one-to-one correspondence between isomorphism classes of principal G-bundles over X and homotopy classes of maps $X \to BG$. (Explicitly, we have a "universal" G-bundle $EG \to BG$, and a map $f\colon X \to BG$ defines a G-bundle over X by pulling back: let E be the fiber product $X \times_{BG} EG$.)

Therefore, computing the cohomology of the classifying space gives information about the classification of principal G-bundles over an arbitrary space. Namely, an element $u \in H^i(BG, R)$ gives a *characteristic class* for G-bundles: for any G-bundle E over a space X, we get an element $u(E) \in H^i(X, R)$, by pulling back u via the map $X \to BG$ corresponding to E.

A homomorphism $G \to H$ of topological groups determines a homotopy class of continuous maps $BG \to BH$. For example, we can view this as the obvious map $(EG \times EH)/G \to EH/H = BH$. As a result, given a commutative ring R, a homomorphism $G \to H$ determines a "pullback map" on group cohomology:

$$H^*(BH, R) \to H^*(BG, R)$$

Example The classifying space of the group $\mathbf{Z}/2$ can be viewed as the infinite real projective space $\mathbf{RP}^\infty = \cup_{n\geq 0}\mathbf{RP}^n$. Its cohomology with coefficients in the field $\mathbf{F}_2 = \mathbf{Z}/2$ is a polynomial ring,

$$H^*(B\mathbf{Z}/2, \mathbf{F}_2) = \mathbf{F}_2[x],$$

where x has degree 1. On the finite-dimensional approximation $\mathbf{RP}^n$ to $B\mathbf{Z}/2$, x restricts to the class of a hyperplane $\mathbf{RP}^{n-1} \subset \mathbf{RP}^n$.

Example The classifying space of the general linear group $GL(n, \mathbf{C})$ can be viewed as the Grassmannian Gr(n, ∞) of n-dimensional complex linear subspaces in $\mathbf{C}^\infty$. The cohomology of this classifying space is a polynomial ring,

$$H^*(BGL(n, \mathbf{C}), \mathbf{Z}) = \mathbf{Z}[c_1, \ldots, c_n].$$

A standard reference for this calculation is Milnor-Stasheff [106, theorem 14.5]. (We determine the Chow ring of $BGL(n)$ in Theorem 2.13, by a method that also works for cohomology.) These generators $c_1, \ldots, c_n$ are called Chern classes. They have degrees $|c_i| = 2i$, meaning that $c_i \in H^{2i}(BGL(n, \mathbf{C}), \mathbf{Z})$.

There is an equivalence of categories between rank-n complex vector bundles V over a space X and principal $GL(n, \mathbf{C})$-bundles E over X; given E, we define $V = (E \times \mathbf{C}^n)/GL(n, \mathbf{C})$. Therefore, the Chern classes give invariants for complex vector bundles V over any space X, $c_i(V) \in H^{2i}(X, \mathbf{Z})$.

Note that $GL(n, \mathbf{C})$ deformation retracts onto the unitary group $U(n)$. (For a matrix in $GL(n, \mathbf{C})$, the columns form a basis for $\mathbf{C}^n$. The Gram-Schmidt process shows how to move them continuously to an orthonormal basis for $\mathbf{C}^n$, which can be identified with an element of $U(n)$.) It follows that the resulting continuous map $BU(n) \to BGL(n, \mathbf{C})$ is a homotopy equivalence. So the previous calculation can be restated as $H^*(BU(n), \mathbf{Z}) = \mathbf{Z}[c_1, \ldots, c_n]$.

As a result, for any compact Lie group G (e.g., a finite group), any complex representation $G \to U(n)$ has Chern classes $c_i \in H^i(BG, \mathbf{Z})$ for $i = 1, \ldots, n$, defined by the pullback map $H^*(BU(n), \mathbf{Z}) \to H^*(BG, \mathbf{Z})$. We can also say that a representation of G determines a vector bundle on BG, and these are the Chern classes of that bundle.

Although we won't need this, it is interesting to note that for compact Lie groups G and H, a continuous map $BG \to BH$ need not be homotopic to one coming from a homomorphism $G \to H$. Sullivan gave the first example: for any odd positive integer a, there is an "unstable Adams operation" $\psi^a \colon BSU(2) \to BSU(2)$ that induces multiplication by a^2 on $H^4(BSU(2), \mathbf{Z}) \cong \mathbf{Z}$ [128, corollaries 5.10, 5.11]. Only the map ψ^1 (the identity map) comes from a group homomorphism $SU(2) \to SU(2)$.

1.2 Equivariant cohomology and basic calculations

Let G be a topological group acting on a topological space X. The *(Borel) equivariant cohomology* of X with respect to G is $H^i_G(X, R) = H^i((X \times EG)/G, R)$. That is, we make the action of G free without changing the homotopy type of X, and then take the quotient by G. In particular, if G acts freely on X, then equivariant cohomology is simply the cohomology of the quotient space, $H^i_G(X) = H^i(X/G)$. At the other extreme, we write H^i_G for the G-equivariant cohomology of a point (with a given coefficient ring, which we usually take to be the field $\mathbf{F}_p = \mathbf{Z}/p$ for a prime number p):

$$H^i_G := H^i_G(\text{point}, \mathbf{F}_p) = H^i(BG, \mathbf{F}_p).$$

Evens and Venkov proved the finite generation of the cohomology ring of a finite group. We give Venkov's elegant proof using equivariant cohomology, which works more generally for compact Lie groups [12, vol. 2, theorem 3.10.1]. Venkov's method helped to inspire Quillen's work on group cohomology and the later developments described in this book.

Theorem 1.1 *Let G be a compact Lie group and R a noetherian ring. Then $H^*(BG, R)$ is a finitely generated R-algebra. For any closed subgroup H of G, $H^*(BH, R)$ is a finitely generated module over $H^*(BG, R)$.*

Here the map $BH \to BG$ gives a ring homomorphism $H^*(BG, R) \to H^*(BH, R)$, and so we can view $H^*(BH, R)$ as a module over $H^*(BG, R)$.

Proof Every compact Lie group G has a faithful complex representation, giving an imbedding of G into $U(n)$ for some n [20, theorem III.4.1]. Since $H^*(BU(n), R) = R[c_1, \ldots, c_n]$ is a finitely generated R-algebra, the first statement of the theorem follows if we can show that $H^*(BG, R)$ is a finitely generated module over the ring of Chern classes $H^*(BU(n), R)$. (This will also imply the second statement of the theorem: for $H \subset G \subset U(n)$, $H^*(BH, R)$ is a finitely generated module over $R[c_1, \ldots, c_n]$ and hence over $H^*(BG, R)$.)

The Leray-Serre spectral sequence of the fibration $U(n)/G \to BG \to BU(n)$ has the form

$$E_2^{ij} = H^i(BU(n), H^j(U(n)/G, R)) \Rightarrow H^{i+j}(BG, R).$$

Since $U(n)/G$ is a closed manifold, its cohomology groups are finitely generated and are zero in degrees greater than the dimension of $U(n)/G$ (which is $n^2 - \dim(G)$). So the E_2 term of the spectral sequence has finitely many rows, each of which is a finitely generated module over $H^*(BU(n), R)$. Since the ring $H^*(BU(n), R)$ is noetherian, every submodule of a finitely generated module over $H^*(BU(n), R)$ is finitely generated, and hence any quotient of a submodule is finitely generated. The differentials in the spectral sequence are linear over $H^*(BU(n), R)$, and so the E_∞ term of the spectral sequence also has finitely many rows, each of which is a finitely generated module over $H^*(BU(n), R)$. Since $H^*(BG, R)$ is filtered with these rows as quotients, $H^*(BG, R)$ is a finitely generated module over $H^*(BU(n), R)$. □

The cohomology of abelian groups is easy to compute. To state the result, write $R\langle x_1, \ldots, x_n\rangle$ for the free graded-commutative algebra over a commutative ring R. This is a graded ring, with given degrees $|x_i| \in \mathbf{Z}$ for the generators, which is the tensor product of the polynomial ring on the generators of even degree with the exterior algebra on the generators of odd degree. We use this notation only for rings R containing $1/2$. (The point is that the cohomology ring of a topological space with coefficients in any commutative ring R is graded-commutative in the sense that $xy = (-1)^{|x||y|}yx$, but only when R contains $1/2$ does this imply that $x^2 = 0$ for x of odd degree. The $\mathbf{F}_2$-cohomology ring of a topological space is commutative in the naive sense.)

Theorem 1.2 *The cohomology ring of $B(S^1)^n$ with any coefficient ring R is the polynomial ring $R[y_1, \dots, y_n]$ with $|y_i| = 2$.*

The integral cohomology ring of $B(\mathbf{Z}/n)$ for a positive integer n is $\mathbf{Z}[y]/(ny)$ with $|y| = 2$. The generator y can be viewed as the first Chern class of a 1-dimensional complex representation $\mathbf{Z}/n \subset U(1)$.

The $\mathbf{F}_2$-cohomology ring of $B(\mathbf{Z}/2)$ is the polynomial ring $\mathbf{F}_2[x]$ with $|x| = 1$. The $\mathbf{F}_2$-cohomology ring of $B(\mathbf{Z}/2^r)$ for $r \geq 2$ is $\mathbf{F}_2[x, y]/(x^2)$ with $|x| = 1$ and $|y| = 2$.

Finally, for an odd prime number p and any $r \geq 1$, the $\mathbf{F}_p$-cohomology ring of $B(\mathbf{Z}/p^r)$ is $\mathbf{F}_p\langle x, y\rangle$ with $|x| = 1$ and $|y| = 2$.

These results can be proved by viewing BS^1 as the infinite projective space $\mathbf{CP}^\infty$ and viewing $B\mathbf{Z}/n$ as the principal S^1-bundle over $\mathbf{CP}^\infty$ whose first Chern class is n times a generator of $H^2(\mathbf{CP}^\infty, \mathbf{Z})$. Or one can give an algebraic proof, as in [1, section II.4]. These results determine the cohomology of BG for any abelian compact Lie group G using the Künneth formula, since $B(G \times H) = BG \times BH$.

For any topological space X, the *Bockstein* $\beta\colon H^i(X, \mathbf{Z}/p) \to H^{i+1}(X, \mathbf{Z})$ is the boundary map associated to the short exact sequence of coefficient groups

$$0 \to \mathbf{Z} \xrightarrow[p]{} \mathbf{Z} \to \mathbf{Z}/p \to 0.$$

The resulting long exact sequence shows that the Bockstein vanishes on integral classes. The composition $H^i(X, \mathbf{Z}/p) \xrightarrow[\beta]{} H^{i+1}(X, \mathbf{Z}) \to H^{i+1}(X, \mathbf{Z}/p)$ is also called the Bockstein. Because the Bockstein vanishes on integral classes, $\beta^2 = 0$. The Bockstein is a derivation on the mod p cohomology ring of any space, in the sense that $\beta(xy) = \beta(x)y + (-1)^{|x|}x\,\beta(y)$ for x, y in $H^*(X, \mathbf{Z}/p)$ [68, section 3.E]. We also note that $\beta x = x^2$ for all $x \in H^1(X, \mathbf{Z}/2)$.

The Bockstein on mod p cohomology is a convenient way to encode some information about integral cohomology. For that reason, we record the Bockstein on the mod p cohomology of the cyclic group $\mathbf{Z}/p^r$: in the preceding notation, $\beta y = 0$ since y is an integral class, and βx is equal to y for $r = 1$ (where we write $y = x^2$ for the group $\mathbf{Z}/2$) and to zero for $r > 1$.

1.3 Algebraic definition of group cohomology

We now present the purely algebraic definition of the cohomology of a discrete group. This is good to know, but it is not used in the rest of the book.

The algebraic definition of group cohomology is one answer to the question of how the algebraic structure of a group determines its cohomology ring. It does not answer all the questions. For example, what special properties does the mod p cohomology ring of a finite simple group have? Or a finite p-group?

To give the definition, let G be a discrete group. We can identify modules over the group ring $\mathbf{Z}G$ with abelian groups on which G acts by automorphisms. Consider the functor from $\mathbf{Z}G$-modules M to abelian groups given by the invariants $M^G := \{x \in M : gx = x \text{ for all } g \in G\}$. This is a left exact functor, meaning that a short exact sequence $0 \to A \to B \to C \to 0$ of $\mathbf{Z}G$-modules determines an exact sequence:

$$0 \to A^G \to B^G \to C^G.$$

We can therefore consider the right-derived functors of M^G, which are called the *cohomology* of G with coefficients in M, $H^i(G, M)$. In particular, $H^0(G, M) = M^G$, and a short exact sequence of $\mathbf{Z}G$-modules gives a long exact sequence of cohomology groups:

$$0 \to H^0(G, A) \to H^0(G, B) \to H^0(G, C) \to H^1(G, A) \to \cdots.$$

Moreover, this notion of group cohomology agrees with the topological definition: for any $\mathbf{Z}G$-module M, $H^*(G, M)$ is isomorphic to the cohomology of the topological space BG with coefficients in the locally constant sheaf associated to M. In particular, if G acts trivially on M, then this is the usual notion of cohomology of the space BG with coefficients in the abelian group M.

We recall how right-derived functors are defined: choose a resolution

$$0 \to M \to I_0 \to I_1 \to \cdots$$

of M by injective $\mathbf{Z}G$-modules and define $H^*(G, M)$ to be the cohomology of the chain complex:

$$0 \to I_0^G \to I_1^G \to \cdots.$$

We can fit group cohomology into a bigger picture by observing that $M^G = \text{Hom}_{\mathbf{Z}G}(\mathbf{Z}, M)$ for any $\mathbf{Z}G$-module M, where G acts trivially on $\mathbf{Z}$. The derived functors of Hom are called Ext, and so we have:

$$H^i(G, M) \cong \text{Ext}^i_{\mathbf{Z}G}(\mathbf{Z}, M)$$

[149]. Ext can also be viewed as the left-derived functor of Hom in the first variable, and so group cohomology can be computed using either a projective

resolution of the $\mathbf{Z}G$-module $\mathbf{Z}$ or an injective resolution of M. A useful variant is that if M is a representation of G over a field k, then

$$H^i(G, M) \cong \mathrm{Ext}^i_{kG}(k, M).$$

For example, this definition makes it clear that for a finite group G and prime number p, there is an algorithm to compute any given cohomology group $H^i(G, \mathbf{F}_p)$. It suffices to work out the first $i+1$ steps of a free resolution of $\mathbf{F}_p$ as an $\mathbf{F}_pG$-module,

$$F_{i+1} \to \cdots \to F_0 \to \mathbf{F}_p \to 0,$$

which amounts to doing linear algebra over $\mathbf{F}_p$. Then $H^i(G, \mathbf{F}_p)$ is the cohomology of the chain complex

$$\mathrm{Hom}_{\mathbf{F}_pG}(F_{i-1}, \mathbf{F}_p) \to \mathrm{Hom}_{\mathbf{F}_pG}(F_i, \mathbf{F}_p) \to \mathrm{Hom}_{\mathbf{F}_pG}(F_{i+1}, \mathbf{F}_p).$$

There is a standard free resolution of $\mathbf{Z}$ as a $\mathbf{Z}G$-module that works for any group G [12, vol. 1, section 3.4], but it is usually too big for computations. Rather, the programs that compute the cohomology of finite groups construct a minimal resolution as far as is needed [51, 52].

Example Let $G = \mathbf{Z}/p$ for a prime number p. Write g for a generator of the group G. Let tr be the element $1 + g + \cdots + g^{p-1}$, called the *trace*, in the group ring $\mathbf{F}_pG$. Then $\mathbf{F}_p$ has a free resolution as an $\mathbf{F}_pG$-module that is periodic, of the form

$$\cdots \to \mathbf{F}_pG \xrightarrow[1-g]{} \mathbf{F}_pG \xrightarrow[\mathrm{tr}]{} \mathbf{F}_pG \xrightarrow[1-g]{} \mathbf{F}_pG \to \mathbf{F}_p \to 0.$$

Taking Hom over $\mathbf{F}_pG$ from this resolution to $\mathbf{F}_p$, all the differentials become zero. It follows that $H^i(G, \mathbf{F}_p) \cong \mathbf{F}_p$ for every $i \geq 0$, in agreement with Theorem 1.2.

The low-dimensional cohomology groups have simple interpretations. For any group G and abelian group A, $H^1(G, A)$ can be identified with the abelian group of homomorphisms $G \to A$. Also, $H^2(G, A)$ is the group of isomorphism classes of central extensions of G by A [24, theorem 3.12]. By definition, an *extension* of G by A is a group E with normal subgroup A and a specified isomorphism $E/A \cong G$. It is *central* if all elements of the subgroup A commute with all elements of E.

2

The Chow Ring of a Classifying Space

The Chow groups of an algebraic variety are an analog of homology groups, with generators and relations given in terms of algebraic subvarieties. In this chapter we define Chow groups and state their main formal properties, including a version of homotopy invariance. Using those properties, we define the Chow ring of the classifying space of an algebraic group, a central topic of this book. More generally, we give Edidin and Graham's definition of the equivariant Chow ring of a variety with group action. The chapter ends with a discussion of some open problems about Chow rings of classifying spaces. Examples suggest that the Chow ring of the classifying space of a group is simpler, and closer to representation theory, than the cohomology ring is. But we know much less about general properties of the Chow ring, such as finite generation.

We state the formal properties of Chow groups without proof, using Fulton's book as a reference [43]. Building on that, we develop equivariant Chow groups in more detail. We refer to the papers [138] and [38] for some results, but we do the basic calculations of equivariant Chow groups.

2.1 The Chow group of algebraic cycles

Let us define Chow groups, following Fulton [43]. We work in the category of separated schemes of finite type over a field k. A *variety* over k is a reduced irreducible scheme (which is separated and of finite type over k, by our assumptions). An i-dimensional *algebraic cycle* on a scheme X over k is a finite $\mathbf{Z}$-linear combination of closed subvarieties of dimension i. The subgroup of algebraic cycles *rationally equivalent* to zero is generated by the elements $\sum_D \text{ord}_D(f)D$, for every $(i+1)$-dimensional closed subvariety W of X and every nonzero rational function f on W. The sum runs over all codimension-1 subvarieties D of W, and $\text{ord}_D(f)$ is the order of vanishing of f along D

[43, section 1.2]. The *Chow group* $CH_i(X)$ is the group of i-dimensional cycles modulo rational equivalence.

For a scheme X over the complex numbers, we can give the set $X(\mathbf{C})$ of complex points the classical (Euclidean) topology, instead of the Zariski topology. For X over the complex numbers, there is a natural "cycle map" from the Chow groups of X to the Borel-Moore homology of the associated topological space [43, proposition 19.1.1], $CH_i(X) \to H_{2i}^{BM}(X, \mathbf{Z})$. The Borel-Moore homology of a locally compact space is also known as homology with closed support. The numbering is explained by the fact that a subvariety of complex dimension i has real dimension $2i$. The definition of the cycle map uses the fact that a complex manifold has a natural orientation.

The cycle map is far from being an isomorphism in general. For example, if X is a smooth complex projective curve, then the Chow group of 0-cycles, $CH_0(X)$, maps onto $H_0(X, \mathbf{Z}) = \mathbf{Z}$ with kernel the group of complex points of the Jacobian of the curve. The Jacobian is an abelian variety of dimension equal to the genus of X, and so $CH_0(X)$ is an uncountable abelian group when X has genus at least 1.

A proper morphism $f: X \to Y$ of schemes over a field k determines a pushforward map on Chow groups, $f_*: CH_i(X) \to CH_i(Y)$. A flat morphism $f: X \to Y$ with fibers of dimension r determines a pullback map, $f^*: CH_i(Y) \to CH_{i+r}(X)$. (The morphism f is allowed to have some fibers empty.) Both types of homomorphism occur in the basic exact sequence for Chow groups, as follows [43, proposition 1.8].

Lemma 2.1 *Let X be a separated scheme of finite type over a field k. Let Z be a closed subscheme. Then the proper pushforward and flat pullback maps fit into an exact sequence*

$$CH_i(Z) \to CH_i(X) \to CH_i(X - Z) \to 0.$$

For X a complex scheme, the basic exact sequence for Chow groups maps to the long exact sequence of Borel-Moore homology groups:

$$\cdots \to H_{2i}^{BM}(Z, \mathbf{Z}) \to H_{2i}^{BM}(X, \mathbf{Z}) \to H_{2i}^{BM}(X - Z, \mathbf{Z}) \to H_{2i-1}^{BM}(Z, \mathbf{Z}) \to \cdots.$$

Note the differences between the two sequences. In the exact sequence of Chow groups, we do not say anything about the kernel of $CH_i Z \to CH_i X$. Indeed, the exact sequence of Chow groups can be extended to the left, but that involves a generalization of Chow groups known as motivic homology groups (or, equivalently, higher Chow groups); see Section 6.2. But Chow groups are simpler in one way than ordinary homology: the restriction map to an open subset is always surjective on Chow groups. Geometrically, this is because the closure in X of a subvariety of $X - Z$ is a subvariety of X. This phenomenon lies behind various ways in which Chow groups behave more simply than ordinary homology.

Chow groups are homotopy invariant in the following sense. An *affine bundle* $E \to B$ is a morphism that is locally a product with fibers A^n. We do not assume anything about the structure group of the fibration. So the total space of a vector bundle is an affine bundle, but affine bundles are more general.

Lemma 2.2 *For an affine bundle $E \to B$ with fibers of dimension n, the pullback $CH_i B \to CH_{i+n} E$ is an isomorphism.*

Proof One natural approach uses motivic homology groups, a generalization of Chow groups. Namely, flat pullback gives an isomorphism from the motivic homology of any k-scheme B to the motivic homology of $B \times A^n$ [14, theorem 2.1]. The lemma follows via the localization sequence for motivic homology [15]. (We state the localization sequence for smooth k-schemes as Theorem 6.8.) □

For a smooth scheme X of dimension n over a field k, we write $CH^i(X)$ for the Chow group of codimension-i cycles, $CH^i(X) = CH_{n-i}(X)$. Intersection of cycles makes the Chow groups of a smooth scheme into a commutative ring, $CH^i(X) \times CH^j(X) \to CH^{i+j}(X)$. Fulton and MacPherson's approach to constructing this product first reduces the problem to that of intersecting a cycle on $X \times X$ with the diagonal, and then defines the latter intersection by deformation to the normal cone [43, chapter 6]. Any morphism $f: X \to Y$ of smooth schemes over k determines a pullback map $f^*: CH^*Y \to CH^*X$, which is a homomorphism of graded rings. (When f is flat, this coincides with the flat pullback map $f^*: CH_*Y \to CH_*X$.) For a smooth complex scheme X of dimension n, Poincaré duality is an isomorphism $H^i(X, \mathbf{Z}) \cong H^{BM}_{2n-i}(X, \mathbf{Z})$. So we have a cycle map $CH^*X \to H^*(X, \mathbf{Z})$, and this is a ring homomorphism, sending CH^i into H^{2i}.

Note that homotopy invariance of Chow rings (Lemma 2.2) does not mean that two smooth complex varieties that are homotopy equivalent as topological spaces (in the classical topology) have isomorphic Chow rings. For example, an elliptic curve X over $\mathbf{C}$ is homotopy equivalent, as a topological space, to $Y = (A^1 - 0)^2$. But $CH^1 Y$ is zero by the basic exact sequence of Chow groups (Lemma 2.1), whereas the abelian group $CH^1 X$ is an extension of $\mathbf{Z}$ by the group $X(\mathbf{C}) \cong (S^1)^2$ [67, example II.6.10.2, example IV.1.3.7].

A vector bundle E on a smooth scheme X has Chern classes $c_i E \in CH^i X$, with the same formal properties as in topology. We record the Chow ring of a projective bundle, which is given by the same formula as the cohomology ring of a projective bundle [43, remark 3.2.4, theorem 3.3]:

Lemma 2.3 *Let X be a smooth scheme over a field. Let E be a vector bundle of rank n on X. Let $\pi : P(E) \to X$ be the projective bundle of codimension-1* *linear subspaces of E. Then*

$$CH^*P(E) \cong CH^*X[u]/(u^n - c_1(E)u^{n-1} + \cdots + (-1)^n c_n(E)),$$

where u is the first Chern class of the quotient line bundle $O(1)$ *of* π^*E.

It is also useful to know the Chow ring of the total space of a line bundle minus the zero section. (In the terminology of Section 2.2, this is the Chow ring of a principal G_m-bundle.) Lemma 2.4 follows from homotopy invariance of Chow groups (Lemma 2.2) together with the basic exact sequence for Chow groups (Lemma 2.1). Alternatively, this is [43, example 2.6.3]:

Lemma 2.4 *Let X be a smooth scheme over a field k. Let L be a line bundle over X. View L as a smooth scheme over k with a morphism* $L \to X$ *and zero section* $X \subset L$. *Then the Chow ring of L minus the zero section is*

$$CH^*(L - X) \cong CH^*X/(c_1 L).$$

By definition, Chow groups contain a huge amount of algebro-geometric information, but they are very hard to compute for general varieties. Some of the main problems in algebraic geometry, such as the Hodge conjecture, are attempts to understand Chow groups.

2.2 The Chow ring of a classifying space

The Chow ring of the classifying space of an algebraic group was defined in [138] and independently by Morel and Voevodsky [107, proposition 2.6]. Edidin and Graham generalized the definition to define the equivariant Chow ring and (more generally) equivariant motivic cohomology [38].

Among many applications of equivariant Chow groups, we mention Brosnan's construction of Steenrod operations on mod p Chow groups [21]. In the more general setting of motivic cohomology, Voevodsky constructed Steenrod operations as a crucial part of his proof with Rost of the Bloch-Kato conjecture. We summarize the properties of Voevodsky's Steenrod operations in section 6.3. Voevodsky's construction includes a computation of the motivic cohomology of the symmetric groups [144, 146].

A *group scheme* over a field k is a scheme G over k together with morphisms $G \times_k G \to G$ (multiplication), $G \to G$ (inverse), and $\mathrm{Spec}(k) \to G$ (identity) over k that satisfy the axioms of a group (associativity, identity, and inverse). Some basic examples of group schemes are the *additive group* G_a over k, meaning the curve A^1 with group operation being addition, and the *multiplicative group* G_m over k, meaning the curve $A^1 - 0$ with group operation being multiplication.

Let G be a group scheme of finite type over a field k. We say that an action of G on a separated scheme X of finite type over k is *free* if the morphism $G \times_k X \to X \times_k X$ given by $(g, x) \mapsto (x, gx)$ is an isomorphism to a closed subscheme. Given a free action, we say that a scheme Y is the *quotient* X/G if Y has finite type over k and we are given a flat surjective morphism $f: X \to Y$ over k such that f is constant on G-orbits and the natural map $G \times_k X \to X \times_Y X$ is an isomorphism. In this situation, we also say that $X \to Y$ is a *principal* G*-bundle.*

Note that the map $X \to X/G$ need not be locally a product in the Zariski topology. For example, the group $\mathbf{Z}/2$ acts freely on $A^1 - 0$ over $\mathbf{C}$ by $x \mapsto -x$, and $(A^1 - 0)/(\mathbf{Z}/2)$ is isomorphic to $A^1 - 0$, with the quotient map $f: A^1 - 0 \to A^1 - 0$ being $x \mapsto x^2$. This principal $\mathbf{Z}/2$-bundle is not Zariski locally trivial. Indeed, for any nonempty open set $U \subset A^1 - 0$ (so U is the complement of a finite set), $f^{-1}(U)$ is again the complement of a finite subset in $A^1 - 0$, and so it is connected, in particular not isomorphic to $U \times (\mathbf{Z}/2)$. Nonetheless, any principal G-bundle is locally trivial in the "fppf topology"; that is a restatement of the definition. A readable introduction to group schemes and principal bundles is Waterhouse [147].

One important result about principal G-bundles $X \to Y$ is that there is an equivalence of categories between G-equivariant vector bundles on X and vector bundles on Y. What makes this subtle is that vector bundles are Zariski locally trivial by definition, whereas the morphism $X \to Y$ is in general not Zariski locally trivial. Nonetheless, this equivalence has a straightforward algebraic proof, part of Grothendieck's theory of faithfully flat descent [147, section 17.2]. (The case of descent used here is also known as Hilbert's Theorem 90.)

Every affine group scheme G of finite type over a field k has a faithful representation [147, theorem 3.4]. That is, G is isomorphic to a closed subgroup scheme of $GL(n)$ over k for some natural number n.

For any affine group scheme G of finite type over a field k and any $i \geq 0$, we define the Chow group $CH^i BG$ to be $CH^i(V - S)/G$ for any representation V of G over k and any G-invariant closed subset S of V such that G acts freely on $V - S$, the quotient $(V - S)/G$ exists as a scheme, and S has codimension greater than i. This definition of $CH^i BG$ is independent of the choices of V and S, by Theorem 2.5. Moreover, the theorem gives a well-defined ring $CH^* BG$.

The point is that, for G an algebraic group over the complex numbers, BG is typically an infinite-dimensional topological space, whereas algebraic varieties in the usual sense have finite dimension. But the spaces $V - S$ come closer and closer to being contractible (in the topological sense) as the codimension of S increases. So a direct limit of the spaces $(V - S)/G$ as the codimension of S increases is homotopy equivalent to BG. This suggested the definition of the Chow ring of BG, for an algebraic group G over any field. In

Morel-Voevodsky's A^1-homotopy category, one can take direct limits of algebraic varieties, and so they were able to define BG as an object in their category [107, proposition 2.6].

In the original papers, G was assumed to be an algebraic group, which is usually understood to mean a smooth group scheme of finite type over k. (Every group scheme of finite type over a field k of characteristic zero is smooth over k [147], but for k of characteristic $p > 0$ there are non-smooth group schemes over k, such as $\mu_p = \{x \in G_m : x^p = 1\}$ and $\alpha_p = \{x \in G_a : x^p = 0\}$. To see that μ_p and α_p are not smooth over k, note that they have dimension 0, but the derivative of the defining equation is px^{p-1}, which is zero because $p = 0$ in k; so the Zariski tangent spaces of μ_p and α_p at the identity have dimension 1.) The assumption of smoothness is not necessary, however. The only observation one needs to define CH^*BG for G not smooth is that if $E \to B$ is a principal G-bundle with E smooth over k, then B is also smooth over k. (Apply [99, theorem 23.7] over the algebraic closure of k.) For example, $(V - S)/G$ is smooth over k even if G is not. We do not make much use of this extra generality.

Theorem 2.5 ([138, theorem 1.1]) *Let G be an affine group scheme of finite type over a field k and let i be a natural number. Let V be a representation of G over k, and let S be a G-invariant closed subset of V such that G acts freely on $V - S$, the quotient $(V - S)/G$ exists as a scheme, and S has codimension greater than i in V. Then the Chow ring of $(V - S)/G$ in degrees at most i depends only on G, not on V and S.*

Proof To prove the independence of S (given that S has codimension greater than i in V), let S' be a larger G-invariant closed subset, still with codimension greater than i. Since G acts freely on $V - S$, $(S' - S)/G$ has codimension greater than i in $(V - S)/G$. So the restriction map $CH^j(V - S)/G \to CH^j(V - S')/G$ is an isomorphism for $j \leq i$ by the basic exact sequence for Chow groups (Lemma 2.1). So the Chow ring of $(V - S)/G$ is independent of S in the range we consider.

The independence of V follows from the double fibration method, used by Bogomolov and others in invariant theory [17]. That is, consider two representations V and W of G such that G acts freely outside subsets S_V and S_W of codimension greater than i and such that the quotients $(V - S_V)/G$ and $(W - S_W)/G$ exist as varieties. Then consider the direct sum $V \oplus W$. The quotient $((V - S_V) \times W)/G$ exists as a variety, being a vector bundle over $(V - S_V)/G$ (constructed by faithfully flat descent, as discussed earlier in this section). Likewise the quotient $(V \times (W - S_W))/G$ exists as a variety, being a vector bundle over $(W - S_W)/G$. Independence of S (applied to the representation $V \oplus W$) shows that these two total spaces of vector bundles have the same Chow ring in degrees at most i. By homotopy invariance of Chow

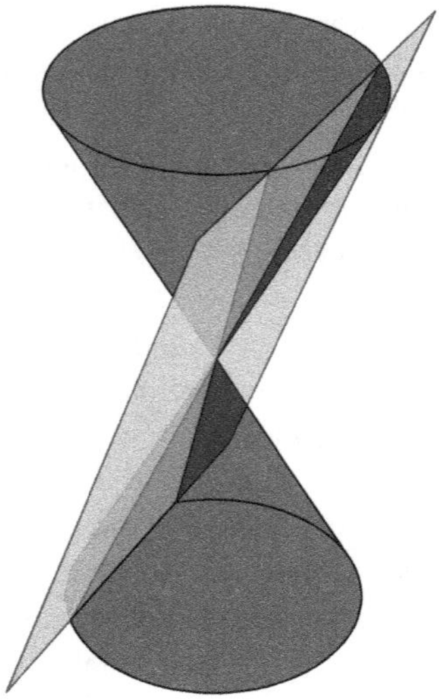

Figure 2.1 $A^2/(\mathbf{Z}/2)$, the quadric cone, intersected with a plane.

rings (Lemma 2.2), $(V - S_V)/G$ and $(W - S_W)/G$ have the same Chow ring in degrees at most i. □

Remark 2.6 Here is a simple example of Chow groups of classifying spaces that one can visualize; this example appears in Hartshorne [67, example II.6.5.2]. Consider the group $G = \mathbf{Z}/2$ as an algebraic group over $\mathbf{C}$. We will shortly compute that CH^1BG is isomorphic to $\mathbf{Z}/2$; here I just want to explain the geometric meaning of that calculation. Consider a 2-dimensional complex vector space V on which the generator of G acts by -1. Then G acts freely on V outside the origin S. Since S has codimension 2 in V, we have $CH^1BG = CH^1(V - S)/G$. Thus the computation of CH^1BG means that $CH^1(V - S)/G$ is isomorphic to $\mathbf{Z}/2$. Equivalently, the Chow group $CH_1(V/G)$ is isomorphic to $\mathbf{Z}/2$; removing the singular point from the geometric quotient V/G does not change this Chow group, by the basic exact sequence (Lemma 2.1). Since $CH_1(V/G)$ is the Chow group of codimension-1 cycles on the surface V/G, it is also called the divisor class group of V/G.

Here V/G is the affine quadric cone $\{(x, y, z) \in A^3 : xz = y^2\}$ over $\mathbf{C}$, with the morphism $V \to V/G$ given by $(u, v) \mapsto (u^2, uv, v^2)$ (Figure 2.1). The generator A of $CH_1(V/G)$ is the class of any line through the origin in this cone. The fact that $2A = 0$ can be seen geometrically by intersecting a plane through the origin in A^3 with the cone V/G; you always get a sum of two lines, or (if the plane is tangent to the cone) 2 times a line. (A plane is the divisor of a rational function on A^3, namely a linear function, and so its intersection with V/G (with multiplicities) is zero in $CH_1(V/G)$.)

Figure 2.2 A^2/Q_8, the D_4 surface singularity.

Another example one can visualize is the quotient of $V = A^2_{\mathbf{C}}$ by the quaternion group $G = Q_8$, which is isomorphic to the "D_4 singularity" $z^2 = x^2y - y^3$ in $A^3_{\mathbf{C}}$ (Figure 2.2). The full Chow ring $CH^*(BG)/2$ is computed in Lemma 13.1. Here $CH^1BG \cong CH_1(V/G)$ is isomorphic to $\mathbf{Z}/2 \oplus \mathbf{Z}/2$, generated by the three lines A, B, C through the origin in V/G that are visible in the figure. These classes satisfy $2A = 2B = 2C = 0$ in $CH_1(V/G)$ (by intersecting V/G with a tangent plane along any of the three lines) and $A + B + C = 0$ (by intersecting V/G with a plane containing the three lines).

Remark 2.7 ([138, remark 1.4]) If G is a finite group, then the geometric quotient V/G exists as an affine variety for all representations V of G [108, amplification 1.3]. So $(V - S)/G$ is a quasi-projective variety for all closed subsets $S \subset V$ such that G acts freely on $V - S$.

For any affine group scheme G over a field k and any positive integer s, there is a representation V of G and a closed subset $S \subset V$ of codimension at least s such that G acts freely on $V - S$ and $(V - S)/G$ exists as a quasi-projective variety. To see this, let W be any faithful representation W of G, and let $n = \dim(W)$. Let $V = \mathrm{Hom}(A^{N+n}, W) \cong W^{\oplus N+n}$ for N large. Let S be the closed subset in V of non-surjective linear maps $A^{N+n} \to W$. Then S has codimension $N + 1$ in V, as one easily counts. Also, $(V - S)/G$ exists as a quasi-projective variety. Indeed, $(V - S)/GL(n)$ is the Grassmannian $\text{Gr}(N, N+n)$ of N-dimensional linear subspaces of A^{N+n}, which we can view as the homogeneous space $GL(N+n)/((GL(N) \times GL(n)) \ltimes \text{Hom}(A^n, A^N))$. Therefore $(V - S)/G$ is the homogeneous space $GL(N+n)/((GL(N) \times G) \ltimes \text{Hom}(A^n, A^N))$. Any quotient of a linear algebraic group by a closed subgroup scheme exists as a quasi-projective scheme [147, pp. 122–123]. So $(V - S)/G$ is a quasi-projective variety.

Moreover, the natural map $GL(N+n)/(GL(N) \times G) \to (V - S)/G$ is an affine bundle. By homotopy invariance of Chow groups (Lemma 2.2), it follows that

$$CH^i BG \cong CH^i \; GL(N+n)/(GL(N) \times G)$$

for all $i \leq N$.

One reason to be interested in the Chow ring of BG is that it is equal to the ring of characteristic classes for principal G-bundles over smooth k-schemes, in the following sense [138, theorem 1.3].

Theorem 2.8 *Let G be an affine group scheme of finite type over a field k. Then the group $CH^i BG$ defined above is naturally identified with the set of assignments α to every smooth k-scheme X with a principal G-bundle E over X of an element $\alpha(E) \in CH^i X$, such that for any morphism $f: Y \to X$ over k we have $\alpha(f^* E) = f^*(\alpha(E))$. The ring structure on $CH^* BG$ is the obvious one on the set of characteristic classes.*

2.3 The equivariant Chow ring

We now consider a generalization. For an affine group scheme G over a field k that acts on a smooth k-scheme X, the *equivariant Chow ring* $CH^*_G X$ is defined by $CH^i_G X = CH^i(X \times (V - S))/G$ for any representation V of G over k and any closed G-invariant subset S of V such that G acts freely on $V - S$, the quotient $(X \times (V - S))/G$ exists as a scheme, and S has codimension greater than i. Again, $CH^*_G X$ is independent of the choices of V and S [38]. The paper [38] defines equivariant Chow groups for any G-scheme X, but since we take X to be smooth, the groups $CH^*_G X$ form a graded ring. (Briefly, the point is that the quotients $(X \times (V - S))/G$ are smooth when X is smooth.)

The condition that $(X \times (V - S))/G$ is a scheme does not pose a difficulty. First, under mild assumptions on X and G, there are many pairs (V, S) for which $(X \times (V - S))/G$ is a scheme and S has codimension as big as we like, by Remark 2.7 and [38, proposition 23]. Second, $(X \times (V - S))/G$ always exists as an algebraic space, and once one defines Chow groups for algebraic spaces, one can define $CH^i_G X$ using any (V, S) such that G acts freely on

$V - S$ and S has codimension greater than i. This is the preferred approach in Edidin-Graham [38, section 6.1].

Kresch extended the definition of equivariant Chow groups to define Chow groups for any algebraic stack of finite type over a field [81].

Since equivariant Chow groups are defined as the Chow groups of auxiliary varieties, the formal properties of Chow groups imply analogous properties for equivariant Chow groups, of which we mention the main ones. First, for a smooth scheme X with G-action over the complex numbers, we have a cycle map

$$CH^i_G X \to H^{2i}_G(X, \mathbf{Z})$$

and in particular

$$CH^i BG \to H^{2i}(BG, \mathbf{Z}).$$

Next, a proper morphism $f\colon X \to Y$ of smooth G-schemes over a field k determines a pushforward map on equivariant Chow groups, $f_*\colon CH^i_G(X) \to CH_G^{i+\dim(Y)-\dim(X)}(Y)$. Any morphism $f\colon X \to Y$ of smooth G-schemes determines a pullback map, $f^*\colon CH^i_G Y \to CH^i_G X$, and f^* is a ring homomorphism. Both types of homomorphism occur in the basic exact sequence for equivariant Chow groups, as follows.

Lemma 2.9 *Let G be an affine group scheme of finite type over a field k that acts on a smooth k-scheme X and preserves a smooth closed k-subscheme Y of codimension r. Then the proper pushforward and flat pullback homomorphisms give an exact sequence*

$$CH^{i-r}_G Y \to CH^i_G X \to CH^i_G(X - Y) \to 0.$$

There are also homomorphisms relating equivariant Chow rings for different groups. (These are analogous to the formal properties of equivariant cohomology.) First, for any homomorphism $G \to H$ of k-group schemes and any H-scheme X over k, we have a ring homomorphism $CH^*_H X \to CH^*_G X$. In particular, this gives a pullback homomorphism $CH^* BH \to CH^* BG$. (We can view that as the pullback map on Chow groups associated to a morphism of smooth varieties that approximates $(EG \times EH)/G \to EH/H = BH$. Here we use the notation EG for a contractible space with free G-action, when G is a topological group; we think of the smooth varieties $V - S$ as approximating EG when G is a k-group scheme.) There is also a "transfer" map in the other direction, discussed in Section 2.5.

When we talk about the Chow ring of BG for a finite group G, we are thinking of G as an algebraic group over some field. One common choice of base field is the complex numbers.

For an affine group scheme G over a field k, any representation $G \to GL(n)$ over k determines a rank-n vector bundle over BG (that is, over the finite-dimensional varieties $(V - S)/G$ that approximate BG), by the equivalence between principal $GL(n)$-bundles and vector bundles. It follows that an n-dimensional representation V of G over k has Chern classes $c_i \in CH^i BG$ for $1 \leq i \leq n$, the Chern classes of the corresponding vector bundle.

2.4 Basic computations

In this section, we compute the Chow ring of the classifying space for abelian groups and the general linear group $GL(n)$. We prove some partial results on the Künneth formula for Chow rings of classifying spaces; the general Chow Künneth formula is an open problem, to which we return in Chapter 17. Finally, we explain that Chow rings of classifying spaces are easy to compute with rational coefficients.

Theorem 2.10 *Let k be a field. The Chow ring of $B(G_m)^r$ is the polynomial ring $\mathbf{Z}[y_1, \ldots, y_r]$ with $|y_i| = 1$.*

Let k be a field, and let n be a positive integer. Let μ_n be the group scheme of nth roots of unity, the kernel of the nth power map on G_m over k. Then the Chow ring of $B\mu_n$ is isomorphic to $\mathbf{Z}[y]/(ny)$ with $|y| = 1$. The generator y is the first Chern class of the natural 1-dimensional representation $\mu_n \subset GL(1)$ over k.

If n is invertible in k and k contains a primitive nth root of unity, then μ_n is isomorphic to $\mathbf{Z}/n$ as an algebraic group over k, and so the Chow ring of $B(\mathbf{Z}/n)_k$ is isomorphic to $\mathbf{Z}[y]/(ny)$ with $|y| = 1$.

Proof Let V_a be the direct sum of a copies of the natural 1-dimensional representation of the multiplicative group G_m. Then G_m acts freely on $V_a - 0$, and $(V_a - 0)/G_m \cong \mathbf{P}^{a-1}$. Since the point 0 has codimension a in V_a, the Chow ring of BG_m is defined to agree with the Chow ring of $\mathbf{P}^{a-1}$ in degrees at most $a - 1$. We know that $CH^*(\mathbf{P}^{a-1}) \cong \mathbf{Z}[y]/(y^a)$, with $|y| = 1$, for example by Lemma 2.3. It follows that $CH^*(BG_m) \cong \mathbf{Z}[y]$. Likewise, $B(G_m)^r$ is approximated by products of r projective spaces, whose Chow ring is given by the projective bundle formula (Lemma 2.3). It follows that $CH^*B(G_m)^r \cong \mathbf{Z}[y_1, \ldots, y_r]$.

We can view V_a as a representation of $\mu_n \subset G_m$. The quotient $(V_a - 0)/(\mu_n)$ is the principal G_m-bundle over $\mathbf{P}^{a-1}$ whose first Chern class is n times the generator of $CH^1\mathbf{P}^{a-1} \cong \mathbf{Z}$. We can view this G_m-bundle as the line bundle $O(n)$ minus the zero section. By Lemma 2.4,

$$\begin{aligned} CH^*(V_a - 0)/(\mu_n) &\cong CH^*(\mathbf{P}^{a-1})/(c_1(O(n))) \\ &= \mathbf{Z}[y]/(y^a, ny), \end{aligned}$$

where $|y| = 1$. It follows that $CH^*B\mu_n \cong \mathbf{Z}[y]/(ny)$. □

For smooth varieties X and Y over a field k, the product map $CH^*X \otimes_{\mathbf{Z}} CH^*(Y) \to CH^*(X \times Y)$ need not be an isomorphism. For example, if X is an elliptic curve, then the class of the diagonal in $CH^1(X \times X)$ is not in the image of $CH^*X \otimes_{\mathbf{Z}} CH^*X$. It is an open question whether the Künneth formula (meaning that this product map is an isomorphism) holds for the Chow ring of the product of two finite groups, viewed as groups over $\mathbf{C}$. That does hold in many cases such as abelian groups over $\mathbf{C}$ [138, section 6]. The following lemma is a bit more general. Lemma 2.12 is already interesting when the base field is the complex numbers, but we assume as little as we can about the base field.

Definition 2.11 For a subgroup G of the symmetric group S_n and a group H, the *wreath product* $G \wr H$ is the semidirect product group $G \ltimes H^n$, where G acts on H^n by permuting the factors.

Lemma 2.12 *Let G be a group scheme over a field k that satisfies one of the following assumptions. Then the product map*

$$CH^*BG \otimes_{\mathbf{Z}} CH^*X \to CH^*(BG \times X)$$

is an isomorphism for all smooth schemes X of finite type over k. It follows that for these groups G, $CH^*BG \otimes_{\mathbf{Z}} CH^*BH \to CH^*B(G \times H)$ is an isomorphism for all affine group schemes H of finite type over k.

(i) *G is the multiplicative group G_m.*
(ii) *G is a finite abelian group of exponent e viewed as an algebraic group over k, e is invertible in k, and k contains the eth roots of unity.*
(iii) *G is an iterated wreath product $\mathbf{Z}/p \wr \cdots \wr \mathbf{Z}/p \wr G_m$ over k, p is invertible in k, and k contains the pth roots of unity.*
(iv) *G is an iterated wreath product $\mathbf{Z}/p \wr \cdots \wr \mathbf{Z}/p \wr A$ for a finite abelian group A of exponent e, viewed as an algebraic group over k. Also, p and e are invertible in k and k contains the pth and eth roots of unity.*

Proof The assumptions imply that BG can be approximated by smooth linear varieties over k in the sense of [140], by the proof of [138, lemma 9.1]. This implies the Chow Künneth formula for $BG \times X$ with X arbitrary, by the discussion after [138, lemma 6.1]. □

For example, let G be an elementary abelian p-group $(\mathbf{Z}/p)^n$, considered as an algebraic group over the complex numbers. Then Lemma 2.12 implies that $CH^*(BG)/p \cong \mathbf{F}_p[y_1, \ldots, y_n]$ with $|y_i| = 1$. Notice that this is simpler than the cohomology of BG. For example, for p odd, $H^*(BG, \mathbf{F}_p)$ is the free graded-commutative algebra

$$\mathbf{F}_p\langle x_1, \ldots, x_n, y_1, \ldots, y_n \rangle,$$

where $|x_i| = 1$ and $|y_i| = 2$. In this case, the ring homomorphism $CH^*(BG)/p \to H^*(BG, \mathbf{F}_p)$ (which takes CH^i to H^{2i}) is injective, and the image is the polynomial subring $\mathbf{F}_p[y_1, \ldots, y_n]$ of the cohomology ring.

Another way in which the Chow ring is simpler is that the Chow ring of $(\mathbf{Z}/p)^n$ is generated by Chern classes of complex representations, whereas that fails for the cohomology ring (whether we use $\mathbf{Z}$ or $\mathbf{F}_p$ coefficients). The even-degree subring $H^{\text{ev}}(B(\mathbf{Z}/p)^n, \mathbf{Z})$ is generated by Chern classes for $n \leq 2$, but not for $n \geq 3$. Group cohomology with $\mathbf{F}_2$ coefficients has a tighter relation to representation theory provided by Stiefel-Whitney classes of real representations. Indeed, the $\mathbf{F}_2$-cohomology ring of any finite abelian group is generated by Stiefel-Whitney classes; this has some analogy with the relation between the Chow ring and Chern classes. To avoid undue optimism, note that Gunawardena-Kahn-Thomas exhibited a group G of order 2^5 such that $H^2(BG, \mathbf{F}_2)$ is not generated by Stiefel-Whitney classes or even by transfers of Stiefel-Whitney classes [66, pp. 337–338].

Here is a fundamental calculation. Let $GL(n)$ denote the general linear group over any field k.

Theorem 2.13 *The Chow ring of $BGL(n)$ is isomorphic to $\mathbf{Z}[c_1, \ldots, c_n]$, with $|c_i| = i$. The generators are called the Chern classes.*

Proof There is a natural rank-n vector bundle V on $BGL(n)$ (that is, on the finite-dimensional smooth varieties $U/GL(n)$ that approximate $BGL(n)$), by the equivalence between principal $GL(n)$-bundles and vector bundles. Taking the Chern classes of V gives a ring homomorphism $\mathbf{Z}[c_1, \ldots, c_n] \to CH^*BGL(n)$, which we want to show is an isomorphism.

For any vector bundle E of rank n on a smooth scheme X over k, let $\text{Fl}(E) \to X$ denote the bundle of flags $0 \subset E_1 \subset E_2 \cdots \subset E_n = E$ in E, where E_i is a linear subspace of dimension i for all i. We can view this flag bundle as an iterated projective bundle. (Explicitly, $\text{Fl}(E)$ is a bundle over the projective bundle $\pi: \mathbf{P}(E) \to X$ with fiber $\text{Fl}(E_{n-1})$, where E_{n-1} is the kernel of the surjection $\pi^*E \to O(1)$.) By induction on the rank of E, the projective bundle formula (Lemma 2.3) gives that

$$CH^*\text{Fl}(E) \cong CH^*X[y_1, \ldots, y_n]/(e_i(y_1, \ldots, y_n) = c_i E),$$

where y_i is the first Chern class of the bundle E_i/E_{i-1} on $\text{Fl}(E)$, and e_i denotes the ith elementary symmetric function, for $i = 1, \ldots, n$.

Let B be the group of upper-triangular matrices in $GL(n)$. Then $GL(n)/B$ is the variety of flags in the vector space A^n over k. We have a fibration

$$GL(n)/B \to BB \to BGL(n)$$

that makes BB the flag bundle $\text{Fl}(V)$ over $BGL(n)$. (To be rigorous, U/B is the flag bundle $\text{Fl}(V)$ over $U/GL(n)$, for the finite-dimensional approximations

$U/GL(n)$ of $BGL(n)$.) It follows that

$$CH^*BB \cong CH^*BGL(n)[y_1, \ldots, y_n]/(e_i(y_1, \ldots, y_n) = c_i V).$$

On the other hand, the Borel subgroup B of $GL(n)$ is the semidirect product $(G_m)^n \ltimes N$ where N is the group of strictly upper-triangular matrices. The unipotent group N is an iterated extension of copies of the additive group G_a. Using homotopy invariance of Chow groups (Lemma 2.2), one deduces that CH^*BB is isomorphic to $CH^*B(G_m)^n = \mathbf{Z}[y_1, \ldots, y_n]$.

The homomorphism $\mathbf{Z}[c_1, \ldots, c_n] \to BGL(n)$ gives a homomorphism φ from

$$\mathbf{Z}[c_1, \ldots, c_n, y_1, \ldots, y_n]/(e_i(y_1, \ldots, y_n) = c_i)$$

to

$$CH^*BGL(n)[y_1, \ldots, y_n]/(e_i(y_1, \ldots, y_n) = c_i V) \cong CH^*BB.$$

The composition of φ with the isomorphism of the latter ring to $\mathbf{Z}[y_1, \ldots, y_n]$ is clearly an isomorphism; so φ is an isomorphism. But φ is the direct sum of $n!$ copies of the homomorphism $\mathbf{Z}[c_1, \ldots, c_n] \to CH^*BGL(n)$ (indexed by the monomials $y_1^{a_1} \cdots y_n^{a_n}$ with $0 \leq a_i \leq i-1$, for example). It follows that $\mathbf{Z}[c_1, \ldots, c_n] \to CH^*BGL(n)$ is an isomorphism. □

Finally, the following result shows that Chow rings of classifying spaces are easy to compute after tensoring with the rationals. So we will be concerned mostly with integral or mod p calculations.

Theorem 2.14 *For any affine algebraic group G over $\mathbf{C}$, the natural map*

$$CH^*BG \otimes \mathbf{Q} \to H^*(BG, \mathbf{Q})$$

is an isomorphism.

Proof We use the notion of a unipotent group scheme from Section 2.7. By definition, a smooth connected affine group G over a field k is *reductive* if every smooth connected unipotent subgroup of $G_{\overline{k}}$ is trivial. Over a perfect field k, every smooth connected affine k-group G has a unique maximal normal unipotent subgroup U, called the *unipotent radical* of G, and the quotient group G/U is reductive [18, section 11.21].

Let k be the complex numbers. The theorem was proved by Edidin and Graham for complex reductive groups, as part of a more general statement about equivariant Chow groups [38, proposition 6]. That implies the same statement for any connected group G, using homotopy invariance of Chow rings (Lemma 2.2) to show that G has the same equivariant Chow ring as the quotient of G by its unipotent radical. The result follows for arbitrary complex algebraic groups by reducing to the connected case using transfers as in Section 2.5. (That is, one shows that $CH^*(BG) \otimes \mathbf{Q} = (CH^*(BG_0) \otimes \mathbf{Q})^{G/G_0}$, where G_0 denotes the identity component of G, and similarly for rational cohomology.)

In more detail, the surjectivity of $CH^*BG \to H^*(BG, \mathbf{Q})$ for any complex algebraic group G follows from the fact that $H^*(BG, \mathbf{Q})$ is generated by Chern classes of complex representations of G, for every complex algebraic group G [87, proof of theorem 1]. The injectivity, in the main case of a complex reductive group G, follows from analyzing the fibration $BB \to BG$ with fiber the flag manifold G/B, where B is a Borel subgroup of G. This argument also gives more precise results about the relation between CH^*BG and $H^*(BG, \mathbf{Z})$ for G reductive in terms of the "torsion index" of G [139, theorem 1.3]. (We define the torsion index in section 16.1.) □

2.5 Transfer

For H a closed subgroup scheme of finite index in an affine k-group scheme G, there is an abelian group homomorphism $\mathrm{tr}_H^G \colon CH^iBH \to CH^iBG$ called *transfer*. (Indeed, for U/G a finite-dimensional approximation to BG, the morphism $U/H \to U/G$ is finite since H has finite index in G, and transfer is proper pushforward on Chow groups.) There is also a transfer homomorphism on cohomology, $\mathrm{tr}_H^G \colon H^i(BH, R) \to H^i(BG, R)$, when H is a closed subgroup of finite index in a compact Lie group G.

We list some of the formal properties of the transfer on Chow groups. According to the additive properties (in particular the double coset formula (iii)), for a finite group G viewed as an algebraic group over a field k, the assignment $H \mapsto CH^*BH$ for subgroups H of G is a *Mackey functor*. According to the multiplicative properties (in particular, that CH^*BH is a commutative ring for each subgroup H of G, together with the projection formula (i)), $H \mapsto CH^*BH$ is a *Green functor* [148]. Group cohomology $H \mapsto H^{\mathrm{ev}}(BH, R)$ for subgroups H of G is also a Green functor [24, proposition III.9.5]. For a subgroup $H \subset G$, write res_H^G for the restriction map $CH^*BG \to CH^*BH$. For a subgroup $H \subset G$ and an element $g \in G$, write $x \mapsto gx$ for the isomorphism $CH^*BH \to CH^*B(gHg^{-1})$ given by conjugation by g.

Lemma 2.15

(i) *(Projection formula) Let H be a closed subgroup scheme of finite index in an affine k-group scheme G. Then the transfer $\mathrm{tr}_H^G \colon CH^*BH \to CH^*BG$ is a homomorphism of CH^*BG-modules, where CH^*BH is viewed as a CH^*BG-module by pullback.*

(ii) *We have $\mathrm{tr}_H^G(1) = [G\colon H]$. (For general group schemes, the index $[G\colon H]$ means the dimension of the k-vector space of regular functions on G/H.)*

(iii) *(Double coset formula) Let K and H be subgroups of a finite group G, viewed as an algebraic group over a field k. Then*

$$\mathrm{res}^G_K \mathrm{tr}^G_H x = \sum_{g \in K\backslash G/H} \mathrm{tr}^K_{K\cap gHg^{-1}} \mathrm{res}^{gHg^{-1}}_{K\cap gHg^{-1}} gx$$

for x in CH^*BH.

(iv) *Let H be a normal subgroup of a finite group G, viewed as an algebraic group over a field k. Then*

$$\mathrm{res}^G_H \mathrm{tr}^G_H x = \sum_{g \in G/H} gx$$

for x in CH^*BH.

Proof These follow from the properties of proper pushforward on Chow groups of smooth varieties, since each Chow group $CH^i BG$ is defined as $CH^i(V - S)/G$ for a suitable smooth variety $(V - S)/G$. In more detail, (i) is proved in [43, example 8.1.7]. To prove (ii), use that the pushforward map on Chow groups for finite morphisms commutes with flat pullback [43, proposition 1.7]. Thus, to compute the degree of the finite map $U/H \to U/G$, it suffices to compute the degree of the pulled back map $U \times G/H \to U$, which is easy by definition of the algebraic cycle associated to a subscheme [43, lemma 1.7.1, section 1.5].

We use the following lemma, a special case of [43, proposition 6.6(c)].

Lemma 2.16 *Let $Y \to Z$ be a finite etale morphism of smooth schemes over a field k, and let $X \to Z$ be any morphism of smooth schemes over k. Consider the fiber product*

$$\begin{array}{ccc} X \times_Z Y & \longrightarrow & Y \\ \downarrow & & \downarrow \\ X & \longrightarrow & Z \end{array}$$

Then pushforward commutes with pullback, as homomorphisms $CH^i Y \to CH^i X$.

Lemma 2.16 gives (iii) by considering the pullback of the finite morphism $BH \to BG$ along the morphism $BK \to BG$. (iv) is a special case of (iii). ☐

Using these properties, the basic applications of the transfer in group cohomology also work for Chow rings. For example, from properties (i) and (ii), the transfer and restriction maps satisfy $\mathrm{tr}^G_H(x|_H) = [G\colon H]x$ for any $x \in CH^*BG$. Applying this to the trivial group $H = 1 \subset G$, we deduce that the abelian group $CH^i BG$ is killed by $|G|$ for any finite k-group scheme G and any $i > 0$.

Also, let G be a finite group, p a prime number, and consider the Chow ring modulo p, $CH_G^* = CH^*(BG)/p$. For P a Sylow p-subgroup of G, we have $\mathrm{tr}_P^G \mathrm{res}_P^G(x) = [G\colon P]x$ in CH_G^*, where $[G\colon P]$ is a unit in $\mathbf{F}_p$. It follows that CH_G^* is a summand of CH_P^*. By the same argument, H_G^* is a summand of H_P^*. For this reason, many questions about Chow rings or cohomology of finite groups can be reduced to questions about p-groups.

Cartan and Eilenberg gave an explicit description of H_G^* as the subring of "stable elements" in H_P^* [24, theorem III.10.3]. The same statement holds for any *cohomological* Mackey functor (meaning one that satisfies property (ii) above) taking values in $\mathbf{Z}_{(p)}$-modules, such as mod p Chow rings [148, corollary 3.7, proposition 7.2]. Namely, an element x in CH_P^i is in the image of CH_G^i if and only if for every subgroup H of P and every element $g \in G$ such that gHg^{-1} is contained in P, $g(x|_H) = x|_{gHg^{-1}}$.

2.6 Becker-Gottlieb transfer for Chow groups

Becker and Gottlieb defined a transfer map on cohomology for any closed subgroup of a compact Lie group, not necessarily of finite index [94]. I extended Becker-Gottlieb transfer to Chow groups, and it was written up by Vezzosi [142, theorem 2.1]. We present the construction in this section. More generally, Becker-Gottlieb transfer can be viewed as a stable map in Morel-Voevodsky's A^1-homotopy category [107]. Using that machinery, it may be possible to generalize Theorem 2.17 to fields of any characteristic. (The proof as written requires a smooth G-equivariant compactification of $G/N(T)$ by a divisor with simple normal crossings, for G a reductive group over a field and T a maximal torus in G. No such compactification seems to be known explicitly, even for $G = GL(n)$.)

By definition, a *torus* over a field k is a k-group scheme that becomes isomorphic to $(G_m)^n$ over the algebraic closure of k, for some natural number n. A torus is *split* if it is isomorphic to $(G_m)^n$ over k.

Theorem 2.17 *Let G be a smooth affine group scheme over a field k of characteristic zero such that the identity component G^0 is reductive. Let T be a maximal torus in G and $N(T)$ its normalizer in G. Then the restriction map*

$$CH^*BG \to CH^*BN(T)$$

is split injective, as a map of CH^*BG-modules.

Proof Let X be a smooth projective variety over a field k, and let $D = \cup_{i=1}^r D_i$ be a divisor with simple normal crossings on X. The vector bundle $\Omega^1(\log D)$ on X of *1-forms with log poles along D* can be defined as the sheaf of rational 1-forms α such that both α and $d\alpha$ have at most simple poles along D. In etale local coordinates where $D = \{x_1 \cdots x_m = 0\} \subset A^n$, $\Omega^1_X(\log D)$ has a basis of sections consisting of $dx_1/x_1, \ldots, dx_m/x_m, dx_{m+1}, \ldots, dx_n$. We refer to [40, chapter 2] for the various exact sequences involving $\Omega^1(\log D)$. In particular, for each component D_i of D, we have the residue exact sequence

$$0 \to \Omega^1_{D_i}(\log D_i \cap (\cup_{j \neq i} D_j)) \to \Omega^1_X(\log D)|_{D_i} \to O_{D_i} \to 0.$$

The *logarithmic tangent bundle* $TX(-\log D)$ is defined as the dual bundle $(\Omega^1(\log D))^*$ on X. Equivalently, $TX(-\log D)$ is the sheaf of vector fields on X that are tangent to D on the smooth locus of D.

Since we are considering varieties over a field k of characteristic zero, we can define the Euler characteristic of a variety by reducing to the case where k is a subfield of $\mathbf{C}$ and using ordinary cohomology (say, with compact support and rational coefficients): $\chi(X) = \sum(-1)^i \dim_{\mathbf{Q}} H^i_c(X, \mathbf{Q})$. One could also define the Euler characteristic using l-adic cohomology. For a smooth proper n-fold over k, the Euler characteristic $\chi(X)$ is equal to the degree of the top Chern class of the tangent bundle $c_n(TX) \in CH^n X$ [43, example 8.1.12], [104, theorem 12.3]. Using the exact sequences for the logarithmic tangent bundle and the additivity properties of the Euler characteristic, it follows that for any smooth proper variety X with a simple normal crossing divisor D, $\chi(X - D) = \int_X c_n(TX(-\log D))$.

Now let $g\colon U \to B$ be a smooth morphism between smooth k-schemes that admits a smooth relative compactification $f\colon X \to B$. That is, U is open in X, f is a smooth proper morphism, $X - U$ is a divisor with simple normal crossings $\cup_i D_i$, and all intersections $D_I = \cap_{i \in I} D_i$ are smooth over B. Let n be the dimension of the fibers. Define a modified pushforward $g_\sharp\colon CH^j U \to CH^j B$, *Becker-Gottlieb transfer on Chow groups*, by

$$g_\sharp(x) = f_*(\widetilde{x} c_n(T_{X/B}(-\log D)))$$

for any lift $\widetilde{x}$ of x to $CH^j X$. Here the relative logarithmic tangent bundle $T_{X/B}(-\log D)$ is a vector bundle of rank n on X. By the basic exact sequence of Chow groups (Lemma 2.1), to show that $g_\sharp$ is well-defined (independent of the lift $\widetilde{x}$), it suffices to show that the formula gives zero for the pushforward to X of a cycle on D_i for some i. That holds because $c_n(T_{X/B}(-\log D))$ restricts to zero in $CH^n D_i$. Indeed, the rank-n vector bundle $T_{X/B}(-\log D)$ restricted to D_i contains a trivial line sub-bundle, by the dual of the residue exact sequence for $\Omega^1_{X/B}(\log D)$.

The transfer $g_\sharp\colon CH^* U \to CH^* B$ is $CH^* B$-linear. Write F for any fiber of g. Then we have

$$g_\sharp \circ g^* = \chi(F)$$

by the projection formula, since $c_n(T_{X/B}(-\log D))$ restricts on each fiber of $X \to B$ to a zero-cycle of degree $\chi(F)$. A priori, the homomorphism $g_\sharp$ may depend on the compactification X of U, but that does not matter for our purpose.

Let G be a smooth affine group scheme over a field k such that the identity component G^0 is reductive. Let T be a maximal k-torus in G. Grothendieck showed that T remains a maximal torus over the algebraic closure $\overline{k}$ [18, theorem 18.2]. Also, all maximal tori in $G^0_{\overline{k}}$ are conjugate by elements of $G^0(\overline{k})$ [18, corollary 11.3]. It follows that the normalizer $N(T)$ in G meets every connected component of $G_{\overline{k}}$. (Indeed, given a point x in $G(\overline{k})$, xTx^{-1} is a maximal torus in G^0, and so there is an element g in $G^0(\overline{k})$ with $gxTx^{-1}g^{-1} = T$. Then gx is a $\overline{k}$-point of $N(T)$ in the connected component of x.) Therefore, the morphism $G^0/N_{G^0}(T) \to G/N_G(T)$ is an isomorphism.

For a reductive group H (which is connected, by definition) with maximal torus T over a field k, the scheme $X = H/N_H(T)$ has Euler characteristic 1. Indeed, this is a geometric statement, meaning that we can replace k by an extension field. So we can assume that T is contained in a Borel subgroup B of H. The Bruhat decomposition expresses H/B as the disjoint union of cells (affine spaces) indexed by the Weyl group $W = N(T)/T$ [18, theorem 14.12]. So $\chi(H/B) = |W|$. By homotopy invariance of ordinary cohomology, it follows that $\chi(H/T) = |W|$. The Euler characteristic is multiplicative under finite etale morphisms since k has characteristic zero, and so $\chi(H/N(T)) = \chi(H/T)/|W| = 1$. (Alternatively, one can show directly that $H/N(T)$ has the $\mathbf{Q}_l$-cohomology of a point; that works for k of any characteristic.)

Two paragraphs back, we showed that $G^0/N_{G^0}(T) = G/N_G(T)$. By the previous paragraph, it follows that $\chi(G/N_G(T)) = 1$.

We apply the transfer map $g_\sharp$ to the fibration $G/N(T) \to BN(T) \to BG$. To be precise, consider the fibration $G/N(T) \to U/N(T) \to U/G$ over a finite-dimensional approximation U/G to BG. Since k has characteristic zero, the smooth variety $F = G/N(T)$ has a smooth G-equivariant compactification $\overline{F}$ with complement a divisor with simple normal crossings, by equivariant resolution of singularities [79, proposition 3.9.1]. Let $\overline{X}$ be the $\overline{F}$-bundle over U/G associated to the G-action on $\overline{F}$. Then the beginning of this proof gives a homomorphism $g_\sharp\colon CH^*BN(T) \to CH^*BG$ such that $g_\sharp \circ g^* = 1$, using that $\chi(G/N(T)) = 1$. So CH^*BG is a summand of $CH^*BN(T)$ as a CH^*BG-module. □

2.7 Groups in characteristic p

Morel and Voevodsky observed that for G a finite etale group scheme of order a power of p over a field k of characteristic p, the classifying space BG is A^1-homotopy equivalent to Spec k [107, proposition 3.3]. (By definition, G is etale over k if it is smooth of dimension zero.) We now prove a slight generalization. We make no further use of this, but it justifies concentrating on p-groups over fields of characteristic not p in the rest of the book. It might be interesting to explore the Chow ring and motivic cohomology for finite connected group schemes over a field k of characteristic p that are not unipotent, such as the kernel of the Frobenius homomorphism in $GL(n)$ or in other reductive groups. We have already seen that the Chow ring of $B\mu_p$ over k is nontrivial (Theorem 2.10).

Lemma 2.18 *Let G be a unipotent group scheme over a field k. Then BG is A^1-homotopy equivalent to* Spec k. *In particular, $CH^i BG = 0$ for $i > 0$.*

By definition, a group scheme over a field k is *unipotent* if it is isomorphic to a closed subgroup scheme of the group of strictly upper triangular matrices in $GL(n)$ over k for some n. Every finite etale group scheme of order a power of p over a field of characteristic p is unipotent [32, définition XVII.1.3, théorème XVII.3.5(ii)].

Proof Embed G as a closed subgroup scheme of the group U of strictly upper triangular matrices in $GL(n)$ for some n. Then U is a *split* unipotent group over k, meaning that it is an iterated extension of copies of the additive group G_a over k. (Every smooth connected unipotent group over a perfect field is split, although we don't need that fact [18, theorem 15.4].)

Let G be embedded in a split unipotent group U over k. I claim that we can identify the classifying space BG with the finite-dimensional variety U/G. (For our purpose, that just means that $CH^*BG \cong CH^*(U/G)$, although in fact the argument shows that BG and U/G are isomorphic in Morel-Voevodsky's A^1-homotopy category.) Namely, for any approximation $(V-S)/G$ to BG as in the definition, consider the variety $(U \times (V-S))/G$, which fibers over both U/G and $(V-S)/G$. The fibration over U/G is a vector bundle minus a subset of codimension equal to the codimension of S in V, which we take to be large. The fibration over $(V-S)/G$ is a principal U-bundle. Such a bundle is Zariski locally trivial since U is split unipotent (by reducing to the case of the additive group), and the fiber U is isomorphic to affine space as a variety. It follows that $(U \times (V-S))/G \to (V-S)/G$ is an A^1-homotopy equivalence, and in particular induces an isomorphism on Chow groups. So CH^*BG is isomorphic to $CH^*(U/G)$.

Let G be any closed subgroup scheme of a split unipotent group U over k. Then Rosenlicht showed that the variety U/G is isomorphic to affine space of some dimension over k [116, theorem 5]. The proof is by induction on the dimension of U. The statement is clear if $U = 1$. Otherwise, U contains a central subgroup Z isomorphic to G_a over k. Then the morphism

$$U/G \to (U/Z)/(G/(G \cap Z))$$

is a principal $Z/(G \cap Z)$-bundle. By induction on the dimension of U, $(U/Z)/(G/(G \cap Z))$ is isomorphic to affine space over k. Any quotient group scheme of $Z = G_a$ is either trivial or isomorphic to G_a [18, theorem 15.4]. Principal G_a-bundles on affine space A^n over k are classified by the group $H^1(A^n, O) = 0$, and so they are trivial. So U/G is isomorphic to affine space over k. We deduce that BG is A^1-homotopy equivalent to Spec k. In particular, $CH^i BG = 0$ for $i > 0$. □

2.8 Wreath products and the symmetric groups

For the symmetric groups, we have essentially complete information on the cohomology ring and Chow ring. The results rely on the special structure of the Sylow p-subgroups of the symmetric groups; they are products of iterated wreath products of cyclic groups. In this section, we state the basic results on the cohomology and Chow ring of a wreath product (Definition 2.11), referring to [138] for the proofs.

Nakaoka and Quillen gave an explicit description of the cohomology of a wreath product group $\mathbf{Z}/p \wr H = \mathbf{Z}/p \ltimes H^p$ [1, theorems IV.1.7 and IV.4.3]. For our purposes, the most important fact is the following result of Quillen's [111, proposition 3.1]:

Theorem 2.19 *Let H be any group, and let $G = \mathbf{Z}/p \wr H$. Then the restriction map on $\mathbf{F}_p$-cohomology,*

$$H^*_{\mathbf{Z}/p \wr H} \to H^*_{H^p} \times H^*_{\mathbf{Z}/p \times H},$$

is injective.

By definition, the $\mathbf{F}_p$-cohomology ring of a group G is *detected* on subgroups $H_1, \ldots, H_r$ if the restriction homomorphism $H^*_G \to \prod_{i=1}^r H^*_{H_i}$ is injective. Likewise for mod p Chow rings.

Corollary 2.20 *Let p be a prime number and n a positive integer. Then the $\mathbf{F}_p$-cohomology of the symmetric group S_n is detected on elementary abelian p-subgroups.*

Proof By transfer, it suffices to prove this after replacing S_n by a Sylow p-subgroup. For $n = a_j p^j + \cdots + a_1 p + a_0$ with $0 \leq a_j \leq p - 1$, a Sylow p-subgroup of S_n is the product over all j of a_j copies of a Sylow p-subgroup of S_{p^j}, which in turn is the j-fold wreath product $\mathbf{Z}/p \wr \cdots \wr \mathbf{Z}/p$. Then the statement follows by induction from Theorem 2.19. □

We now turn to Chow rings. The Chow ring of a wreath product $\mathbf{Z}/p \wr H$ is even simpler than the cohomology ring, assuming certain good properties of

H. In particular, we have a simple description of generators for the Chow ring, as well as a detection theorem analogous to Theorem 2.19.

For an algebraic group G over a field k, a *transferred Euler class* in CH_G^* is an $\mathbf{F}_p$-linear combination of transfers of Euler classes (top Chern classes) of representations of finite-index subgroups over k.

In the following lemma, for p invertible in k, we consider the cycle map from the mod p Chow group CH_G^i to etale cohomology $H_{\text{et}}^{2i}(BG_{k_s}, \mathbf{F}_p(i))$, where k_s denotes a separable closure of k. For $k = \mathbf{C}$, this etale cohomology group coincides with the usual cohomology of BG. In fact, a choice of pth root of unity in k_s always determines an identification of these etale cohomology groups with the group cohomology of G; see Lemma 8.3.

Lemma 2.21 *Let p be a prime number, and let G be a group scheme over a field k that satisfies one of the following assumptions. Then the mod p Chow ring CH_G^* is detected on elementary abelian subgroups. Also, $CH_G^i \to H_{\text{et}}^{2i}(BG_{k_s}, \mathbf{F}_p(i))$ is injective for all i. Finally, CH_G^* consists of transferred Euler classes.*

(i) *G is the multiplicative group G_m.*
(ii) *G is a finite abelian group of exponent e viewed as an algebraic group over k, e is invertible in k, and k contains the eth roots of unity.*
(iii) *G is an iterated wreath product $\mathbf{Z}/p \wr \cdots \wr \mathbf{Z}/p \wr G_m$ over k, p is invertible in k, and k contains the pth roots of unity.*
(iv) *G is an iterated wreath product $\mathbf{Z}/p \wr \cdots \wr \mathbf{Z}/p \wr A$ for a finite abelian group A of exponent e, viewed as an algebraic group over k. Also, p and e are invertible in k and k contains the pth and eth roots of unity.*

Proof Let k be a field of characteristic not p that contains the pth roots of unity. Let H be an affine group scheme over k such that BH can be approximated by smooth quasi-projective varieties that can be cut into open subsets of affine spaces over k, and $CH_H^i \to H_{\text{et}}^{2i}(BH, \mathbf{F}_p(i))$ is injective for all i. Then [138, section 9] shows that $\mathbf{Z}/p \wr H$ satisfies the same assumptions. That paper also gives an explicit additive basis for $CH^*_{\mathbf{Z}/p\wr H}$ in terms of CH_H^*. In particular, the description shows that if CH_H^* is generated by transferred Euler classes, then so is $CH^*_{\mathbf{Z}/p\wr H}$ [138, section 11]. This gives the results we want on generation of the Chow ring by transferred Euler classes, and injectivity of the mod p cycle map. The description also shows that $CH^*_{\mathbf{Z}/p\wr H}$ maps onto the invariants $(CH^*_{H^p})^{\mathbf{Z}/p}$.

Under the same assumptions on H, [138, section 9] also shows that $CH^*_{\mathbf{Z}/p\wr H}$ is detected on the subgroups H^p and $\mathbf{Z}/p \times H$. Combined with Lemma 2.12 on the Chow Künneth formula, it follows by induction that CH_G^* is detected on elementary abelian subgroups for the groups in the theorem. □

Corollary 2.22 *Let n be a positive integer, p a prime number, and k a field. Suppose that p is invertible in k and that k contains the pth roots of unity. Let G be either the symmetric group S_n or the wreath product $S_n \wr G_m$, viewed as a group scheme over k. Then the mod p Chow ring CH^*_G consists of transferred Euler classes, and is detected on elementary abelian p-subgroups.*

Proof By transfer, it suffices to prove this after replacing S_n by a Sylow p-subgroup. For $n = a_j p^j + \cdots + a_1 p + a_0$ with $0 \leq a_j \leq p-1$, a Sylow p-subgroup of S_n is the product over all j of a_j copies of a Sylow p-subgroup of S_{p^j}, which in turn is the j-fold wreath product $\mathbf{Z}/p \wr \cdots \wr \mathbf{Z}/p$. Then the statement follows by induction from Lemma 2.21 (applied to the group $\mathbf{Z}/p \wr \cdots \wr \mathbf{Z}/p$ or $\mathbf{Z}/p \wr \cdots \wr \mathbf{Z}/p \wr G_m$) together with Lemma 2.12 (on the Chow Künneth formula). □

2.9 General linear groups over finite fields

This section gives the calculations by Quillen and Guillot of the cohomology and Chow ring of the general linear group over finite fields. These are important calculations, but we do not use them elsewhere in the book. Adem and Milgram summarize what is known about the cohomology of other finite groups of Lie type [1, chapter VII].

Let q be a power of a prime number p, and let l be a prime number different from p. Then Quillen computed the mod l cohomology ring of the finite group $GL(n, \mathbf{F}_q)$, as follows [115, theorem 4, remark after theorem 1]. See also Benson [12, vol. 2, theorem 2.9.3] for a summary of Quillen's argument. This result led to Quillen's calculation of the algebraic K-theory of finite fields. It turns out to be a simple example; for example, the mod l cohomology of $GL(n, \mathbf{F}_q)$ for $l \neq p$ is detected on abelian subgroups, which is far from true for finite groups in general. The mod p cohomology of $GL(n, \mathbf{F}_q)$ is far more complicated and is largely unknown.

The main reason for the simplicity of the mod l cohomology of $GL(n, \mathbf{F}_q)$ is that an l-Sylow subgroup of $GL(n, \mathbf{F}_q)$ with $l \neq p$ is a product of iterated wreath products of abelian groups when $l \neq 2$ (and has a similar description when $l = 2$). Compare Theorem 2.19.

Theorem 2.23 *Let q be a power of a prime number p. Let l be a prime number different from p, r the multiplicative order of q modulo l, and $m = \lfloor n/r \rfloor$. For l odd, the $\mathbf{F}_l$-cohomology ring of $GL(n, \mathbf{F}_q)$ is the free graded-commutative algebra*

$$\mathbf{F}_l\langle e_r, e_{2r}, \ldots e_{mr}, c_r, c_{2r}, \ldots, c_{mr}\rangle,$$

where $|e_{ir}| = 2ir - 1$ *and* $|c_{ir}| = 2ir$. *If* $l = 2$ *and* $q \equiv 1 \pmod 4$, *then*

$$H^*_{GL(n,\mathbf{F}_q)} \cong \mathbf{F}_2[e_1, e_2, \ldots e_n, c_1, c_2, \ldots, c_n]/(e_i^2 = 0),$$

where $|e_i| = 2i - 1$ *and* $|c_i| = 2i$. *If* $l = 2$ *and* $q \equiv 3 \pmod 4$, *then*

$$H^*_{GL(n,\mathbf{F}_q)} \cong \mathbf{F}_2[e_1, e_2, \ldots e_n, c_1, c_2, \ldots, c_n]/\left(e_i^2 = \sum_{a=0}^{i-1} c_a c_{2i-1-a}\right),$$

where $c_0 = 1$ *and* $c_i = 0$ *for* $i > n$.

The standard representation of $GL(n, \mathbf{F}_q)$ on $(\mathbf{F}_q)^n$ has a natural lift to a virtual complex representation of $GL(n, \mathbf{F}_q)$, called the Brauer lift ρ; see Serre [124, theorem 43] or Benson [12, vol. 1, section 5.9]. The classes c_i in Theorem 2.23 are the Chern classes of ρ.

Guillot computed the mod l Chow ring of the finite group $GL(n, \mathbf{F}_q)$, viewed as an algebraic group over $\mathbf{C}$, as follows [60, theorem 4.7]. Again, this turns out to be a simple example: the Chow ring injects into the cohomology ring, and is detected on elementary abelian subgroups.

Theorem 2.24 *Let* q *be a power of a prime number* p. *Let* l *be an odd prime number different from* p, r *the multiplicative order of* q *modulo* l, *and* $m = \lfloor n/r \rfloor$. *Let* c_i *be the* i*th Chern class of the Brauer lift* ρ *of the standard representation of* $GL(n, \mathbf{F}_q)$. *Then the mod* l *Chow ring of* $GL(n, \mathbf{F}_q)$ *is the polynomial ring*

$$\mathbf{F}_l[c_r, c_{2r}, \ldots, c_{mr}].$$

2.10 Questions about the Chow ring of a finite group

In contrast to cohomology, there is no algorithm to compute Chow groups, and they can be big. In particular, Schoen gave examples of smooth projective 3-folds X over $\overline{\mathbf{Q}}$ and prime numbers p such that $CH^2(X)/p \cong CH^2(X_{\mathbf{C}})/p$ is infinite [118]. There are other varieties for which the subgroup of $CH^3(X)$ killed by p is infinite [119]. It is an open question whether the Chow groups $CH^i BG$ of the classifying space of an algebraic group are finitely generated abelian groups. For a finite group G, which we generally view as an algebraic group over the complex numbers, $CH^i BG$ is killed by $|G|$ for $i > 0$, and so the question is whether $CH^i BG$ is finite for $i > 0$; that is true in all the known computations. The results of this book shed some light: we can reduce to checking finiteness of $CH^i BG$ for small values of i, and in some cases that is enough to solve the problem (Corollary 10.5 and Theorem 11.1).

For the examples of groups G over $\mathbf{C}$ we have seen in this chapter, the homomorphism from the Chow ring of BG to the cohomology ring was injective, but that fails in general. In particular, for every prime number p, there is a p-group G such that $CH^2(BG)/p \to H^4(BG, \mathbf{F}_p)$ is not injective. (Then the product $H = G \times \mathbf{Z}/p$ has $CH^3BH \to H^6(BH, \mathbf{Z})$ not injective.) An example is the extraspecial 2-group $G = 2_+^{1+4}$, as shown in [137, section 5] (which we summarize in the proof of Theorem 15.13). We give more complicated examples at odd primes in Theorem 15.7.

A major tool in the cohomology theory of finite groups is the Hochschild-Serre spectral sequence $H^*(Q, H^*(N, \mathbf{F}_p)) \Rightarrow H^*(G, \mathbf{F}_p)$ for a group extension $1 \to N \to G \to Q \to 1$. No analog is known for Chow rings. Indeed, there seems to be no way to define the Chow groups of a group G with coefficients in a nontrivial G-module. Rost found what seems to be the right notion of Chow groups with coefficients, but the possible coefficients are more complicated objects known as cycle modules [117, 63].

Although the fibration $BN \to BG \to BQ$ makes sense in algebraic geometry (using finite-dimensional approximations), almost nothing is known about the Chow groups of a fibration in general. The difficulty is that the fibration is locally trivial in the etale topology, but not in the Zariski topology. More broadly, the whole problem of computing Chow rings of classifying spaces can be considered as a problem of "etale descent" for Chow groups: given a finite group G acting freely on a scheme X, how are $CH^*(X)$ and $CH^*(X/G)$ related? With rational coefficients we have the simple answer that $CH^*(X/G) \otimes \mathbf{Q} = (CH^*(X) \otimes \mathbf{Q})^G$ (proved using transfer maps). But not much is known integrally, or modulo a prime number.

Later (Theorem 8.10) we will prove Yagita's theorem that the Chow ring of a finite group is qualitatively similar to the cohomology ring, in the following sense. For a fixed prime number p, write $CH_G^* = CH^*(BG)/p$ and $H_G^* = H^*(BG, \mathbf{F}_p)$. Then the cycle map $CH_G^* \to H_G^*$ is an F-isomorphism. That is: every element of the kernel of $CH_G^* \to H_G^*$ is nilpotent, and for every element x of H_G^*, there is an $r \geq 0$ such that x^{p^r} is in the image of the cycle map. It follows that the "variety" of CH_G^* is the same as that of H_G^*, meaning that the morphism $\operatorname{Spec} H_G^{\text{ev}} \to \operatorname{Spec} CH_G^*$ is a universal homeomorphism. (That is, it remains a homeomorphism after any extension of the base field $\mathbf{F}_p$.) This holds even though it is not known whether CH_G^* is finitely generated as an $\mathbf{F}_p$-algebra. Conceivably CH_G^* could contain an enormous square-zero ideal.

Examples suggest that the Chow ring of a group is more closely tied to representation theory than the cohomology ring is. The following result is some justification for that idea [138, corollary 3.2]. The geometric filtration of the representation ring $R(G)$ is defined in the proof.

Theorem 2.25 *Let G be an affine group scheme of finite type over a field k, p a prime number. Then the mod p Chow ring CH_G^* in degrees $\leq p$ is generated* *by Chern classes of representations of G over k. Also, the natural surjection $CH^iBG \to \mathrm{gr}^i_{\mathrm{geom}}R(G)$ is an isomorphism p-locally for $i \leq p$.*

Proof Thomason defined equivariant G-theory $G_i^G X$ of a G-scheme X over k as the Quillen K-theory of the abelian category of G-equivariant coherent sheaves on X, for $i \geq 0$ [136]. In particular, $G_i^G k$ is the K-theory of the abelian category of representations of G, and so $G_0^G k = R(G)$. For X smooth over k, the natural map $K_i^G X \to G_i^G X$ is an isomorphism, where $K_i^G X$ is the Quillen K-theory of the exact category of G-equivariant vector bundles on X.

Thomason proved homotopy invariance of equivariant G-theory, which means in particular that $G_i^G k \cong G_i^G V$ for every representation V of G. Moreover, for any closed G-invariant subset S of a G-scheme X, Thomason showed that every G-equivariant coherent sheaf on $X - S$ is the restriction of a G-equivariant coherent sheaf on V [136, corollary 2.4]. So we have a surjection

$$G_0^G X \twoheadrightarrow G_0^G(X - S).$$

For a smooth variety X over a field k, the *geometric filtration* of the algebraic K-group K_0X means the filtration by codimension of support [57]. That is, an element of K_0X belongs to $F^r_{\mathrm{geom}}K_0X$ if it restricts to zero in $K^0(X - S)$ for some closed subset S of codimension at least r. Equivalently, identifying K_0X with G_0X (since X is smooth over k), an element of K_0X belongs to $F^r_{\mathrm{geom}}K_0X$ if it can be represented by a coherent sheaf whose support has codimension at least r.

We define the geometric filtration of $R(G)$ as follows. For any natural number r, let V be a representation of G with a G-invariant closed subset S of codimension greater than r such that G acts freely on $V - S$ with quotient a scheme over k. Let $F^r_{\mathrm{geom}}R(G)$ be the subgroup of $R(G)$ of elements that restrict to zero in $K_0^G(V - S') = K_0(V - S')/G$ for some closed G-invariant closed subset $S \subset S' \subset V$ of codimension at least r in V. This is independent of the choices of V and S, by the same argument as for Theorem 2.5 showing that CH^*BG is well-defined. Moreover, by Thomason's surjection earlier in the proof, $R(G) = G_0^G V \to K_0(V - S)/G$ is surjective, and therefore the natural map

$$\mathrm{gr}^i_{\mathrm{geom}}R(G) \to \mathrm{gr}^i_{\mathrm{geom}}K_0(V - S)/G$$

is an isomorphism for $i \leq r$.

Let $X = (V - S)/G$; it remains to relate K_0X with the Chow groups of X. There is a natural map from CH^iX to $\mathrm{gr}^i_{\mathrm{geom}}K_0X$, taking a subvariety Z of codimension i to the class of the coherent sheaf O_Z. The map $CH^iX \to \mathrm{gr}^i_{\mathrm{geom}}K_0X$ is surjective, and the ith Chern class gives a map back such that the composition

$$CH^i X \to \mathrm{gr}^i_{\mathrm{geom}} K_0 X \underset{c_i}{\to} CH^i X$$

is multiplication by $(-1)^{i-1}(i-1)!$, by Riemann-Roch without denominators [43, example 15.3.1], [57]. It follows that the surjection from $CH^i X$ to $\mathrm{gr}^i_{\mathrm{geom}} K_0 X$ becomes an isomorphism after inverting $(i-1)!$, and the subgroup of $CH^i X$ generated by Chern classes of elements of $K_0 X$ contains $(i-1)! CH^i X$. This is what we want. □

In particular, CH^1 is easy to compute.

Lemma 2.26 *Let G be an affine group scheme of finite type over a field k. Then the first Chern class gives an isomorphism*

$$c_1 \colon \mathrm{Hom}_k(G, G_m) \to CH^1 BG.$$

Proof The homomorphism c_1 is surjective by Theorem 2.25, using that the first Chern class of any representation V of G is equal to $c_1(\det(V))$.

To show that c_1 is injective, let $\alpha \colon G \to G_m$ be a representation such that $c_1 \alpha = 0$ in $CH^1 BG$. Let V be a representation of G with a closed G-invariant subset S of codimension at least 2 such that G acts freely on $V - S$ with quotient a scheme. Then the pullback of α by $V \to \mathrm{Spec}(k)$ is a G-equivariant line bundle L_α on V. Since $(V-S)/G$ is smooth over k, we can identify $CH^1((V-S)/G)$ with the group $\mathrm{Pic}((V-S)/G)$ of isomorphism classes of line bundles on $(V-S)/G$ [67, corollary II.6.16]. So the restriction of L_α to $V - S$ is G-equivariantly trivial. That is, L_α has a nowhere-vanishing G-equivariant section s on $V - S$. Since S has codimension at least 2, s extends to a nowhere-vanishing section of L_α over V. This section is G-equivariant because its restriction to $V - S$ is G-equivariant. By restricting s to the origin in V, it follows that the homomorphism $\alpha \colon G \to G_m$ is trivial. □

There is also an explicit description of $CH^2 BG$ in terms of cohomology, Lemma 15.1.

3

Depth and Regularity

In this chapter, we define some fundamental concepts of commutative algebra: depth and Castelnuovo-Mumford regularity. The depth of a ring (say, a finitely generated commutative graded algebra) is the maximum length of a regular sequence of elements of positive degree. For some purposes, depth is a good measure of how well-behaved a ring is. The rings with maximal depth, known as Cohen-Macaulay rings, are the rings that are *free* finitely generated modules over some polynomial subring. Regularity is a quantitative measure of the complexity of a graded ring in terms of the degrees of generators, relations, and so on. It is not related to the notion of a regular local ring.

Depth and regularity have extremely good formal properties. That allowed Symonds to prove strong bounds for the degrees of generators of the cohomology ring of a finite group by studying the a priori harder problem of bounding the regularity, as we see in Chapter 4.

The chapter ends with Duflot's theorem, which gives a lower bound for the depth of the cohomology ring of a group, and an analog for the Chow ring.

3.1 Depth and regularity in terms of local cohomology

In this section, we define depth and Castelnuovo-Mumford regularity for modules over a graded ring. Our definitions are in terms of local cohomology, because that makes the formal properties of these invariants easy to prove. But the reason these invariants are important in the rest of the book is their interpretation in terms of generators and relations for a module, which we prove as Theorem 3.14. Our exposition partly follows Benson [13] and Symonds [131].

Something new in our treatment is that in order to deal with Chow rings, which are not known to be finitely generated algebras, we have to consider non-noetherian rings and modules that are not finitely generated. The Chow ring of the classifying space of an algebraic group is at least generated in a bounded set of degrees as a module over a graded polynomial ring on finitely many generators (Theorem 5.2).

Let k be a field and let $R = k \oplus R_1 \oplus R_2 \oplus \cdots$ be a commutative graded k-algebra. Let $\mathfrak{m}$ be the maximal ideal $R^{>0}$ in R. For the rest of this section, we define an R-module to be a graded R-module $M = \oplus_{l \in \mathbf{Z}} M_l$. Ideals in R are understood to be homogeneous, and homomorphisms of R-modules are understood to mean homomorphisms that preserve the grading.

For an ideal I contained in $\mathfrak{m}$, let y_i be a set of homogeneous generators of positive degree for I, indexed by some totally ordered set S. Define the *local cohomology* groups $H_I^*(M)$ as the cohomology of the following cochain complex, called the Cech complex, with M placed in degree zero:

$$0 \to M \to \oplus_{l \in \mathbf{Z}} \prod_i M[1/y_i]_l \to \oplus_{l \in \mathbf{Z}} \prod_{i<j} M[1/y_i y_j]_l \to \cdots .$$

Here the subscript l refers to the lth graded piece of these modules. The boundary maps in the Cech complex are given by $(-1)^r$ times the obvious homomorphism from $M[1/y_0 \ldots \widehat{y_r} \cdots y_m]$ to $M[1/y_0 \ldots y_m]$. Each group $H_I^m(M)$ is clearly a graded R-module, $H_I^m(M) = \oplus_{l \in \mathbf{Z}} H_I^m(M)_l$. For a finitely generated ideal $I = (y_1, \ldots, y_n)$, we can write the Cech complex more simply as:

$$0 \to M \to \prod_i M[1/y_i] \to \prod_{i<j} M[1/y_i y_j] \to \cdots .$$

For any ideal I in a commutative ring R, the definition implies that $H_I^0(M)$ is the I-torsion submodule of M, defined as $\{x \in M : (\forall f \in I)(\exists m \geq 0)\ f^m x = 0\}$. An introduction to local cohomology in the classical setting of noetherian rings is [73]. Local cohomology was defined by Grothendieck for arbitrary commutative rings [59, exposé II].

To see that the local cohomology groups $H_I^*(M)$ are independent of the choice of generators for an ideal I, we use the following geometric interpretation. Let Proj(R) be the scheme associated to the graded ring R, and let $\widetilde{M}$ be the quasicoherent sheaf on Proj(R) associated to M [55, definition II.2.5.3]. For an integer l, write $M(l)$ for the graded module M with degrees lowered by l. Let U be the complement of the closed subset of Proj(R) defined by the ideal I. Then local cohomology $H_I^*(M)$ is isomorphic to the cohomology of the complex with M placed in degree 0 and only the first boundary map nonzero:

$$0 \to M \to \oplus_{l \in \mathbf{Z}} H^0(U, \widetilde{M(l)}) \xrightarrow{0} \oplus_{l \in \mathbf{Z}} H^1(U, \widetilde{M(l)}) \xrightarrow{0} \cdots .$$

Indeed, this is identified with the cohomology of the Cech complex when we compute the cohomology of U using its open cover by the affine open subsets $\{y_i \neq 0\}$ [56, proposition III.1.4.1], [55, proposition II.2.5.2]. From this interpretation, we see that local cohomology $H_I^*(M)$, as a graded R-module, does not depend on the choice of generators of I. In fact, it depends only on the radical $\text{rad}(I) = \{x \in R : x^m \in I \text{ for some } m > 0\}$ of I, because two ideals with the same radical define the same closed subset of $\text{Proj}(R)$.

Example For R a commutative graded ring, it follows that

$$H^0_{\mathfrak{m}}(R) = \ker\left(R \to \oplus_{l\in\mathbf{Z}} H^0(\text{Proj}(R), O(l))\right),$$
$$H^1_{\mathfrak{m}}(R) = \text{coker}\left(R \to \oplus_{l\in\mathbf{Z}} H^0(\text{Proj}(R), O(l))\right),$$

and

$$H^i_{\mathfrak{m}}(R) = \oplus_{l\in\mathbf{Z}} H^{i-1}(\text{Proj}(R), O(l))$$

for $i > 1$. For R a graded polynomial ring $k[y_1, \ldots, y_n]$ with the degrees $|y_j|$ positive, $H^i_{\mathfrak{m}}(R)$ is nonzero only if $i = n$, either by this geometric interpretation or by the algebraic definition of local cohomology. Moreover,

$$H^n_{\mathfrak{m}}(R) = k[y_1^{\pm 1}, \ldots, y_n^{\pm 1}]/(y_1^{i_1} \cdots y_n^{i_n} = 0 \text{ if some } i_j \geq 0)$$

[73, example 7.16].

Remark 3.1 The grading on each local cohomology group $H^i_{\mathfrak{m}}(M)$ encodes information about the degrees of generators and relations for an R-module M. The precise relation is described in the main result of this chapter, Theorem 3.14. Here we show the relation in simple examples.

Let R be the polynomial ring $k[x]$ with x in degree 1. (Then $\text{Proj}(R)$ is a point, and local cohomology is easy to compute by either the algebraic or the geometric interpretation.) Let M be the free R-module on one generator in degree a. Then $H^i_{\mathfrak{m}}(M)$ is zero for $i \neq 1$, and $H^1_{\mathfrak{m}}(M)$ is the $k[x]$-module $k[x, x^{-1}]/x^a\, k[x]$. The vanishing of $H^0_{\mathfrak{m}}(M)$ is equivalent to the freeness of M as an R-module, and in this case the highest degree of an element needed to generate M is 1 plus the highest degree (namely, $a - 1$) occurring in $H^1_{\mathfrak{m}}(M)$, as Theorem 3.14 explains.

Next, let M be the R-module generated by an element e in degree a modulo the relation $x^b e = 0$, for a positive integer b. Then $H^0_{\mathfrak{m}}(M) \cong M$ is nonzero in degrees $a, a+1, \ldots, a+b-1$, while $H^1_{\mathfrak{m}}(M) = 0$. The non-vanishing of $H^0_{\mathfrak{m}}(M)$ is equivalent to the non-freeness of M as an R-module, and the relation $x^b e = 0$ for M in degree $a + b$ is responsible for $a + b - 1$ being the highest degree occurring in $H^0_{\mathfrak{m}}(M)$, as Theorem 3.14 explains.

A short exact sequence $0 \to M_1 \to M_2 \to M_3 \to 0$ of R-modules gives a long exact sequence of local cohomology groups

$$0 \to H^0_I(M_1) \to H^0_I(M_2) \to H^0_I(M_3) \to H^1_I(M_1) \to \cdots,$$

since the Cech complexes form a short exact sequence of chain complexes (or by the properties of sheaf cohomology).

For noetherian rings, Grothendieck gave another interpretation of local cohomology [59, lemma II.8]. To stick with our situation, we state this only for graded rings. Note that local cohomology $H_I^*(M)$ as we have defined it (for R and M graded) coincides with local cohomology in the ungraded sense (as considered in some references) when R is a finitely generated k-algebra. Indeed, I is a finitely generated ideal in that case, and so both types of local cohomology are computed by the same Cech complex.

Lemma 3.2 *Let I be an ideal in a finitely generated graded k-algebra R. Then local cohomology is a direct limit of Ext groups,*

$$H_I^i(M) = \varinjlim_l \operatorname{Ext}_R^i(R/I^l, M).$$

Equivalently, for R noetherian, the groups $H_I^*(M)$ are the right-derived functors of the left exact functor $H_I^0(M)$ on the category of R-modules. This can fail for R not noetherian, even for a finitely generated ideal I. Explicitly, let R be a ring with unbounded y-torsion for some element $y \in R$, such as $R = k[y, z_1, z_2, z_3, \ldots]/(yz_1, y^2z_2, y^3z_3, \ldots)$. Then any injective R-module J containing the module $\{x \in R : yx = 0\}$ has $H_{(y)}^1(J) \neq 0$ by Grothendieck [59, lemma II.9], whereas the derived functors $E_{(y)}^i(M) := \varinjlim_l \operatorname{Ext}_R^i(R/\mathfrak{m}^l, M)$ of $H_{(y)}^0(M)$ on the category of R-modules have $E_{(y)}^1(J) = 0$, since J is injective.

Theorem 3.3 (Independence Theorem for local cohomology) *If $f : R' \to R$ is a homomorphism of commutative graded rings, $I' \subset R'$ is an ideal contained in the maximal ideal $(R')^{>0}$, and M is an R-module, then f induces an isomorphism*

$$H_{I'}^i(R', M) \to H_{I'R}^i(R, M),$$

where we view M as an R'-module via f.

Proof This is immediate from the Cech interpretation: for $I' = (y_i : I \in S)$, both local cohomology groups are the cohomology of the same Cech complex

$$0 \to M \to \oplus_{l \in \mathbf{Z}} \prod_i M[1/y_i]_l \to \oplus_{l \in \mathbf{Z}} \prod_{i<j} M[1/y_i y_j]_l \to \cdots . \qquad \square$$

The Independence Theorem is a key advantage of the definition of local cohomology in terms of the Cech complex (or, equivalently, in terms of sheaf cohomology).

One last general property of local cohomology is that each local cohomology group $H_I^i(M)$ is an I-torsion module. That is, for every $y \in I$ and every α in $H_I^i(M)$, there is a natural number m such that $y^m \alpha = 0$. Indeed, let I be generated by y together with some set of elements T. Then $H_I^*(M)$ is computed by the Cech complex of M with respect to the set $\{y\} \cup T$, which we can view as the complex $0 \to R \to R[1/y] \to 0$ tensored with the Cech complex of M with respect to T. Therefore, if we tensor the Cech complex over R with $R[1/y]$, we get some chain complex tensored with $0 \to R[1/y] \to R[1/y] \to 0$, and so the cohomology becomes zero. Since $R[1/y]$ is a flat R-module, it follows that $H_I^*(M) \otimes_R R[1/y] = 0$. Equivalently, every element of $H_I^*(M)$ is killed by a power of y, as we want.

We make the following definition for later use.

Definition 3.4 Let R be a finitely generated graded algebra over a field k, assumed to be finite over its center $Z(R)$. A *system of parameters* in R is a sequence $y_1, \ldots, y_n$ of elements of $Z(R)$ such that R is finite over the polynomial ring $k[y_1, \ldots, y_n]$ and n is equal to the dimension of $Z(R)$. Equivalently, R is finite over $k[y_1, \ldots, y_n]$ and $k[y_1, \ldots, y_n]$ injects into R.

We now define depth in terms of local cohomology. Under extra finiteness assumptions, we give several other interpretations, which may be clearer.

Definition 3.5 Let M be a module over a commutative graded ring R. The *depth* of M, depth(R, M), is the supremum of the integers j such that $H^i_{\mathfrak{m}}(M) = 0$ for all $i < j$.

The depth of a nonzero bounded below module over a graded polynomial ring $k[y_1, \ldots, y_n]$ is at most n. We prove this as part of a more precise statement, Theorem 3.14.

Lemma 3.6 *Let R be a commutative graded ring. Then the depth of an R-module M is at least the length n of any M-regular sequence in $\mathfrak{m}$, meaning a sequence $y_1, \ldots, y_n$ in $\mathfrak{m}$ such that y_i is a non-zero-divisor on $M/(y_1, \ldots, y_{i-1})$ for $i = 1, \ldots, n$.*

If R is a finitely generated k-algebra and M is a finitely generated R-module, then the depth of M is equal to the supremum of the lengths of all M-regular sequences in $\mathfrak{m}$.

Proof Let $y_1, \ldots, y_n$ be an M-regular sequence in $\mathfrak{m}$. By definition, we have an exact sequence of R-modules

$$0 \to M/(y_1, \ldots, y_{i-1}) \xrightarrow[y_i]{} M/(y_1, \ldots, y_{i-1}) \to M/(y_1, \ldots, y_i) \to 0$$

for each i. The resulting long exact sequence of local cohomology has the form:

$$\to H^{n-i-1}_{\mathfrak{m}}(M/(y_1, \ldots, y_i)) \to H^{n-i}_{\mathfrak{m}}(M/(y_1, \ldots, y_{i-1}))$$
$$\xrightarrow[y_i]{} H^{n-i}_{\mathfrak{m}}(M/(y_1, \ldots, y_{i-1})) \to$$

We show that $H^j_{\mathfrak{m}}(M/(y_1, \ldots, y_i)) = 0$ for $j < n - i$ by descending induction on i. This is clear for $i = n$. Suppose it is true for a given value of i. The exact sequence shows that multiplication by y_i is injective on $H^j_{\mathfrak{m}}(M/(y_1, \ldots, y_{i-1}))$ for $j < n - i + 1$. We showed that this local cohomology group is $\mathfrak{m}$-torsion; in particular, every element of this group is killed by some power of y_i. Therefore, $H^j_{\mathfrak{m}}(M/(y_1, \ldots, y_{i-1})) = 0$ for $j < n - i + 1$, completing the induction. For $i = 0$, we conclude that $H^j_{\mathfrak{m}}(M) = 0$ for $j < n$. That is, M has depth at least n by our definition.

This lower bound for the depth is an equality for finitely generated modules M over a noetherian ring R. A proof is given in [73, theorem 9.1]. □

For modules that are not finitely generated, the definition of depth in terms of local cohomology has better properties than the maximum length of a regular sequence. For example, let k be an algebraically closed field, $R = k[x, y]$, and $M = R/(x) \oplus \bigoplus_{a \in k} R/(ax + y)$. Then M is an infinite direct sum of modules of depth 1, and so it has depth 1 in our sense. Indeed, local cohomology with respect to the finitely generated ideal $\mathfrak{m} = (x, y)$ commutes with arbitrary direct sums, by the Cech complex. But there is no M-regular sequence of length 1 in $\mathfrak{m}$.

Definition 3.7 Let R be a commutative graded ring. Let M be a graded R-module. Let $a_i(R, M)$ denote the maximum degree of a nonzero element of $H^i_{\mathfrak{m}}(R, M)$ (possibly ∞ if unbounded or $-\infty$ if $H^i_{\mathfrak{m}}(R, M) = 0$). The (Castelnuovo-Mumford) regularity of M over R is

$$\operatorname{reg}(R, M) = \sup_i \{a_i(R, M) + i\}.$$

Example Let R be a graded polynomial ring $R = k[y_1, \ldots, y_n]$. By the computation stated earlier, the local cohomology $H^i_{\mathfrak{m}}(R)$ is zero except when $i = n$, and the top-degree subspace of $H^n_{\mathfrak{m}}(R)$ is the k-vector space spanned by $y_1^{-1} \cdots y_n^{-1}$. It follows that the regularity of R as an R-module is equal to $-\sigma(R)$, where we define:

Definition 3.8 For a graded polynomial ring $R = k[y_1, \ldots, y_n]$, let $\sigma(R) = \sum_i (|y_i| - 1)$.

We state some simple properties of regularity for later use.

Lemma 3.9 *Let R be a commutative graded ring and M a graded R-module.*

(i) *Shifting M up in degree by an integer a increases the regularity of M by a.*
(ii) *If M is bounded above, then* $\operatorname{reg}(R, M)$ *is equal to the top degree in which M is nonzero.*
(iii) *For a short exact sequence $0 \to A \to B \to C \to 0$ of R-modules,* $\operatorname{reg}(R, B) \le \max(\operatorname{reg}(R, A), \operatorname{reg}(R, C))$.

Proof The definition of regularity implies (i). The Cech complex shows that a bounded above module M has $H^0_{\mathfrak{m}}(M) = M$ and $H^i_{\mathfrak{m}}(M) = 0$ for $i > 0$, which gives (ii) by the definition of regularity. The long exact sequence of local cohomology implies (iii). □

For a commutative graded ring R, we define the depth and regularity of R to mean the depth and regularity of R as a module over itself.

Lemma 3.10 *Let $R = k[y_1, \dots, y_n]$ be a graded polynomial ring with $|y_i| > 0$ for all i, and let S be a commutative graded ring with a homomorphism $R \to S$. Suppose that S is generated as an R-module by a set of elements of bounded degree. Then* $\operatorname{depth}(S) = \operatorname{depth}(R, S)$ *(the depth of S as an R-module) and* $\operatorname{reg}(S) = \operatorname{reg}(R, S)$.

Proof We are given that there is an $m \geq 0$ such that S is generated as an R-module by elements of degree less than m. Therefore, $S^{\geq m}$ maps to zero in $S/(y_1, \dots, y_n)$, and so the ideal $(S^{>0})^m$ is contained in $(y_1, \dots, y_n)$. In particular, $S^{>0}$ is the radical of the finitely generated ideal $R^{>0}S = (y_1, \dots, y_n)$.

By the Independence Theorem (Theorem 3.3), we have $H^*_{R^{>0}}(R, S) \cong H^*_{R^{>0}S}(S, S)$, compatibly with the gradings on these groups. Since $S^{>0}$ is the radical of $R^{>0}S$, it follows that $H^*_{R^{>0}}(R, S) \cong H^*_{S^{>0}}(S, S)$. Depth and regularity are defined in terms of local cohomology, and so we have $\operatorname{depth}(R, S) = \operatorname{depth}(S)$ and $\operatorname{reg}(S) = \operatorname{reg}(R, S)$. □

3.2 Depth and regularity in terms of generators and relations

The formal properties of depth and regularity are easiest to prove using local cohomology. But when R is a graded polynomial ring, we want to relate depth and regularity to simpler invariants, such as the degrees of generators and relations of a module. For that purpose, we now reformulate depth and regularity in terms of a projective resolution of a module. For finitely generated modules, this can be done using Grothendieck's local duality theorem [73, theorem 11.29], but we want to avoid the assumption of finite generation.

Let $R = k[y_1, \dots, y_n]$ be a graded polynomial ring. Let M be an R-module that is bounded below (meaning that $M_i = 0$ for i less than some integer i_0). Clearly M has a resolution by free modules, which are understood to be graded:

$$\cdots \to F_1 \to F_0 \to M \to 0.$$

We say that a free resolution is *minimal* if the associated k-linear maps $F_i/\mathfrak{m}F_i \to F_{i-1}/\mathfrak{m}F_{i-1}$ are zero. (We repeat that $\mathfrak{m}$ denotes the ideal $R^{>0}$.) By Eilenberg, M has a unique minimal resolution up to isomorphism [37, proposition 15, proposition 7]. By definition of the minimal resolution, the group

$\mathrm{Tor}_i^R(k, M)$ is canonically identified with $F_i/\mathfrak{m}F_i$, the space of generators of F_i as a free module.

We use the Koszul resolution [149, corollary 4.5.5]:

Lemma 3.11 *Let R be the polynomial ring $R = k[y_1, \ldots, y_n]$ over a commutative ring k. Consider k as an R-module with y_i acting by zero on k for all i. Then k has a free resolution of the form:*

$$0 \to R^{\oplus\binom{n}{n}} \to \cdots \to R^{\oplus\binom{n}{1}} \to R^{\oplus\binom{n}{0}} \to k.$$

Explicitly, let $e(I)$ be basis elements for $R^{\oplus\binom{n}{a}}$ indexed by the a-element subsets I of $\{1, \ldots, n\}$. Then the differential is

$$d(e(i_1 \cdots i_a)) = \sum_{j=1}^{a} (-1)^j y_j e(i_1 \cdots \widehat{i_j} \cdots i_a)$$

for $i_1 < \ldots < i_a$.

The Koszul resolution immediately implies the Hilbert syzygy theorem in our setting [149, theorem 4.3.8].

Theorem 3.12 (Hilbert syzygy theorem) *Let $R = k[y_1, \ldots, y_n]$ be a graded polynomial ring over a field k. Let M be a bounded below R-module. Then M has a free resolution of length at most n.*

Proof By the Koszul resolution (Lemma 3.11), $\mathrm{Tor}_i^R(k, M)$ is zero for all $i > n$. By the discussion of minimal resolutions above, it follows that the minimal resolution of M has length at most n. □

For later use, we give the following characterization of regular sequences.

Lemma 3.13 *Let M be a bounded below module over a commutative graded k-algebra R. Let $y_1, \ldots, y_n \in \mathfrak{m}$. The following are equivalent.*

(i) *$y_1, \ldots, y_n$ is an M-regular sequence.*
(ii) *M is a flat $k[y_1, \ldots, y_n]$-module.*
(iii) *M is a free $k[y_1, \ldots, y_n]$-module.*

Proof Let S be the graded polynomial ring $k[y_1, \ldots, y_n]$; then M is an S-module. Suppose that $y_1, \ldots, y_n$ is an M-regular sequence. Consider the short exact sequence of S-modules

$$0 \to S/(y_1, \ldots, y_{i-1}) \xrightarrow[y_i]{} S/(y_1, \ldots, y_{i-1}) \to S/(y_1, \ldots, y_i) \to 0$$

for $i = 1, \ldots, n$. This gives a long exact sequence of Tor groups,

$$\mathrm{Tor}_1^S(S/(y_1, \ldots, y_{i-1}), M) \to \mathrm{Tor}_1^S(S/(y_1, \ldots, y_i), M) \\ \to M/(y_1, \ldots, y_{i-1}) \xrightarrow[y_i]{} M/(y_1, \ldots, y_{i-1}).$$

It follows by induction on i that $\mathrm{Tor}_j^S(S/(y_1, \ldots, y_i), M) = 0$ for $i = 0, \ldots, n$ and $j > 0$, this being trivial for $i = 0$. Taking $i = n$, we find that $\mathrm{Tor}_j^S(k, M) = 0$ for $j > 0$.

It follows that the minimal resolution of M has length 0; that is, M is a free S-module. Conversely, if M is a free S-module, then $y_1, \ldots, y_n$ is clearly an M-regular sequence. Finally, if M is a flat S-module, then $\mathrm{Tor}_j^S(k, M) = 0$ for $j > 0$. Again, it follows using the minimal resolution that M is a free S-module. □

Let M be a bounded below module over a graded polynomial ring R. Let $\cdots \to F_1 \to F_0 \to M \to 0$ be the minimal resolution of M. By the isomorphism $\mathrm{Tor}_i(k, M) \cong F_i/\mathfrak{m}F_i$, the projective dimension $\mathrm{pd}(R, M)$ (the shortest length of a projective resolution of M) is equal to the largest i such that $F_i \neq 0$ in the minimal resolution of M.

For a bounded below R-module M, let $\rho_i(R, M)$ be the maximum degree of a nonzero element of $F_i/\mathfrak{m}F_i$ (possibly ∞ or $-\infty$) in the minimal resolution of M; equivalently, $\rho_i(R, M)$ is the maximum degree of a generator of F_i. Define

$$\mathrm{Preg}(R, M) = \sup_i (\rho_i(R, M) - i) - \sigma(R),$$

where $\sigma(R)$ was defined in Definition 3.8. In view of Theorem 3.14, Preg does not need a name of its own; it is simply another way to compute the regularity as defined earlier, in the case of bounded below modules. In the common situation where all the y_i have degree 1, we have $\sigma(R) = 0$, and so this definition of regularity can be written without mentioning $\sigma(R)$.

Theorem 3.14 *Let $R = k[y_1, \ldots, y_n]$ be a graded polynomial ring. Let M be a nonzero graded R-module that is bounded below. Then* $\mathrm{reg}(R, M) = \mathrm{Preg}(R, M)$ *and* $\mathrm{depth}(R, M) + \mathrm{pd}(M) = n$.

We are interested in $\mathrm{Preg}(R, M)$ because it gives information about generators and relations of the R-module M, by its definition. In particular, the R-module M is generated by elements of degree at most $\mathrm{Preg}(R, M)$.

Proof We first prove the formula for depth, due to Auslander-Buchsbaum for M finitely generated [73, theorem 8.13]. We first show that $\mathrm{depth}(R, M) \geq n - \mathrm{pd}(M)$, that is, that $H^i_{\mathfrak{m}}(M) = 0$ for $i < n - \mathrm{pd}(M)$. For $\mathrm{pd}(M) = 0$, M is a nonzero free graded R-module, and so this follows from the calculation that $H^i_{\mathfrak{m}}(R)$ is nonzero if and only if $i = n$. For any $\mathrm{pd}(M)$, the upper bound for depth follows by induction, using that a short exact sequence $0 \to M_1 \to M_2 \to M_3 \to 0$ of R-modules gives a long exact sequence of local cohomology:

$$\to H^i_{\mathfrak{m}}(M_1) \to H^i_{\mathfrak{m}}(M_2) \to H^i_{\mathfrak{m}}(M_3) \to H^{i+1}_{\mathfrak{m}}(M_1) \to .$$

Conversely, suppose M is nonzero and has projective dimension d. By the Hilbert syzygy theorem (Theorem 3.12), d is at most n. (In particular, d is finite.) Next, the minimal resolution of M has $F_d \neq 0$. By the discussion of projective dimension before Lemma 3.13, it follows that $\mathrm{Tor}_d^R(k, M)$ is not zero.

By Cartan and Eilenberg, there is a canonical isomorphism

$$\mathrm{Tor}_i^R(k, M) \cong \mathrm{Ext}_R^{n-i}(k, M)$$

for every R-module M and every integer i [27, exercise VIII.7]. Indeed, if we compute these groups using the Koszul resolution of k as an R-module (Lemma 3.11), then the two groups are the cohomology of the same chain complex (with a shift in degrees). Both groups are graded, with the part of Tor_i in degree j corresponding to the part of Ext^{n-i} in degree $j - \sum |y_i| = j - \sigma(R) - n$.

Therefore, $\mathrm{Ext}_R^j(k, M)$ is 0 for $j < n - d$ and not zero for $j = n - d$. For each $l \geq 0$, the exact sequence $0 \to \mathfrak{m}^l/\mathfrak{m}^{l+1} \to R/\mathfrak{m}^{l+1} \to R/\mathfrak{m}^l \to 0$ of R-modules induces a long exact sequence of Ext groups,

$$\mathrm{Ext}_R^{n-d-1}(\mathfrak{m}^l/\mathfrak{m}^{l+1}, M) \to \mathrm{Ext}_R^{n-d}(R/\mathfrak{m}^l, M) \to \mathrm{Ext}_R^{n-d}(R/\mathfrak{m}^{l+1}, M).$$

We read off that the homomorphisms $\mathrm{Ext}^{n-d}(k, M) \to \mathrm{Ext}^{n-d}(R/\mathfrak{m}^2, M) \to \cdots$ are all injective. Since R is noetherian, Lemma 3.2 gives that

$$H_{\mathfrak{m}}^i(M) = \varinjlim_l \mathrm{Ext}_R^i(R/\mathfrak{m}^l, M).$$

Therefore, we have $H_{\mathfrak{m}}^{n-d}(M) \neq 0$. This completes the proof that $\mathrm{depth}(R, M) + \mathrm{pd}(M) = n$.

Next, let us show that $\mathrm{reg}(R, M) \leq \mathrm{Preg}(R, M)$. We know that $\mathrm{Tor}_i(k, M)$ is zero in degrees greater than $\mathrm{Preg}(R, M) + i + \sigma(R)$, for all i. By Cartan-Eilenberg's isomorphism above, $\mathrm{Ext}_R^{n-i}(k, M)$ is zero in degrees greater than $\mathrm{Preg}(R, M) + i - n$, for all i. By the long exact sequence of Ext groups above, it follows that $H_{\mathfrak{m}}^{n-i}(M)$ is zero in degrees greater than $\mathrm{Preg}(R, M) + i - n$, for all i. This means that $\mathrm{reg}(R, M) \leq \mathrm{Preg}(R, M)$.

Conversely, suppose that M is nonzero and has $\mathrm{Preg}(R, M) = r$. First suppose that r is finite. Then $\mathrm{Tor}_i^R(k, M)$ is zero in degrees greater than $r + i + \sigma(R)$ for all i, and there is a j such that $\mathrm{Tor}_j^R(k, M)$ is nonzero in degree $r + j + \sigma(R)$. Equivalently, $\mathrm{Ext}_R^{n-i}(k, M)$ is zero in degrees greater than $r + i - n$ for all i, and $\mathrm{Ext}_R^{n-j}(k, M)$ is nonzero in degree $r + j - n$. In particular, $\mathrm{Ext}_R^{n-j-1}(k, M)$ is zero in degrees greater than $r + j + 1 - n$, and so $\mathrm{Ext}_R^{n-j-1}(\mathfrak{m}^l/\mathfrak{m}^{l+1}, M)$ is zero in degrees greater than $r + j + 1 - n - l$ (hence in degree $r + j - n$) for every positive integer l. (This uses that the generators $y_1, \ldots, y_n$ of the ideal $\mathfrak{m}$ all have degree at least 1.) Considering the exact sequence

$$\mathrm{Ext}_R^{n-j-1}(\mathfrak{m}^l/\mathfrak{m}^{l+1}, M) \to \mathrm{Ext}_R^{n-j}(R/\mathfrak{m}^l, M) \to \mathrm{Ext}_R^{n-j}(R/\mathfrak{m}^{l+1}, M)$$

in graded degree $r+j-n$, we find that the maps $\operatorname{Ext}_R^{n-j}(k, M) \to \operatorname{Ext}_R^{n-j}(R/\mathfrak{m}^2, M) \to \cdots$ are all injective in degree $r+j-n$. It follows that the direct limit, $H_{\mathfrak{m}}^{n-j}(M)$, is nonzero in degree $r+j-n$. Thus $\operatorname{reg}(R, M) \geq r$. That completes the proof that $\operatorname{reg}(R, M) = \operatorname{Preg}(R, M)$ when $\operatorname{Preg}(R, M)$ is finite.

Finally, suppose that $\operatorname{Preg}(R, M)$ is infinite; we will show that $\operatorname{reg}(R, M)$ is infinite. The proof is similar to that for the finite case. Our assumption means that there is a j such that $\operatorname{Tor}_j^R(k, M)$ is nonzero in arbitrary high degrees; note that j must be in the set $\{0, 1, \ldots, n\}$. Let j be the maximum number with this property. Then $\operatorname{Ext}_R^{n-j}(k, M)$ is nonzero in arbitrarily high degrees, and $n-j$ is minimal with this property. Then there is an integer r such that $\operatorname{Ext}_R^{n-j-1}(k, M)$ is zero in degrees greater than $r+j+1-n$ (to be parallel with the notation in the previous case of the proof). Then $\operatorname{Ext}_R^{n-j-1}(\mathfrak{m}^l/\mathfrak{m}^{l+1}, M)$ is zero in degrees greater than $r+j+1-n-l$ (hence in degrees greater than $r+j-n$) for every positive integer l. By the exact sequence

$$\operatorname{Ext}_R^{n-j-1}(\mathfrak{m}^l/\mathfrak{m}^{l+1}, M) \to \operatorname{Ext}_R^{n-j}(R/\mathfrak{m}^l, M) \to \operatorname{Ext}_R^{n-j}(R/\mathfrak{m}^{l+1}, M),$$

the homomorphisms $\operatorname{Ext}_R^{n-j}(k, M) \to \operatorname{Ext}_R^{n-j}(R/\mathfrak{m}^2, M) \to \cdots$ are all injective in degrees greater than $r+j-n$. It follows that the direct limit, $H_{\mathfrak{m}}^{n-j}(M)$, is nonzero in arbitrarily high degrees. So $\operatorname{reg}(R, M) = \infty$, as we want. □

Definition 3.15 Let R be a graded k-algebra (not necessarily commutative). We say that R is *Cohen-Macaulay* if the local cohomology $H_{\mathfrak{m}}^i(R)$ is concentrated in one degree, where $\mathfrak{m}$ is the maximal ideal of $Z(R)$ and we consider R as a module over $Z(R)$.

This is a standard definition in the case of finitely generated commutative k-algebras. We define the Cohen-Macaulay property in this generality because we want to consider it for the $\mathbf{F}_p$-cohomology ring of a finite group, which is only graded-commutative, and also to the mod p Chow ring of a finite group, which is commutative but is only known to be generated in bounded degrees over a polynomial subring (Theorem 5.2). In particular, the Chow ring of a finite group is not known to be noetherian. (For rings not generated in bounded degrees over a polynomial subring, Definition 3.15 is probably not very meaningful.)

Lemma 3.16 *Let R be a graded k-algebra, not necessarily commutative. If R is a free module with generators in bounded degrees over a graded polynomial ring S contained in the center of R, then R is Cohen-Macaulay.*

Proof Write $S = k[x_1, \ldots, x_n]$, with maximal ideal $\mathfrak{m}_S$, and let $\mathfrak{m}$ be the maximal ideal of $Z(R)$. The ideals $\mathfrak{m}_S Z(R)$ and $\mathfrak{m}$ in $Z(R)$ have the same radical, by our assumption, and so R has the same local cohomology with respect to these two ideals. By the Independence Theorem, it follows that $H^i_{\mathfrak{m}_S} R \cong H^i_{\mathfrak{m}} R$ for all i. Since R is a free S-module, we have $H^i_{\mathfrak{m}_S} R = 0$ for $i \neq n$. Therefore, R is Cohen-Macaulay in the sense of Definition 3.15. □

3.3 Duflot's lower bound for depth

In this section, we prove Duflot's lower bound for the depth of the cohomology ring of a finite group [34], and generalize it to the Chow ring. Duflot's theorem has inspired a lot of work on group cohomology over the past 30 years, including Theorem 9.1 in this book.

Depth is defined in Definition 3.5 for a module over a commutative graded ring. The depth of the cohomology ring $H^*_G = H^*(BG, \mathbf{F}_p)$ is understood to mean its depth over the even-degree subring H^{ev}_G, which is commutative. For $p = 2$, the whole ring H^*_G is commutative, and this definition coincides with the depth of H^*_G as a module over itself, by Lemma 3.10.

Theorem 3.17 *Let G be a finite group, and let p be a prime number. Let S be a Sylow p-subgroup of G and let $C = Z(S)[p]$ be the p-torsion subgroup of the center of S. Let $\zeta_1, \ldots, \zeta_s$ be elements of $H^*_G = H^*(BG, \mathbf{F}_p)$ (of even degree if p is odd) such that the restrictions of $\zeta_1, \ldots, \zeta_s$ form a regular sequence in H^*_C. Then $\zeta_1, \ldots, \zeta_s$ is an H^*_G-regular sequence.*

Proof We follow Carlson's exposition [26, theorem 12.3.3] of Broto-Henn's proof [22]. We are given that H^*_C is a flat (or equivalently, free) $\mathbf{F}_p[\zeta_1, \ldots, \zeta_s]$-module, by Lemma 3.13. Suppose we can show that the cohomology of a Sylow p-subgroup H^*_S is a flat $\mathbf{F}_p[\zeta_1, \ldots, \zeta_s]$-module. Then it follows that H^*_G is a flat $\mathbf{F}_p[\zeta_1, \ldots, \zeta_s]$-module, because H^*_G is a summand of H^*_S as an H^*_G-module via transfer (Section 2.5). Thus we have reduced to proving the theorem for G a p-group.

Lemma 3.18 *Let G be a p-group. Let C be an elementary abelian subgroup of the center of G. Then H^*_G is an H^*_C-comodule.*

Proof Since C is central in G, we have a group homomorphism $C \times G \to G$ defined by $(c, g) \mapsto cg$. The pullback homomorphism $H^*_G \to H^*_{C\times G} = H^*_C \otimes_{\mathbf{F}_p} H^*_G$ makes H^*_G a comodule over the Hopf algebra H^*_C. □

We now prove Theorem 3.17, where we have arranged for G to be a p-group and $C = Z(G)[p]$. Write

$$\alpha\colon H^*_G \to H^*_C \otimes_{\mathbf{F}_p} H^*_G \cong H^*_{C\times G}$$

for the pullback map associated to the homomorphism $C \times G \to G$, $(c, g) \mapsto cg$. The inclusion $G \hookrightarrow C \times G$ by $g \mapsto (1, g)$ gives a restriction map $H^*_{C\times G} \to$

H_G^*. In terms of the identification $H_{C\times G}^* = H_C^* \otimes_{\mathbf{F}_p} H_G^*$, this restriction map is the obvious ring homomorphism

$$\beta\colon H_C^* \otimes_{\mathbf{F}_p} H_G^* \to H_G^*$$

sending $H_C^{>0}$ to zero. Since the composition $G \hookrightarrow C \times G \to G$ is the identity, the composition

$$H_G^* \underset{\alpha}{\to} H_C^* \otimes_{\mathbf{F}_p} H_G^* \underset{\beta}{\to} H_G^*$$

is the identity. Thus H_G^* is a summand of $H_C^* \otimes_{\mathbf{F}_p} H_G^*$ as an H_G^*-module, where we make $H_C^* \otimes_{\mathbf{F}_p} H_G^*$ into an H_G^*-module using α.

In particular, H_G^* is a summand of $H_C^* \otimes_{\mathbf{F}_p} H_G^*$ as an $\mathbf{F}_p[\zeta_1, \ldots, \zeta_s]$-module. We can filter $H_C^* \otimes_{\mathbf{F}_p} H_G^*$ by the $\mathbf{F}_p[\zeta_1, \ldots, \zeta_s]$-submodules $H_C^* \otimes_{\mathbf{F}_p} H_G^{\geq j}$, for $j \geq 0$. The corresponding quotient modules can be identified with $H_C^* \otimes_{\mathbf{F}_p} H_G^j$, with $\zeta_1, \ldots, \zeta_s$ acting by 0 on H_G^j. Since $\zeta_1, \ldots, \zeta_s$ form a regular sequence in H_C^*, H_C^* is a free $\mathbf{F}_p[\zeta_1, \ldots, \zeta_s]$-module by Lemma 3.13, and so these quotients are free $\mathbf{F}_p[\zeta_1, \ldots, \zeta_s]$-modules. So, by induction, $H_C^* \otimes_{\mathbf{F}_p} H_G^*$ is a free $\mathbf{F}_p[\zeta_1, \ldots, \zeta_s]$-module. Since H_G^* is a summand of $H_C^* \otimes_{\mathbf{F}_p} H_G^*$ as a graded $\mathbf{F}_p[\zeta_1, \ldots, \zeta_s]$-module, H_G^* is also a free $\mathbf{F}_p[\zeta_1, \ldots, \zeta_s]$-module, by Lemma 3.13. □

Corollary 3.19 *Let G be a finite group and p a prime number. The depth of $H_G^* = H^*(BG, \mathbf{F}_p)$ is at least the p-rank of the center of a Sylow p-subgroup of G.*

Proof Let S be a Sylow p-subgroup of G, and let $C = Z(G)[p]$, which is isomorphic to $(\mathbf{Z}/p)^c$ for some c. Let V be a faithful complex representation of G, with $n = \dim(V)$. By Theorem 1.1, H_C^* is finite over H_G^*, and in fact over $\mathbf{F}_p[c_1V, \ldots, c_nV]$. By Lemma 3.10, since the ring H_C^* is Cohen-Macaulay, there are elements $\zeta_1, \ldots, \zeta_c$ in $\mathbf{F}_p[c_1V, \ldots, c_nV]$ that restrict to a regular sequence in H_C^*. By Theorem 3.17, $\zeta_1, \ldots, \zeta_c$ is an H_G^*-regular sequence. By Lemma 3.6, the depth of H_G^* is at least c. □

We now generalize Duflot's theorem on depth to the Chow ring.

Theorem 3.20 *Let G be a finite group, p a prime number, k a field of characteristic not p that contains the pth roots of unity. Consider G as an algebraic group over k. Let S be a Sylow p-subgroup of G and let $C = Z(S)[p]$ be the p-torsion subgroup of the center of S. Let $\zeta_1, \ldots, \zeta_s$ be elements of $CH_G^* = CH^*(BG)/p$ such that the restrictions of $\zeta_1, \ldots, \zeta_s$ form a regular sequence in CH_C^*. Then $\zeta_1, \ldots, \zeta_s$ is a regular sequence in CH_G^*.*

The proof is essentially the same as for cohomology. Some details are: The assumption on k implies that $C \cong (\mathbf{Z}/p)^c$ has Chow ring $CH_C^* =$

$\mathbf{F}_p[y_1, \ldots, y_c]$. This is slightly different from H^*_C, but the argument is unaffected by the change. The Künneth formula is not known for the Chow ring of a product of two finite groups in general, but it is true when one factor is abelian of exponent e and k contains the eth roots of unity (Lemma 2.12), which is the case needed to show that CH^*_S is a CH^*_C-comodule. Finally, the $\mathbf{F}_p$-vector spaces CH^i_S are not known to be finite-dimensional, but the filtration argument in the proof of Theorem 3.17 still works.

Corollary 3.21 *Let G be a finite group, p a prime number, k a field of characteristic not p that contains the pth roots of unity. Consider G as an algebraic group over k. The depth of CH^*_G is at least the p-rank of the center of a Sylow p-subgroup of G.*

Proof Let C be the p-torsion subgroup of the center of a Sylow p-subgroup of G. We have $C \cong (\mathbf{Z}/p)^c$ for some c. By the same proof as for Corollary 3.19, there are Chern classes in CH^*_G, $\zeta_1, \ldots, \zeta_c$, which restrict to a regular sequence in $CH^*_C = \mathbf{F}_p[y_1, \ldots, y_c]$. By Theorem 3.20, $\zeta_1, \ldots, \zeta_c$ form a regular sequence in CH^*_G. By Lemma 3.6, the depth of CH^*_G is at least c. □

4

Regularity of Group Cohomology

We now prove Symonds's theorem on the regularity of group cohomology, and its consequences for the degrees of generators and relations.

4.1 Regularity of group cohomology and applications

Theorem 4.1 (Symonds [130]) *Let p be a prime number. Let N be a smooth manifold with finite-dimensional $\mathbf{F}_p$-cohomology. Let G be a compact Lie group acting on N. Then the equivariant cohomology ring $H^*_G(N, \mathbf{F}_p)$ (graded with H^i in degree i) has regularity at most* $\dim(N) - \dim(G)$.

In particular, for a finite group G, the cohomology ring $H^*_G = H^*(BG, \mathbf{F}_p)$ has regularity at most zero. Moreover, Benson and Carlson showed that H^*_G has regularity at least zero, using a spectral sequence they constructed from a system of parameters in H^*_G. The key point is a duality on that spectral sequence, analogous to Poincaré duality [13, theorem 4.2]. Combining the results of Benson-Carlson and Symonds gives:

Corollary 4.2 *For a finite group G and a prime number p, the graded ring $H^*_G = H^*(BG, \mathbf{F}_p)$ has regularity equal to zero.*

The fact that H^*_G always has regularity equal to zero is an impressively sharp result. For example, if G is a finite group such that H^*_G is a finitely generated free module over a polynomial ring $\mathbf{F}_p[y]$ in one variable, then Corollary 4.2 says that the highest-degree module generator has degree *equal* to $|y| - 1$. (This case of Corollary 4.2, where H^*_G is Cohen-Macaulay, was proved earlier by Benson and Carlson [12, vol. 2, theorem 5.18.1].) We see this behavior in the cohomology ring of a cyclic group of order an odd prime p, $H^*_{\mathbf{Z}/p} = \mathbf{F}_p\langle x, y\rangle$ (the free graded-commutative algebra) with $|x| = 1$ and $|y| = 2$, and also in the cohomology of the symmetric group, $H^*_{S_p} = \mathbf{F}_p\langle x, y\rangle$ with $|x| = 2p-3$ and $|y| = 2p-2$.

The most dramatic applications of Theorem 4.1 are the bounds it gives for the generators of the cohomology ring of a finite group, such as the following. No bound was known before Symonds's work.

Corollary 4.3 (Symonds) *Let G be a compact Lie group (for example, a finite group) with a faithful complex representation V of dimension n. Let p be a prime number. Then the cohomology $H^*_G = H^*(BG, \mathbf{F}_p)$ is generated as a module over the polynomial ring $H^*_{U(n)} = \mathbf{F}_p[c_1 V, \ldots, c_n V]$ by elements of degree at most $n^2 - \dim(G)$. It follows that the cohomology ring H^*_G is generated by elements of degree at most* $\max(2n, n^2 - \dim(G))$.

Moreover, the relations among the $\mathbf{F}_p[c_1 V, \ldots, c_n V]$-module generators for H^*_G are in degrees at most $n^2 + 1 - \dim(G)$. It follows that the relations in H^*_G as an $\mathbf{F}_p$-algebra are in degree at most $\max(2n, n^2 + 1 - \dim(G), 2(n^2 - \dim(G)))$.

As Symonds pointed out, Corollary 4.3 implies that the mod p cohomology ring of a nontrivial finite group G, for any prime number p, is generated by elements of degree at most $|G| - 1$. But for most groups of interest, applying Corollary 4.3 directly gives a much better bound than $|G| - 1$. For that reason, this book seeks optimal bounds for the cohomology and Chow ring of a finite group in terms of the dimension of a faithful complex representation, rather than in terms of the order of the group.

Proof (Corollary 4.3) We are given a faithful representation $G \subset U(n)$. By Venkov (Theorem 1.1), H^*_G is a finitely generated module over

$$H^*_{U(n)} = \mathbf{F}_p[c_1 V, \ldots, c_n V].$$

In the notation of Definition 3.8, the graded polynomial ring $H^*_{U(n)}$ has $\sigma(H^*_{U(n)}) = \sum_{i=1}^n (2i-1) = n^2$. By Theorem 4.1, H^*_G has regularity at most $-\dim(G)$ as a module over $H^*_{U(n)}$. By the interpretation of regularity in Theorem 3.14, it follows that H^*_G is generated as a module over $H^*_{U(n)}$ by elements of degree at most $\sigma(H^*_{U(n)}) - \dim(G) = n^2 - \dim(G)$. It also follows that the relations among these module generators are in degree at most $\sigma(H^*_{U(n)}) - \dim(G) + 1 = n^2 - \dim(G) + 1$. This implies the statements we want about generators and relations for H^*_G as an $\mathbf{F}_p$-algebra. □

4.2 Proof of Symonds's theorem

Using the formal properties of regularity, together with some basic arguments on equivariant cohomology used by Quillen, Symonds's proof quickly reduces the problem of bounding the regularity of H_G^* for any compact Lie group G to the problem of bounding the regularity of $H_A^* M$, for an elementary abelian group A acting on a smooth manifold M. The latter problem is handled by a theorem of Duflot on actions of elementary abelian groups (not to be confused with her theorem on depth, Corollary 3.19). We also prove Duflot's theorem.

Proof (Theorem 4.1) Write H_G^* for $H^*(BG, \mathbf{F}_p)$. Choose a faithful representation $G \subset U(n)$. We need the following variant of Venkov's theorem (Theorem 1.1).

Lemma 4.4 *Let N be a smooth manifold such that N has finite-dimensional $\mathbf{F}_p$-cohomology. Let G be a compact Lie group acting on N. Let $G \subset U(n)$ be a faithful representation. Then $H_G^* N$ is a finitely generated module over $H_{U(n)}^* = \mathbf{F}_p[c_1, \ldots, c_n]$.*

Proof We can view the homotopy quotient $N//G = (N \times EG)/G$ as a double quotient $G\backslash(N \times U(n))/U(n)$, where G acts on N and by left multiplication on $U(n)$, and $U(n)$ acts trivially on N and by right multiplication on $U(n)$ (so the two actions commute). Thus we have a fibration

$$(N \times U(n))/G \to N//G \to BU(n).$$

Here $(N \times U(n))/G$ is a finite-dimensional manifold, and it has finite-dimensional $\mathbf{F}_p$-cohomology in each degree using the Leray-Serre spectral sequence

$$H^*(BG, H^*(N \times U(n))) \Rightarrow H^*((N \times U(n))/G).$$

So $(N \times U(n))/G$ has finite-dimensional $\mathbf{F}_p$-cohomology. The same argument as in Venkov's theorem (Theorem 1.1), applied to the fibration above, gives that $H_G^* N = H^*(N//G)$ is a finitely generated $H^* BU(n)$-module. □

Thus we know that $H_G^* N$ is finitely generated as an $\mathbf{F}_p[c_1, \ldots, c_n]$-module, and we want to show that it has regularity at most $\dim(N) - \dim(G)$ as an $\mathbf{F}_p[c_1, \ldots, c_n]$-module. (By the properties of regularity (Lemma 3.10), this property of $H_G^* N$ does not depend on the choice of faithful representation $G \subset U(n)$.)

Let T be the subgroup $(S^1)^n \subset U(n)$ of diagonal matrices, and $S = (\mathbf{Z}/p)^n$ the p-torsion subgroup of T. The cohomology ring of the flag manifold $U(n)/T$ is well known: $H^*(U(n)/T)$ is the quotient of the polynomial ring $\mathbf{F}_p[y_1, \ldots, y_n]$ with $|y_i| = 1$ by the elementary symmetric polynomials $c_1, \ldots, c_n$. For example, this follows from the spectral sequence of the fibration

$$U(n)/T \to BT \to BU(n),$$

or by viewing $U(n)/T$ as an iterated projective bundle as in the proof of Theorem 2.13. Let $L_1, \ldots, L_n$ be the complex line bundles on $U(n)/T$ given by the obvious 1-dimensional representations of T. The quotient manifold $U(n)/S$ is the principal $(S^1)^n$-bundle over $U(n)/T$ corresponding to the pth powers of the line bundles $L_1, \ldots, L_n$. The first Chern classes of these pth power line bundles are zero in $H^2(U(n)/T)$, since we are using $\mathbf{F}_p$ coefficients. So, applying the spectral sequence for the cohomology of an S^1-bundle n times, we find that for p odd,

$$H^*(U(n)/S) \cong H^*(U(n)/T)\langle x_1, \ldots, x_n\rangle,$$

where $|x_i| = 1$ for $i = 1, \ldots, n$. (For $p = 2$, x_i^2 may not be zero in $H^*(U(n)/S)$, but it is enough for what follows that $H^*(U(n)/S)$ is a free module over $H^*(U(n)/T)$ with basis elements $\prod_{i\in I} x_i$ for all subsets $I \subset \{1, \ldots, n\}$.) In particular, the highest degree in which $H^*(U(n)/S)$ is nonzero is $n^2 - n + n = n^2$. (That is also clear from the fact that $U(n)/S$ is a closed orientable manifold of dimension n^2.)

Since $G \subset U(n)$, the group G acts on the quotient manifold $U(n)/S$ and also on the given manifold N.

Lemma 4.5 *$H^*_G(N \times U(n)/S)$ is a free $H^*_G N$-module with top generator in degree n^2.*

This was proved by Quillen [113, lemma 6.5].

Proof First consider $H^*_G(N \times U(n)/T)$, which we can view as the cohomology ring of $Y := G\backslash(EG \times N \times U(n))/T$. This double quotient is a $U(n)/T$-bundle over $N//G = G\backslash(EG \times N)$, so its cohomology ring is the same as that of a flag bundle (a $U(n)/T$-bundle) over $N//G$. Such a bundle is an iterated projective bundle (corresponding to a sequence of vector bundles), as in the proof of Theorem 2.13, and so its cohomology ring is known: it is $H^*(BG)[y_1, \ldots, y_n]/(e_i(y_1, \ldots, y_n) = c_i)$ where $|y_i| = 2$, e_i denotes the ith elementary symmetric function, and c_i is the ith Chern class of the given representation $G \to U(n)$. Next, $H^*_G(N \times U(n)/S)$ is the cohomology ring of $G\backslash(EG \times N \times U(n))/S$, which is a principal $(S^1)^n$-bundle over Y, corresponding to the pth powers of the obvious line bundles $L_1, \ldots, L_n$ on Y (with Chern classes $y_1, \ldots, y_n$). So, for p odd, $H^*_G(N \times U(n)/S)$ is the tensor product of the previous ring with the free graded-commutative algebra on generators $x_1, \ldots, x_n$ in degree 1. So

$$H^*_G(N \times U(n)/S) \cong (H^*_G N)\langle x_1, \ldots, x_n, y_1, \ldots, y_n\rangle/(e_i(y_1, \ldots, y_n) = c_i).$$

(For $p = 2$, we have the usual variation: $H^*_G(N \times U(n)/S)$ is a free module over $H^*_G N[y_1, \ldots, y_n]/(e_i(y_1, \ldots, y_n) = c_i)$ with basis elements $\prod_{i\in I} x_i$ for all subsets I of $\{1, \ldots, n\}$.) Thus $H^*_G(N \times U(n)/S)$ is a free $H^*_G N$-module with basis elements corresponding to a basis for the $\mathbf{F}_p$-vector space $H^*(U(n)/S)$. In particular, the top generator is in degree n^2. □

By Lemma 4.5 and the interpretation of regularity in terms of a minimal resolution (Theorem 3.14),

$$\operatorname{reg}(H_G^*(N \times U(n)/S)) = \operatorname{reg}(H_G^* N) + n^2.$$

So Theorem 4.1 reduces to showing that $\operatorname{reg}(H_G^*(N \times U(n)/S)) \leq \dim(N) + n^2 - \dim(G)$. Here we are viewing $H_G^*(N \times U(n)/S)$ as an $\mathbf{F}_p[c_1, \ldots, c_n]$-module via the homomorphism $\mathbf{F}_p[c_1, \ldots, c_n] \to H_G^*$, but it is isomorphic to $H_S^*(G\backslash(N \times U(n)))$, and so we can also view it as an H_S^*-module. It is generated in bounded degrees over both rings, by Lemma 6.3. Therefore, the properties of regularity imply that it has the same regularity over both rings. Thus it suffices to show that

$$\operatorname{reg}(H_S^*(G\backslash(N \times U(n)))) \leq \dim(N) + n^2 - \dim(G).$$

At this point, the group G disappears from the problem; it suffices to show that $\operatorname{reg}(H_S^* M) \leq \dim(M)$ for every action of an elementary abelian p-group $S = (\mathbf{Z}/p)^n$ on a manifold M with finite-dimensional $\mathbf{F}_p$-cohomology. (Apply this to the manifold $M = G\backslash(N \times U(n))$.) This will follow from Duflot's theorem on actions of elementary abelian groups [35], which we now prove.

For any compact Lie group K acting on a smooth manifold M, the fixed point set M^K is a smooth submanifold. (One can choose a K-invariant Riemannian metric on M, and then the exponential map at a point $p \in M^K$ identifies a neighborhood of zero in $(T_p M)^K$ with a neighborhood of p in M^K.) Filter M by the closed submanifolds M^V for the various subgroups $V \subset S$. Let M_i be the closed subset of points with isotropy group in S of rank at least i, and let $M_{(i)}$ be the open subset of M_i consisting of points with isotropy group of rank equal to i. For each $d \geq 0$, let $M_{(i),d}$ be the union of the connected components of $M_{(i)}$ with codimension d in M. Duflot's theorem is:

Theorem 4.6 *There is a short exact sequence of H_S^*-modules*

$$0 \to \oplus_d H_S^{*-d}(M_{(i),d}) \to H_S^*(M - M_{i+1}) \to H_S^*(M - M_i) \to 0.$$

Proof This is part of the long exact localization sequence for equivariant cohomology. One detail is that in order to write the first group as we have (without twisted coefficients), we need to observe that the normal bundle of $M_{(i)}$ in M is orientable when p is odd. (For $p = 2$, there is no issue about orientability.) That holds because the normal bundle to a connected component of M^V, for a subgroup $V \cong (\mathbf{Z}/p)^i$ of S, is a real representation of V with no trivial summands, and such a representation can be given a complex structure in a canonical way, when p is odd.

All we need to prove is the injectivity of the first homomorphism. (This will imply the surjectivity of the last homomorphism, by the long exact sequence.) It suffices to prove that pushing forward from $\coprod M_{(i),d}$ to $M - M_{i+1}$ and then pulling back gives an injection on equivariant cohomology groups. Since the restriction of cycles on one connected component of $M_{(i)}$ to another component is zero, it suffices to show for each subgroup $V \subset S$ of p-rank i that pushing forward from the subspace $(M - M_{i+1})^V$ fixed by V to $M - M_{i+1}$ and then pulling back gives an injection on equivariant cohomology groups. The result of pushing forward from a smooth closed submanifold and then pulling back is multiplication by the Euler class of the normal bundle [106, theorem 11.3]. (For a complex vector bundle, the Euler class is the top Chern class; when $p = 2$, we define the Euler class of a real vector bundle in $\mathbf{F}_2$-cohomology to be the top Stiefel-Whitney class.) Thus, it suffices to show that the Euler class of the normal bundle to $X := (M - M_{i+1})^V$ in $M - M_{i+1}$ is not a zero divisor in $H_S^*((M - M_{i+1})^V)$.

Since $S \cong (\mathbf{Z}/p)^n$, we can choose a splitting $S = V \times W$ for the subgroup $V \subset S$. Since V acts trivially on X while W acts freely on it, we have an isomorphism

$$H_S^*(X) \cong (H^*(X/W))\langle x_1, \ldots, x_v, y_1, \ldots, y_v \rangle,$$

where $|x_i| = 1$, $|y_i| = 2$, and $y_1, \ldots, y_v$ are the first Chern classes of 1-dimensional representations of $V \cong (\mathbf{Z}/p)^v$. (This is for p odd; for $p = 2$, the elements x_i are polynomial generators in degree 1, and $y_i = x_i^2$.)

Let N be the normal bundle to X in $M - M_{i+1}$, which is an S-equivariant real vector bundle on X, with a canonical complex structure for p odd. Because X is the whole fixed point space in $M - M_{i+1}$ for the subgroup V, the sub-bundle of N fixed by V is 0. Thus, if the isotypic decomposition of N with respect to V is $N = \oplus_\alpha E_\alpha$ for $\alpha \in \mathrm{Hom}(V, S^1) \cong (\mathbf{Z}/p)^n$, then $E_0 = 0$. Note that each E_α is an S-equivariant subbundle of the normal bundle N, because S commutes with its subgroup V. The Euler class $\chi(N)$ is the product of the Euler classes of the subbundles E_α, and so it suffices to show that $\chi(E_\alpha)$ is a non-zero-divisor in $H_S^* X$ for each $\alpha \neq 0$ in $\mathrm{Hom}(V, S^1)$.

Let L_α be the 1-dimensional complex representation of S given by projecting $S = V \times W$ to V and applying the representation α of V. Let $F_\alpha = E_\alpha \otimes L_\alpha^*$. Then F_α is an S-equivariant vector bundle on X on which the subgroup V acts trivially. Therefore, the Chern classes of F_α lie in the subring $H^*(X/W)$ of

$$H_S^*(X) = (H^*(X/W))\langle x_1, \ldots, x_v, y_1, \ldots, y_v \rangle,$$

with the variation for $p = 2$ as above. The first Chern class of L_α is a nonzero $\mathbf{F}_p$-linear combination of $y_1, \ldots, y_v$, and so, after a change of basis for V, we can assume that $c_1(L_\alpha) = y_1$. Write m for the rank of the bundle E_α. Since

$E_\alpha = F_\alpha \otimes L_\alpha$, we have

$$\chi(E_\alpha) = y_1^m + c_1(F_\alpha)y_1^{m-1} + \cdots + c_m(F_\alpha),$$

by the formula for the Chern classes of the tensor product of a complex vector bundle with a complex line bundle. (For $p = 2$, we use the analogous formula for the top Stiefel-Whitney class of the tensor product of a real vector bundle with a real line bundle.) Since this is a monic polynomial, it is a non-zero-divisor in $H_S^*(X)$, as we want. □

For a subgroup V of S, let $M^{(V)}$ be the locally closed submanifold of M of points with stabilizer equal to V. By Theorem 4.6, we have a filtration of $H_S^*(M)$ by finitely many H_S^*-submodules such that the subquotients are isomorphic to $H_S^*(M^{(V)})$ for the subgroups V of S, shifted in degree by the codimension of each connected component of $M^{(V)}$ in M. By the basic properties of regularity in Lemma 3.9, it follows that

$$\operatorname{reg}(H_S^*, H_S^*(M)) \leq \max_{V \subset S}\{\operatorname{reg}(H_S^*, H_S^* M^{(V)}) + \operatorname{codim}(M^{(V)} \subset M)\}.$$

So Theorem 4.1 follows if we can show that for each subgroup V of S, $H_S^*(M^{(V)})$ has regularity at most $\dim(M) - \operatorname{codim}(M^{(V)} \subset M) = \dim(M^{(V)})$ as an H_S^*-module.

Choose a splitting $S = V \times W$ for the subgroup $V \subset S \cong (\mathbf{Z}/p)^n$. The group V acts trivially on $M^{(V)}$ while W acts freely, and so we have

$$H_S^*(M^{(V)}) \cong (H^*(M^{(V)}/W))\langle x_1, \ldots, x_v, y_1, \ldots, y_v\rangle,$$

with the usual variant for $p = 2$. This is a finitely generated free module over the polynomial ring $\mathbf{F}_p[y_1, \ldots, y_v]$, with top generator $x_1 \cdots x_v$ (in degree v) times a top-degree element of $H^*(M^{(V)}/W)$. By Lemma 3.10, the regularity of $H_S^*(M^{(V)})$ over H_S^* is equal to its regularity over $\mathbf{F}_p[y_1, \ldots, y_v]$. In terms of Definition 3.8, the graded polynomial ring $\mathbf{F}_p[y_1, \ldots, y_v]$ has σ equal to v. By Theorem 3.14, the regularity of $H_S^*(M^{(V)})$ is equal to its top degree as a free $\mathbf{F}_p[y_1, \ldots, y_v]$-module, minus v. That is equal to the top degree of the $\mathbf{F}_p$-cohomology of the manifold $M^{(V)}/W$, which is at most the dimension of $M^{(V)}$, as we want. □

5

Generators for the Chow Ring

Although Chow rings of classifying spaces are more mysterious than cohomology rings, one aspect of the Chow ring was understood earlier. In the 1990s, I showed that the Chow ring of a group with a faithful representation of degree n was generated in degrees at most n^2 (Theorem 5.1). The proof was much simpler than the proof of Symonds's 2010 regularity theorem for group cohomology, which gave an analogous bound on generators for group cohomology (Corollary 4.3).

In this chapter, we improve the bound on the degree of generators for the Chow ring by about a factor of 2, to $n(n-1)/2$ (Theorem 5.2). The proof is still simple. Later we will bound the regularity of the Chow ring (Theorem 6.5). That is a stronger result – for example, it gives information about relations as well as generators – but the proof is more complicated.

We give examples to show that the bound $n(n-1)/2$ for generators of the Chow ring is optimal, and that Symonds's bound n^2 for generators of the cohomology ring is at least close to optimal, among arbitrary finite groups.

5.1 Bounding the generators of the Chow ring

The following bound goes back to the 1990s [138, theorem 14.1]. We prove a slight generalization as Lemma 6.3.

Theorem 5.1 *Let G be an affine group scheme over a field k with a faithful representation V of dimension n. Then*

$$CH^*GL(n)/G \cong CH^*BG/(c_1V, \ldots, c_nV).$$

As a result, CH^*BG is generated as a module over the Chern classes $\mathbf{Z}[c_1V, \ldots, c_nV]$ by elements of degree at most $n^2 - \dim(G)$. It follows that the ring CH^*BG is generated by elements of degree at most $\max(n, n^2 - \dim(G))$.

At the time, no bound was known for the generators of the cohomology ring of a finite group. It is remarkable that Symonds found a very similar bound for the cohomology in 2010, Corollary 4.3 in this book. (Note that there is no essential difference between compact Lie groups and complex algebraic groups for homotopy-theoretic purposes. Every complex algebraic group is homotopy equivalent to a maximal compact subgroup, and conversely the inclusion from any compact Lie group to its complexification is a homotopy equivalence.)

It turns out that Theorem 5.1 can be improved by about a factor of two, as follows. The improved bound turns out to be optimal. We concentrate on the case of finite groups.

Theorem 5.2 *Let G be an finite group with a faithful representation V of dimension n over a field k with $|G|$ invertible in k. Then, with G considered as an algebraic group over k, the Chow ring CH^*BG is generated as a module over the Chern classes $\mathbf{Z}[c_1 V, \ldots, c_n V]$ by elements of degree at most $n(n-1)/2$. A fortiori, the ring CH^*BG is generated by elements of degree at most n if $n \leq 2$, and at most $n(n-1)/2$ if $n \geq 3$.*

Proof By Theorem 5.1, it suffices to show that $CH^i(GL(n)/G) = 0$ for $i > n(n-1)/2$. The homogeneous variety $GL(n)/G$ has dimension n^2, and so this is not immediately clear. But the proof turns out to be straightforward.

Let U be the group of strictly upper-triangular matrices in $GL(n)$. The homogeneous space $U\backslash GL(n)$ is a quasi-projective variety [147, pp. 122–123], and so its quotient by the finite group G is a quasi-projective variety [32, 2nd ed., remarque V.5.1].

The group U has dimension $n(n-1)/2$, and has a filtration by normal subgroups with all quotients being the additive group G_a over k. Since $|G|$ is invertible in k, the intersection of U with any conjugate of G is trivial. Equivalently, G acts freely on $U\backslash GL(n)$. So the quotient variety $U\backslash GL(n)/G$ is smooth.

We have a principal U-bundle $GL(n)/G \to U\backslash GL(n)/G$. The pullback map $CH^*(U\backslash GL(n)/G) \to CH^*(GL(n)/G)$ is an isomorphism. Indeed, this follows by viewing this principal U-bundle as the composite of a sequence of principal G_a-bundles. The pullback on Chow rings associated to a principal G_a-bundle over a smooth variety is an isomorphism, by homotopy invariance of Chow groups (Lemma 2.2).

Therefore, we have $CH^i(GL(n)/G) = 0$ for $i > n(n+1)/2$, since $U\backslash GL(n)/G$ has dimension $n(n+1)/2$. We want to go further by n. Let T be the group of diagonal matrices in $GL(n)$, so that $T \cong (G_m)^n$. Since T normalizes U, T acts on the variety $U\backslash GL(n)/G$, by $t(UxG) = UtxG$. This action is not free; indeed, every element of G is conjugate in $GL(n)$ to some element of T.

Nonetheless, the action of T on $U\backslash GL(n)/G$ has finite stabilizer groups, since G is finite. Applying the following easy lemma completes the proof. □

Lemma 5.3 *Let X be a smooth variety over a field k. Let T be a split torus that acts on X with finite stabilizer groups. Then $CH^i X = 0$ for $i > \dim(X) - \dim(T)$.*

Proof The T-action on X has only finitely many different stabilizer group schemes in T. Stratify X according to the possible stabilizer groups. The strata are smooth, because the fixed points of any finite subgroup scheme of T on X form a smooth subscheme. By the basic exact sequence of Chow groups (Lemma 2.1), the lemma reduces to the special case where T acts on X with the same finite stabilizer group A everywhere. Here T/A is a split torus, and so we have reduced to the case where a split torus T acts freely on a smooth variety X. That is, X is a principal $(G_m)^n$-bundle over a smooth variety B, for some n.

It suffices to show that the pullback $CH^*B \to CH^*X$ is surjective, since B has dimension $\dim(X) - \dim(T)$. This surjectivity follows by induction from the case of a principal G_m-bundle $X \to B$; that is, X is the complement of the zero section in the total space of a line bundle L over B. In that case, we have

$$CH^*(X) \cong CH^*(B)/(c_1 L)$$

by Lemma 2.4, which gives the surjectivity. (Alternatively, we could handle a principal $(G_m)^n$-bundle in one step, by viewing it as an open subset of a vector bundle.) □

The same proof, with $GL(n)$ replaced by a product of groups $GL(n_i)$, gives the following sharper bound when the given representation is reducible. This will also follow from a later result, Theorem 6.5, which strengthens Theorem 5.2 to a bound on the regularity of the Chow ring, along the lines of Symonds's results on the cohomology ring.

Theorem 5.4 *Let G be an finite group with a faithful representation V of dimension n over a field k with $|G|$ invertible in k. Write V as a direct sum of irreducible representations V_i of dimension n_i, $1 \leq i \leq s$. Then, with G considered as an algebraic group over k, the Chow ring CH^*BG is generated as a module over the Chern classes $\mathbf{Z}[c_j(V_i) : 1 \leq i \leq s,\ 1 \leq j \leq n_i]$ by elements of degree at most $\sum_i n_i(n_i - 1)/2$. A fortiori, the ring CH^*BG is generated by elements of degree at most $max(n_1, \ldots, n_s, \sum_i n_i(n_i - 1)/2)$.*

For completeness, we mention the analogous bound for cohomology. This is immediate from Symonds's regularity theorem, Theorem 4.1.

Theorem 5.5 *Let G be an finite group with a faithful complex representation V of dimension n. Write V as a direct sum of irreducible representations V_i of dimension n_i, $1 \le i \le s$. Then the cohomology ring $H_G^* = H^*(BG, \mathbf{F}_p)$ is generated as a module over the Chern classes $\mathbf{F}_p[c_j(V_i) : 1 \le i \le s, 1 \le j \le n_i]$ by elements of degree at most $\sum_i n_i^2$. A fortiori, the ring H_G^* is generated by elements of degree at most $max(2n_1, \ldots, 2n_s, \sum_i n_i^2)$.*

5.2 Optimality of the bounds

In this section, we show that the bound $n(n-1)/2$ in Theorem 5.2 for generators of the Chow ring is optimal. We also show that Symonds's bound n^2 for generators of the cohomology ring (Corollary 4.3) is at least close to being optimal.

We first show that the bound $n(n-1)/2$ for generators of CH^*BG as a $\mathbf{Z}[c_1V, \ldots, c_nV]$-module is optimal. Write CH_G^* for $CH^*(BG)/p$, for a fixed prime number p. Let G be the group $(\mathbf{Z}/p)^n$, for any positive integer n and any prime number p. Then G has a faithful complex representation V of dimension n, the direct sum of 1-dimensional representations $V_1, \ldots, V_n$. In this example, CH_G^* is the polynomial ring $\mathbf{F}_p[y_1, \ldots, y_n]$, $|y_i| = 1$, where the y_i are the first Chern classes of the representations $V_1, \ldots, V_n$. The pullback homomorphism $CH_{GL(n)}^* = \mathbf{F}_p[c_1, \ldots, c_n] \to CH_G^*$ sends $c_1, \ldots, c_n$ to the elementary symmetric polynomials in $y_1, \ldots, y_n$.

It is a classical algebraic fact that the ring of all polynomials $R[y_1, \ldots, y_n]$ over any commutative ring R is a free module over the subring of symmetric polynomials, with module generators $y_1^{a_1} \cdots y_n^{a_n}$ for $0 \le a_i \le i-1$ [19, theorem IV.6.1]. In particular, the highest-degree generator $y_2 y_3^2 \cdots y_n^{n-1}$ is in degree $n(n-1)/2$. Thus the bound $n(n-1)/2$ for the degree of module generators of CH_G^* over $\mathbf{F}_p[c_1V, \ldots, c_nV]$ (Theorem 5.2) is optimal, for every positive integer n and every prime number p.

This example is unsatisfying in some respects. First, as a ring, CH_G^* is generated in degree 1, and so this example is not interesting if we just look at CH_G^* as a ring. Second, the representation V is reducible, and so it seems artificial to view it as a module over the Chern classes of the whole representation $V = V_1 \oplus \cdots \oplus V_n$. It is more natural to view CH_G^* as a module over $\mathbf{F}_p[c_1V_1, \ldots, c_1V_n]$; Theorem 5.4 shows that this module is generated in degrees at most $\sum_{i=1}^n 1(1-1)/2 = 0$, which is clearly more useful. But it leaves open the question of how good the bounds of Theorem 5.2 are for V irreducible.

Both of these criticisms are answered by the following example. Let G be the wreath product $A_n \wr \mathbf{Z}/p = A_n \ltimes (\mathbf{Z}/p)^n$, for any $n \ge 3$ and any prime number p. Here A_n denotes the alternating group. Then G has a faithful irreducible complex representation of dimension n. (Indeed, G is contained in the normalizer $S_n \wr G_m$ of a maximal torus in $GL(n)$.)

What can we say about algebra generators for CH_G^*? Consider the restriction map

$$\begin{aligned} CH_G^* &\to (CH^*_{(\mathbf{Z}/p)^n})^{A_n} \\ &\cong \mathbf{F}_p[y_1, \ldots, y_n]^{A_n} \\ &\cong \mathbf{F}_p[c_1, \ldots, c_n] \oplus D \cdot \mathbf{F}_p[c_1, \ldots, c_n]. \end{aligned}$$

Here we define

$$D = \sum_{\sigma \in A_n} y_{\sigma(1)}^{n-1} y_{\sigma(2)}^{n-2} \cdots y_{\sigma(n-1)},$$

which is a polynomial of degree $n(n-1)/2$. This description of the invariants of the alternating group is classical [10, p. 104]. For p odd, we could use the square root of the discriminant, $\Delta = \prod_{i<j}(y_i - y_j)$, in place of D as a generator of the invariants of the alternating group, but D works in any characteristic p.

Let u be the transfer of the class $y_1^{n-1} y_2^{n-2} \cdots y_{n-1}$ from $CH^{n(n-1)/2}_{(\mathbf{Z}/p)^n}$ to $CH_G^{n(n-1)/2}$. Then the restriction of u to $CH^*_{(\mathbf{Z}/p)^n}$ is the sum of all conjugates of $y_1^{n-1} y_2^{n-2} \cdots y_{n-1}$ by the elements of $G/(\mathbf{Z}/p)^n = A_n$ (Lemma 2.15). This sum is exactly the class D. Here D is not in the subring generated by lower-degree A_n-invariants, since they are S_n-invariant while D is not. Therefore u is indecomposable in the ring CH_G^*, and it has degree $n(n-1)/2$.

Thus the group $G = A_n \wr \mathbf{Z}/p$ shows that the degree bound $n(n-1)/2$ from Theorem 5.2 for generators of the mod p Chow ring of a finite group is optimal, for all $n \geq 3$ and all prime numbers p. Moreover, the faithful representation of dimension n is irreducible, in this example.

The same group shows that Symonds's bound n^2 for generators of the mod p cohomology ring of a finite group (Corollary 4.3) is at least close to being optimal. Let $n \geq 3$ and let p be a prime number. For p odd, let G be the wreath product $A_n \wr \mathbf{Z}/p = A_n \ltimes (\mathbf{Z}/p)^n$; for $p = 2$, define G to be the wreath product $A_n \wr \mathbf{Z}/4$ instead. Let N be the normal subgroup $(\mathbf{Z}/p)^n$ for p odd, or $(\mathbf{Z}/4)^n$ for $p = 2$. The group G has a faithful irreducible complex representation of dimension n.

What can we say about generators of the ring $H_G^* := H^*(G, \mathbf{F}_p)$? Consider the restriction map

$$\begin{aligned} H_G^* &\to (H_N^*)^{A_n} \\ &\cong \mathbf{F}_p\langle x_1, \ldots, x_n, y_1, \ldots, y_n\rangle^{A_n}. \end{aligned}$$

Here the elements x_i have degree 1 and generate an exterior algebra, while the elements y_i have degree 2 and generate a polynomial algebra. We needed to define N to be $(\mathbf{Z}/4)^n$ rather than $(\mathbf{Z}/2)^n$ when $p = 2$ in order to have this description of H_N^*. The elements $y_1, \ldots, y_n$ are the first Chern classes of 1-dimensional complex representations of N.

Here A_n permutes the generators x_i and the generators y_i. Therefore, the ring homomorphism from H_N^* to $\mathbf{F}_p[y_1, \ldots, y_n]$ that sends each x_i to zero is A_n-equivariant. So we have a ring homomorphism

$$\begin{aligned} \mathbf{F}_p\langle x_1, \ldots, x_n, y_1, \ldots, y_n\rangle^{A_n} &\to \mathbf{F}_p[y_1, \ldots, y_n]^{A_n} \\ &\cong \mathbf{F}_p[c_1, \ldots, c_n] \oplus D \cdot \mathbf{F}_p[c_1, \ldots, c_n], \end{aligned}$$

where $D = \sum_{\sigma \in A_n} y_{\sigma(1)}^{n-1} y_{\sigma(2)}^{n-2} \cdots y_{\sigma(n-1)}$. Here D is in degree $n(n-1) = n^2 - n$, since the elements y_i are in degree 2 in cohomology.

We constructed an element u in the Chow ring of G, as an explicit transfer, whose restriction to the Chow ring of N is the class D. Therefore, we have a cohomology class u in $H_G^{n^2-n}$ whose restriction to N is the cohomology class D. The element D is indecomposable in the invariant ring $\mathbf{F}_p[y_1, \ldots, y_n]^{A_n}$, because the lower-degree elements are S_n-invariant and D is not. Therefore, u is indecomposable in the cohomology ring of G.

Thus the group $G = A_n \wr \mathbf{Z}/p$, or $G = A_n \wr \mathbf{Z}/4$ for $p = 2$, shows that the degree bound n^2 from Corollary 4.3 for generators of the mod p cohomology ring of a finite group is nearly optimal, for all $n \geq 3$ and all prime numbers p. Namely, we need an algebra generator in degree $n^2 - n$, at least. The faithful representation of dimension n is irreducible in this example.

6

Regularity of the Chow Ring

We prove here that the mod p Chow ring of any finite group scheme G has regularity at most zero (Theorem 6.5), as Symonds proved for the cohomology ring. The result has an assumption on the characteristic of the base field, probably unnecessary. The proof is closely analogous to Symonds's argument, using an analog of Duflot's theorem (on actions of elementary abelian groups) for Chow rings. We also prove analogous bounds for the regularity of motivic cohomology, which includes the Chow ring as a special case.

The regularity theorem implies our earlier bound on degrees of the generators for the Chow ring, Theorem 5.2, but it is more powerful. For example, the regularity theorem also bounds the degrees of relations in the Chow ring. The rest of the book uses the regularity theorems for calculations as well as to prove general results on Chow rings and cohomology rings.

Section 6.2 summarizes the properties of motivic cohomology, a natural generalization of the Chow ring. We use motivic cohomology in the proofs of a few later results about Chow rings, notably Yagita's theorem, which computes the Chow ring of a finite group up to F-isomorphism, Theorem 8.10. Theorem 6.10 proves the bound "regularity ≤ 0" for the motivic cohomology of a finite group, generalizing the case of the Chow ring (Theorem 6.5). Theorem 6.10 is not used in the rest of the book.

6.1 Bounding the regularity of the Chow ring

Definition 6.1 Let R be a commutative graded algebra, or a graded-commutative algebra, over a field k. We assume that $R_0 = k$. Let $\overline{k}$ be an algebraic closure of k. Define $\sigma(R)$ to be the minimum of the numbers $\sum_i(|y_i| - 1)$ over all homomorphisms from graded polynomial rings $T = \overline{k}[y_1, \ldots, y_n]$ to $R \otimes_k \overline{k}$ such that $R \otimes_k \overline{k}$ is generated as a T-module by a set of elements of bounded degree.

For the purpose of this book, we could have defined $\sigma(R)$ without the field extension to $\overline{k}$. But $\sigma(R)$ as defined here is potentially a smaller number because of the field extension. Since our estimates of degrees of generators for group cohomology will involve the number $\sigma(R)$ for suitable rings R, it may be convenient to know that it suffices to find systems of parameters (roughly speaking) for these rings after a field extension.

The definition of $\sigma(R)$ might be considered to be artificial; it does not have a cohomological interpretation, as invariants such as depth or Castelnuovo-Mumford regularity do. Nonetheless, it has one very convenient formal property: if $R \to S$ is a homomorphism of graded rings such that S is generated as an R-module by elements of bounded degree, then $\sigma(S) \leq \sigma(R)$. This is immediate from the definition. (Note that it is crucial for this inequality to allow the polynomial ring in the definition of $\sigma(R)$ to have higher dimension than R.) We deduce the following inequality for Chow rings and cohomology. Fix a prime number p and write $CH^*_G = CH^*(BG)/p$ and $H^*_G = H^*(BG, \mathbf{F}_p)$.

Lemma 6.2 *Let $H \subset G$ be affine group schemes over a field k. Let p be a prime number. Then $\sigma(CH^*_H) \leq \sigma(CH^*_G)$. Likewise, for compact Lie groups $H \subset G$, $\sigma(H^*_H) \leq \sigma(H^*_G)$.*

Proof By the formal property of σ mentioned previously, it suffices to show that CH^*_H is generated over CH^*_G by elements of bounded degree, and likewise for cohomology. Consider a faithful representation V of G of some dimension n. Then CH^*_H is generated in bounded degrees as a module over the Chern classes $\mathbf{F}_p[c_1, \ldots, c_n]$ by Theorem 5.2 (or Theorem 5.1). A fortiori, CH^*_H is generated in bounded degrees as a module over CH^*_G, as we want. The same argument applies to cohomology, since H^*_H is finitely generated as a module over the Chern classes of a faithful complex representation, by Venkov (Theorem 1.1). □

The following lemma, generalizing Theorem 5.1, will be used in the proof of Theorem 6.5.

Lemma 6.3 *Let G be an affine group scheme of finite type over a field k. Let G act on a smooth variety X over k. Then $CH^*_G(X)$ is generated by elements of bounded degree as a CH^*BG-module. More strongly, for a faithful representation $G \subset GL(n)$, $CH^*_G(X)$ is generated by elements of degree at most $\dim(X) + n^2 - \dim(G)$ as a $\mathbf{Z}[c_1, \ldots, c_n]$-module.*

Proof It suffices to prove the second statement, that $CH^*_G(X)$ is generated by elements of degree at most $\dim(X) + n^2 - \dim(G)$ as a $\mathbf{Z}[c_1, \ldots, c_n]$-module. Let $Y = (X \times GL(n))/G$; then $GL(n)$ acts on Y, and $CH^*_G(X) \cong CH^*_{GL(n)}Y$. So it suffices to show that if $GL(n)$ acts on a smooth variety Y, then $CH^*_{GL(n)}Y$ is generated by elements of degree at most $\dim(Y)$. This follows if we can prove the isomorphism

$$CH^*Y \cong CH^*_{GL(n)}(Y)/(c_1, \ldots, c_n),$$

since $CH^iY = 0$ for $i > \dim(Y)$. By considering a finite-dimensional approximation X to $Y//GL(n)$, it suffices to show that $CH^*E = CH^*X/(c_1V, \ldots, c_nV)$ when E is the principal $GL(n)$-bundle over a smooth k-scheme X associated to a vector bundle V.

Let B be the subgroup of upper-triangular matrices in $GL(n)$. Then E/B is the flag bundle $\mathrm{Fl}(V)$ over X, whose Chow ring is computed in the proof of Theorem 2.13:

$$CH^*(E/B) = CH^*X[y_1, \ldots, y_n]/(e_i(y_1, \ldots, y_n) = c_iV),$$

where $|y_i| = 1$ and e_i denotes the ith elementary symmetric function, for $i = 1, \ldots, n$. By the computation of the Chow ring of a principal G_m-bundle (Lemma 2.4) plus homotopy invariance of Chow groups (Lemma 2.2),

$$\begin{aligned} CH^*E &= CH^*(E/B)/(y_1, \ldots, y_n) \\ &= CH^*X/(c_1V, \ldots, c_nV). \end{aligned}$$

□

From now on, fix a prime number p and let CH^*_GX denote the equivariant Chow ring modulo p. Lemmas 6.3 and 3.10 imply that the regularity of the ring CH^*_GY for any smooth G-variety Y is equal to the regularity of CH^*_GY as a module over $\mathbf{F}_p[c_1, \ldots, c_n]$ for any faithful representation $G \subset GL(n)$.

Estimating $\sigma(R)$ is useful when we know the regularity of a ring, for the following reason.

Lemma 6.4 *Let R be a graded algebra over a field k. Suppose that* $\mathrm{reg}(R) \leq 0$. *Suppose that the minimum in the definition of $\sigma(R)$ occurs for a graded polynomial ring $k[y_1, \ldots, y_n]$. Then R is generated as a k-algebra by elements of degree at most* $\max\{|y_i|, \sigma(R)\}$, *modulo relations in degree at most* $\max\{|y_i|, \sigma(R) + 1, 2\sigma(R)\}$. *(Thus, if at least two elements y_i have degree at least 2, then the k-algebra R is generated in degree at most $\sigma(R)$ modulo relations in degree $2\sigma(R)$.)*

Proof It suffices to show that $R \otimes_k \overline{k}$ is generated as a $\overline{k}$-algebra by elements of degree at most $\sigma(R)$. Let $R \otimes_k \overline{k}$ be generated as a module over a graded polynomial ring $T = \overline{k}[y_1, \ldots, y_n]$ by elements of bounded degree. Since R (and hence $R \otimes_k \overline{k}$) has regularity at most zero, $R \otimes_k \overline{k}$ is generated as a T-module by elements z_j of degree at most $\sum_{i=1}^n (|y_i| - 1)$. Therefore the algebra $R \otimes \overline{k}$ is generated by the elements y_i and z_j, which have degree at most $\max\{|y_i|, \sigma(R)\}$.

Since $R \otimes_k \overline{k}$ has regularity at most zero, we also know that the relations among the T-module generators z_j are in degrees at most $1 + \sigma(R)$. It is straightforward to deduce that the relations among the algebra generators y_i and z_j for $R \otimes_k \overline{k}$ are in degrees at most $\max\{|y_i|, \sigma(R)+1, 2\sigma(R)\}$ [131, proposition 2.1]. □

Theorem 6.5 *Let G be a finite group scheme over a field k. Let p be a prime number that is invertible in k. Then the ring $CH^*_G = CH^*(BG)/p$ (graded with CH^i in degree i) has regularity ≤ 0.*

The regularity of the Chow ring of a finite group can be less than 0, in contrast to the cohomology ring (Corollary 4.2). For example, $CH^*_{\mathbf{Z}/p} = \mathbf{F}_p[y]$, where $|y| = 1$, has regularity zero, whereas the symmetric group S_p has mod p Chow ring $\mathbf{F}_p[y]$, where $|y| = p-1$, which has regularity $-(p-2)$. The Chow ring of a p-group often has regularity zero, but Lemma 13.6 shows that the regularity can be less than zero.

To get an idea of what Theorem 6.5 says, note that it immediately implies Theorem 5.2: a finite group G with a faithful k-representation of dimension n has CH^*BG generated in degree at most $n(n-1)/2$ as a $\mathbf{Z}[c_1, \ldots, c_n]$-module, once we know from Theorem 5.1 that this module is generated in a bounded set of degrees. It suffices to bound the generators modulo p for each prime p, since the groups $CH^i(BG)$ are killed by the order of G for $i > 0$. By Lemma 6.2, the inclusion $G \subset GL(n)$ implies that $\sigma(G) \leq \sigma(GL(n)) \leq \sum_{i=1}^n (|c_i| - 1) = \sum_{i=1}^n (i-1) = n(n-1)/2$. By Theorem 6.5, the ring $CH^*_G = CH^*(BG)/p$ has regularity at most 0, and so it is generated as an $\mathbf{F}_p[c_1, \ldots, c_n]$-module by elements of degree at most $n(n-1)/2$, as we wanted.

The original proof of Theorem 5.2 is simpler than the proof of Theorem 6.5. But Theorem 6.5 is more powerful, since it bounds the generators of the Chow ring in terms of any set of elements over which the Chow ring is generated by elements of bounded degree, not just the Chern classes of a faithful representation. Also, Theorem 6.5 bounds the degrees of the relations in the Chow ring, not only the generators.

Proof For clarity, we first prove Theorem 6.5 for G a finite k-group scheme of order invertible in k. At the end, we explain the generalization to more general finite group schemes.

Choose a faithful representation of G over k, $G \subset GL(n)$. We know that CH^*_G is generated in bounded degrees as an $\mathbf{F}_p[c_1, \ldots, c_n]$-module by Theorem 5.2 (or the weaker Theorem 5.1), and we want to show that it has regularity at most 0 as an $\mathbf{F}_p[c_1, \ldots, c_n]$-module. (By the properties of regularity, this property does not depend on the choice of faithful representation.)

Let T be the subgroup $(G_m)^n \subset GL(n)$ of diagonal matrices, and $S = (\mu_p)^n$ the p-torsion subgroup scheme of T. The quotient variety $GL(n)/T$ is an iterated affine-space bundle over the flag manifold $GL(n)/B$, and so its Chow ring is well known: $CH^*(GL(n)/T)/p$ is the quotient of the polynomial ring $\mathbf{F}_p[y_1, \ldots, y_n]$ with $|y_i| = 1$ by the elementary symmetric polynomials $c_1, \ldots, c_n$. This calculation is also immediate from Theorem 5.2. The quotient variety $GL(n)/S$ is a principal $(G_m)^n$-bundle over $GL(n)/T$ corresponding to the pth powers of the line bundles $L_1, \ldots, L_n$ corresponding to the obvious 1-dimensional representations of T. Therefore

$$\begin{aligned} CH^*(GL(n)/S)/p &\cong (CH^*(GL(n)/T)/p)/(c_1(L_1^{\otimes p}), \ldots, c_1(L_n^{\otimes p})) \\ &\cong CH^*(GL(n)/T)/p. \end{aligned}$$

In particular, the highest degree in which $CH^*(GL(n)/S)/p$ is nonzero is $n(n-1)/2$.

Since $G \subset GL(n)$, the group G acts on the quotient variety $GL(n)/S$.

Lemma 6.6 *$CH^*_G GL(n)/S$ is a free CH^*_G-module with top generator in degree $n(n-1)/2$.*

Proof This is a straightforward analog of Quillen's Lemma 4.5. First consider $CH^*_G(GL(n)/T)$, which we can view as the mod p Chow ring of $Y := G\backslash(EG \times GL(n))/T$. (To deal only with finite-dimensional varieties, one can replace EG in this argument by an open subset $W - Z$ of some representation W of G such that G acts freely on $W - Z$ and Z has high codimension in W compared to the Chow groups we consider.) This double quotient is a $GL(n)/T$-bundle over BG, so its Chow ring is the same as that of a flag bundle (a $GL(n)/B$-bundle) over BG. Such a bundle is an iterated projective bundle (corresponding to a vector bundle), and so its Chow ring is known: it is $CH^*_G[y_1, \ldots, y_n]/(e_i(y_1, \ldots, y_n) = c_i)$ where $|y_i| = 1$, e_i denotes the ith elementary symmetric function, and c_i is the ith Chern class of the given representation $G \to GL(n)$. Next, $CH^*_G GL(n)/S$ is the mod p Chow ring of $G\backslash(EG \times GL(n))/S$, which is a principal $(G_m)^n$-bundle over Y, corresponding to the pth powers of the obvious line bundles $L_1, \ldots, L_n$ on Y (with Chern classes $y_1, \ldots, y_n$). So $CH^*_G GL(n)/S$ is the quotient of the previous ring by 0, and so

$$CH^*_G GL(n)/S \cong CH^*_G[y_1, \ldots, y_n]/(e_i(y_1, \ldots, y_n) = c_i).$$

This is a free CH^*_G-module with basis elements corresponding to a basis for the $\mathbf{F}_p$-vector space $CH^*(GL(n)/S)/p$. In particular, the top generator is in degree $n(n-1)/2$. □

By Lemma 6.6,

$$\operatorname{reg}(CH^*_G GL(n)/S) = \operatorname{reg}(CH^*_G) + n(n-1)/2.$$

So Theorem 6.5 reduces to showing that $\operatorname{reg}(CH^*_G GL(n)/S) \leq n(n-1)/2$. Here we are viewing $CH^*_G GL(n)/S$ as a CH^*_G-module, but it is isomorphic to $CH^*_S GL(n)/G$, and so we can also view it as a CH^*_S-module. It is generated in bounded degrees over both rings, by Lemma 6.3. Therefore, the properties of regularity imply that it has the same regularity over both rings. Thus it suffices to show that

$$\operatorname{reg}(CH_S^* GL(n)/G) \le n(n-1)/2.$$

Symonds's proof that H_G^* has regularity ≤ 0 uses Duflot's theorem on actions of elementary abelian groups at this point (Theorem 4.6). We generalize Duflot's theorem to the Chow ring instead of cohomology.

Let U be the group of strictly upper-triangular matrices in $GL(n)$. Since the diagonal torus T normalizes U, T acts on the variety $M := U\backslash GL(n)/G$, by $t(UxG) = UtxG$. Here G acts freely on $U\backslash GL(n)$, since $|G|$ is invertible in k and hence U intersects any conjugate of G only in the identity element. The homogeneous space $U\backslash GL(n)$ is a smooth quasi-projective variety [147, pp. 122–123], and so its quotient M by the finite group scheme G is a quasi-projective variety [32, 2nd ed., remarque V.5.1]. Since the action is free, M is smooth of dimension $n(n+1)/2$ over k. Since $S \cong (\mu_p)^n$ is a subgroup scheme of T, S also acts on M. Since U is a repeated extension of additive groups, homotopy invariance of Chow groups (Lemma 2.2) implies that $CH_S^* GL(n)/G \cong CH_S^* M$. So it suffices to show that $CH_S^* M$ has regularity at most $n(n-1)/2$.

Filter M by the closed subsets M^V for the various subgroup schemes $V \subset S$. They are smooth subschemes of M. Each M^V is mapped into itself by the torus T, because T commutes with S. Let M_i be the closed subset of points with isotropy group in S of rank at least i, and let $M_{(i)}$ be the open subset of M_i consisting of points with isotropy group of rank equal to i. For each $d \ge 0$, let $M_{(i),d}$ be the union of the connected components of $M_{(i)}$ with codimension d in M. The analog for the Chow ring of Duflot's theorem on cohomology is:

Theorem 6.7 *There is a short exact sequence of CH_S^*-modules*

$$0 \to \oplus_d CH_S^{*-d} M_{(i),d} \to CH_S^*(M - M_{i+1}) \to CH_S^*(M - M_i) \to 0.$$

Proof This is the usual exact sequence for equivariant Chow groups, Lemma 2.9, tensored over $\mathbf{Z}$ with $\mathbf{F}_p$. All we need to prove is the injectivity of the first homomorphism.

It suffices to prove that pushing forward from $\coprod M_{(i),d}$ to $M - M_{i+1}$ and then pulling back gives an injection on equivariant Chow groups. Since the restriction of cycles on one connected component of $M_{(i)}$ to another component is zero, it suffices to show for each subgroup scheme $V \subset S$ of p-rank i that pushing forward from the subspace $(M - M_{i+1})^V$ fixed by V to $M - M_{i+1}$ and then pulling back gives an injection on equivariant Chow groups. The result of pushing forward from a smooth closed subvariety and then pulling back is multiplication by the Euler class of the normal bundle [43]. Thus, it suffices to show that the Euler class of the normal bundle to $X := (M - M_{i+1})^V$ in $M - M_{i+1}$ is not a zero divisor in $CH_S^*(M - M_{i+1})^V$.

Since $S \cong (\mu_p)^n$, we can choose a splitting $S = V \times W$ for the subgroup $V \subset S$. Since V acts trivially on X while W acts freely on it, we have an isomorphism

$$CH_S^* X \cong CH^*(X/W)[y_1, \dots, y_v],$$

where $y_1, \dots, y_v$ are the first Chern classes of 1-dimensional representations of $V \cong (\mu_p)^v$.

Let N be the normal bundle to X in $M - M_{i+1}$, which is an S-equivariant vector bundle on X. Because X is the whole fixed point space in $M - M_{i+1}$ for the subgroup V, and p is invertible in the base field k, the sub-bundle of N fixed by V is 0. Thus, if the isotypic decomposition of N with respect to V is $N = \oplus_\alpha E_\alpha$ for $\alpha \in \mathrm{Hom}(V, G_m) \cong (\mathbf{Z}/p)^n$, then $E_0 = 0$. Note that each E_α is an S-equivariant subbundle of the normal bundle N, because S commutes with its subgroup V. The Euler class of N is the product of the Euler classes of the subbundles E_α, and so it suffices to show that $\chi(E_\alpha)$ is a non-zero-divisor in $CH_S^* X$ for each $\alpha \neq 0$ in $\mathrm{Hom}(V, G_m)$.

Let L_α be the 1-dimensional representation of S given by projecting $S = V \times W$ to V and applying the representation α of V. Let $F_\alpha = E_\alpha \otimes L_\alpha^*$. Then F_α is an S-equivariant vector bundle on X on which the subgroup V acts trivially. Therefore, the Chern classes of F_α lie in the subring $CH^*(X/W)/p$ of

$$CH_S^* X = (CH^*(X/W)/p)[y_1, \dots, y_v].$$

The first Chern class of L_α is a nonzero $\mathbf{F}_p$-linear combination of $y_1, \dots, y_v$, and so, after a change of basis for V, we can assume that $c_1(L_\alpha) = y_1$. Write m for the rank of the bundle E_α. Since $E_\alpha = F_\alpha \otimes L_\alpha$, we have

$$\chi(E_\alpha) = y_1^m + c_1(F_\alpha) y_1^{m-1} + \cdots + c_m(F_\alpha),$$

by the formula for the Chern classes of the tensor product of a vector bundle with a line bundle. Since this is a monic polynomial, it is a non-zero-divisor in $CH_S^* X$, as we want. □

Let $M^{(V)}$ be the locally closed smooth subscheme of M where the stabilizer subgroup in S is equal to V. By Theorem 6.7, we have a filtration of $CH_S^* M$ by finitely many CH_S^*-submodules such that the subquotients are isomorphic to $CH_S^* M^{(V)}$ for the subgroups V of S, shifted in degree by the codimension of $M^{(V)}$ in M. By the basic properties of regularity in Lemma 3.9, it follows that

$$\mathrm{reg}(CH^* BS, CH_S^* M) \leq \max_{V \subset S} \{\mathrm{reg}(CH^* BS, CH_S^* M^{(V)}) + \mathrm{codim}(M^{(V)} \subset M)\}.$$

So Theorem 6.5 follows if we can show that for each subgroup V of S, $CH_S^* M^{(V)}$ has regularity at most $n(n-1)/2 - \text{codim}(M^{(V)} \subset M) = \dim(M^{(V)}) - n$ as a CH_S^*-module.

Choose a splitting $S = V \times W$ for the subgroup $V \subset S \cong (\mu_p)^n$. The group V acts trivially on $M^{(V)}$ while W acts freely, and so we have

$$CH_S^* M^{(V)} \cong (CH^*(M^{(V)}/W)/p)[y_1, \ldots, y_v],$$

by the Chow Künneth formula, Lemma 2.12. So the regularity of $CH_S^* M^{(V)}$ as a CH_S^*-module is equal to the regularity of $CH^*(M^{(V)}/W)/p$ as a CH_W^*-module. Since $CH^*(M^{(V)}/W)/p$ is 0 in high degrees, its regularity as a CH_W^*-module is equal to the highest degree in which it is nonzero, by Lemma 3.9. So it suffices to show that $CH^i(M^{(V)}/W)/p$ is zero for $i > \dim(M^{(V)}) - n$.

The torus $T = (G_m)^n$ acts with finite stabilizers on $M = U\backslash GL(n)/G$. Because T commutes with the action of S, T also acts on the locally closed subscheme $M^{(V)}$ and hence on $M^{(V)}/W$, again with finite stabilizers. By Lemma 5.3, $CH^i M^{(V)}/W$ is zero for $i > \dim(M^{(V)}/W) - n = \dim(M^{(V)}) - n$, as we want.

This completes the proof that $\text{reg}(CH_G^*) \leq 0$ for every finite k-group scheme G of order invertible in k. We now prove that $\text{reg}(CH_G^*) \leq 0$ for every finite group scheme G over k.

As in the proof above, choose a faithful representation $G \subset GL(n)$ over k. Let T be the diagonal torus $(G_m)^n$ in $GL(n)$, $S = T[p] \cong (\mu_p)^n$, and U the group of strictly upper-triangular matrices in $GL(n)$. As above, it suffices to show that

$$\text{reg}(CH_S^* GL(n)/G) \leq n(n-1)/2.$$

For any finite extension field E of degree prime to p, CH_G^* is a summand of $CH_{G_E}^*$ using transfers, and so $\text{reg}(CH_G^*) \leq 0$ follows from $\text{reg}(CH_{G_E}^*) \leq 0$. So we can freely make field extensions of degree prime to p. Over the perfect closure k^{perf} of k, the underlying reduced scheme $(G_{k^{\text{perf}}})_{\text{red}}$ is a smooth subgroup scheme [32, section VIA.0.2 and proposition VIA.1.3.1]. So, after replacing k by a finite extension of degree a power of the characteristic (which we assumed is prime to p), we can assume that G_{red} is a smooth k-subgroup scheme. The index of G_{red} in G is a power of the characteristic of k, and so CH_G^* is a summand of $CH_{G_{\text{red}}}^*$ using transfers. So we can assume that G is smooth over k.

The finite etale k-group scheme G is determined by the action of $\text{Gal}(k_s/k)$ on the finite group $F = G(k_s)$. After replacing k by an extension of degree prime to p, we can assume that $\text{Gal}(k_s/k)$ acts through a p-group P on F. Then F has a Galois-invariant Sylow p-subgroup, by the usual Sylow theorem applied to the semidirect product $P \ltimes F$. That is, G has a k-subgroup scheme of order a power of p and index prime to p. Using transfer, we can assume that G has order a power of p. So the order of G is invertible in k. We proved earlier that $\mathrm{reg}(CH_G^*) \leq 0$ in this case. □

6.2 Motivic cohomology

This section briefly introduces motivic cohomology.

Bloch defined motivic homology, at first called higher Chow groups, as a way to extend the basic exact sequence for Chow groups to a long exact sequence [14]. For smooth schemes X over a field k, a suitable renumbering of motivic homology is called motivic cohomology. The motivic cohomology groups $H_M^i(X, \mathbf{Z}(j))$ form a bigraded ring, with Chow groups being the special case $CH^i X = H_M^{2i}(X, \mathbf{Z}(i))$ [100, theorem 19.1]. The index j is traditionally called the "weight." As in topology, motivic cohomology groups can be defined with coefficients in any abelian group A, and they satisfy the universal coefficient theorem. The group $H_M^{2i}(X, A(i))$ is $CH^i(X) \otimes_{\mathbf{Z}} A$. For brevity, we sometimes write $H^i(X, A(j))$ for motivic cohomology (without the subscript M).

Bloch's higher Chow groups $CH^i(X, j)$ coincide with motivic homology for schemes X of finite type over a field k, although they are written with numbering by codimension. In particular, when X is smooth over k, higher Chow groups coincide with motivic cohomology, with the numbering given by:

$$CH^a(X, b) \cong H_M^{2a-b}(X, \mathbf{Z}(a)).$$

We summarize Bloch's definition of higher Chow groups, a straightforward generalization of the definition of Chow groups [14, 100]. For a natural number j, define the algebraic j-simplex Δ^j over a field k to be the hyperplane $x_0 + \cdots + x_j = 1$ in affine $(j+1)$-space. The *faces* of Δ^n are the subspaces of Δ^n defined by setting some of the variables x_m to zero. There are natural morphisms $f_m \colon \Delta^{j-1} \to \Delta^j$ for $m = 0, \ldots, j$ whose images are the faces of codimension i. For an equidimensional k-scheme X of finite type, let $z^i(X, j)$ be the free abelian group on the set of codimension-i subvarieties of $X \times_k \Delta^j$ that intersect each face $X \times_k \Delta^r$ in a codimension-i subset. Then there is a natural chain complex $z^i(X, *)$ of the form

$$\cdots \to z^i(X, 2) \to z^i(X, 1) \to z^i(X, 0) \to 0,$$

where the boundary map on $z^i(X, j)$ is $\sum_{m=0}^{j}(-1)^m f_m^*$. The *higher Chow group* $CH^i(X, j)$ is the homology of this chain complex at $z^i(X, j)$.

The localization sequence for motivic cohomology has the following form [15].

Lemma 6.8 *Let X be a smooth scheme over a field k and $Y \subset X$ a smooth closed subscheme of codimension d. Let A be an abelian group. Then there is a long exact sequence of motivic cohomology groups*

$$\begin{aligned} &\to H_M^{i-2d}(Y, A(j-d)) \to H_M^i(X, A(j)) \to H_M^i(X-Y, A(j)) \\ &\to H_M^{i+1-2d}(Y, A(j-d)) \to \end{aligned}$$

The groups $H_M^i(X, A(j))$ are zero for $i > 2j$, as is immediate from the interpretation of motivic cohomology as higher Chow groups. That implies the surjectivity at the right in the basic exact sequence for Chow groups:

$$\cdots \to H_M^{2i-1}(X-Y, \mathbf{Z}(i)) \to CH^{i-d}Y \to CH^iX \to CH^i(X-Y) \to 0.$$

Let n be a positive integer that is invertible in the field k. Then there is a cycle map from motivic cohomology to etale cohomology,

$$H_M^i(X, \mathbf{Z}/n(j)) \to H_{\text{et}}^i(X, \mathbf{Z}/n(j)) = H_{\text{et}}^i(X, \mu_n^{\otimes j})$$

[100, theorem 10.2]. For $k = \mathbf{C}$, this etale cohomology group can be identified with ordinary cohomology, $H^i(X, \mathbf{Z}/n)$. The Beilinson-Lichtenbaum conjecture (Voevodsky's theorem) is the remarkable fact that the cycle map is an isomorphism in a wide range of bidegrees [145, theorem 6.17]:

Theorem 6.9 *Let X be a smooth scheme of finite type over a field k. Let n be a positive integer that is invertible in k. Then the cycle map*

$$H_M^i(X, \mathbf{Z}/n(j)) \to H_{\text{et}}^i(X, \mathbf{Z}/n(j))$$

is an isomorphism for $i \leq j$ and is injective for $i = j+1$.

There are several elementary vanishing properties of motivic cohomology. In particular, $H_M^i(X, A(j))$ is zero for $j < 0$, and also when $i > j + \dim(X)$, for any smooth scheme X over a field. Both are clear from the interpretation of motivic cohomology as higher Chow groups, since there are no cycles of negative codimension or of negative dimension. Combined with the Beilinson-Lichtenbaum conjecture, the latter vanishing gives a complete description of motivic cohomology with finite coefficients for a field k, meaning the motivic cohomology of Spec k. Namely, every motivic cohomology group $H_M^i(k, \mathbf{Z}/n(j))$ is either zero (if $i > j$) or isomorphic to etale cohomology (if $i \leq j$), for n invertible in k.

The Beilinson-Soulé conjecture asserts that $H_M^i(X, A(j))$ is zero for $i < 0$. Although this remains open, it is true for A finite of order invertible in k, by the Beilinson-Lichtenbaum conjecture.

Edidin and Graham defined the equivariant motivic cohomology of a smooth k-scheme X with an action of an affine group scheme G at the same time as they defined equivariant Chow rings. Namely, equivariant motivic cohomology is defined to be the motivic cohomology of suitable approximations $(X \times (V - S))/G$ to $(X \times EG)/G$. The localization sequence generalizes to G-equivariant motivic cohomology [38, proposition 5].

6.3 Steenrod operations on motivic cohomology

We now state the basic properties of Steenrod operations on motivic cohomology. We use these operations for several proofs in this book, all in later chapters.

Let p be a prime number and k a field in which p is invertible. Voevodsky defined Steenrod operations on mod p motivic cohomology, for a smooth scheme X over k:

$$P^a\colon H^i(X, \mathbf{F}_p(j)) \to H^{i+2(p-1)a}(X, \mathbf{F}_p(j + (p-1)a))$$

for $a \geq 0$ [144, section 9]. We also have the Bockstein

$$\beta\colon H^i(X, \mathbf{F}_p(j)) \to H^{i+1}(X, \mathbf{F}_p(j)).$$

For example, the Steenrod operations P^a send mod p Chow groups into themselves (since $CH^j(X)/p = H^{2j}(X, \mathbf{F}_p(j))$), while the Bockstein is zero on mod p Chow groups (since $H^{2j+1}(X, \mathbf{F}_p(j)) = 0$). As in topology, Steenrod operations arise from the failure of multiplication on motivic cohomology to come from a commutative operation at the level of cycles. The proof of Lemma 8.8 makes this point in more detail.

Steenrod operations commute with pullback for arbitrary morphisms of smooth k-schemes. The operation P^0 is the identity. The Bockstein is a derivation, meaning that $\beta(xy) = \beta(x)y + (-1)^i x\beta(y)$ for x in $H^i(X, \mathbf{F}_p(j))$. We have the Cartan formula $P^a(xy) = \sum_{b=0}^{a} P^b(x)P^{a-b}(y)$, assuming that k contains the 4th roots of unity when $p = 2$. The Cartan formula holds without that extra assumption when x and y are in the mod p Chow ring [144, proposition 9.7]. Finally, the Steenrod operations satisfy the "unstable" properties that

$$P^a x = 0$$

for x in $H^i(X, \mathbf{F}_p(j))$ when $a > i - j$ and $a \geq j$, and

$$P^a x = x^p$$

for x in $CH^a(X)/p = H^{2a}(X, \mathbf{F}_p(a))$ [144, lemmas 9.8 and 9.9].

6.4 Regularity of motivic cohomology

In this section, Theorem 6.10 proves the bound "regularity ≤ 0" for the motivic cohomology of a classifying space over an algebraically closed field, generalizing the case of the Chow ring (Theorem 6.5). As an application, Corollary 6.16 shows that the motivic cohomology of BG maps isomorphically to ordinary cohomology in a large range of bidegrees, more than is true for smooth varieties in general.

This section is not used in the rest of the book.

Fix a prime number p. For an affine group scheme G and any integer j, define

$$M_j(G) = \oplus_i H_M^{2i-j}(BG, \mathbf{F}_p(i)).$$

We view $M_j(G)$ as a graded abelian group, graded by the degree of H^*; thus $M_j(G)$ is concentrated in degrees $\equiv j \pmod 2$. In particular, $M_0(G)$ is the Chow ring $CH_G^* = CH^*(BG)/p$, and $M_j(G)$ is zero for $j < 0$. For each j, $M_j(G)$ is a module over the Chow ring CH_G^*. In particular, given a faithful representation $G \to GL(n)$, $M_j(G)$ is a module over the Chern classes, $\mathbf{F}_p[c_1, \ldots, c_n]$. Note that $|c_j| = 2j$ in this context.

Theorem 6.10 *Let G be a finite group scheme over a field k, p a prime number invertible in k, and j a natural number. Suppose that the order of G is invertible in k. Then, for any faithful representation $G \to GL(n)$, $M_j(G)$ is generated by elements of degree at most $n^2 + j$ as a module over $\mathbf{F}_p[c_1, \ldots, c_n]$, and at most n^2 if k is algebraically closed. Moreover, $M_j(G)$ has regularity at most j as an $\mathbf{F}_p[c_1, \ldots, c_n]$-module, and at most 0 if k is algebraically closed.*

Theorem 6.10 implies, say for k algebraically closed, that for any elements $y_1, \ldots, y_m$ of CH_G^* such that CH_G^* is generated by elements of bounded degree as an $\mathbf{F}_p[y_1, \ldots, y_m]$-module, $M_j(G)$ is generated by elements of degree at most $\sum_i (2|y_i| - 1)$ as an $\mathbf{F}_p[y_1, \ldots, y_m]$-module, where $|y_i|$ denotes the degree of y_i in the Chow ring.

For later use, we give a case where the Künneth formula holds for motivic cohomology.

Lemma 6.11 *Let G be an elementary abelian group $(\mathbf{Z}/p)^v$ for some prime number p. Let k be a field of characteristic not p that contains a primitive pth root of unity ζ_p. Let u be the corresponding element of $H_M^0(k, \mathbf{F}_p(1)) = \mu_p(k)$. Let X be a smooth scheme over a field k. Then*

$$H_M^*(X \times BG, \mathbf{F}_p(*)) \cong H_M^*(X, \mathbf{F}_p(*))\langle x_1, \ldots, x_v, y_1, \ldots, y_v \rangle,$$

where each element x_i is in $H^1_M(BG, \mathbf{F}_p(1))$ and $y_i = \beta x_i$ is in $H^2_M(BG, \mathbf{F}_p(1))$. The notation means a free graded-commutative algebra if p is odd, and the commutative algebra with relations $x_i^2 = u y_i$ if $p = 2$.

Proof It suffices by induction to describe the motivic cohomology of $X \times B\mu_p$. Here $B\mu_p$ can be approximated by the quotient variety $(A^{N+1} - 0)/\mu_p$ for N large, and this variety is the complement of the zero section in the line bundle $O(p)$ over $\mathbf{P}^N$. The lemma follows from the motivic cohomology of a projective bundle [14] together with the localization sequence (Lemma 6.8). □

Proof (Theorem 6.10) We simultaneously prove that $M_j(G)$ is generated by a set of elements of bounded degree as an $\mathbf{F}_p[c_1, \ldots, c_n]$-module, and that $M_j(G)$ has regularity at most 0 as an $\mathbf{F}_p[c_1, \ldots, c_n]$-module. The two statements together imply that $M_j(G)$ is generated by elements of degree at most n^2 as an $\mathbf{F}_p[c_1, \ldots, c_n]$-module.

As in the proof of Theorem 6.5, let T be the subgroup $(G_m)^n \subset GL(n)$ of diagonal matrices, and $S = (\mu_p)^n$ the p-torsion subgroup scheme of T.

Lemma 6.12 *As a CH^*_G-module, $M_{j+n,G}(GL(n)/S)$ contains $M_{j,G}$ shifted up by degree n^2 as a summand.*

Proof First consider $H^*_G(GL(n)/T, \mathbf{F}_p(*))$, which we can view as the motivic cohomology of $Y := G\backslash(EG \times GL(n))/T$. Then Y is a $GL(n)/T$-bundle over BG, and so its motivic cohomology is that of a flag bundle (a $GL(n)/B$-bundle) over BG associated to a vector bundle. This motivic cohomology is well known:

$$H^*(Y, \mathbf{F}_p(*)) = H^*(BG, \mathbf{F}_p(*))[y_1, \ldots, y_n]/(e_i(y_1, \ldots, y_n) = c_i),$$

where $|y_i| = (2, 1)$ and c_i in $H^{2i}(BG, \mathbf{F}_p(i))$ is the ith Chern class of the given representation $G \to GL(n)$.

Next, $G\backslash(EG \times GL(n))/S$ is a $(G_m)^n$-bundle over Y corresponding to the pth powers of the obvious line bundles $L_1, \ldots, L_n$ on Y (with Chern classes $y_1, \ldots, y_n$). The Chern classes of these pth powers are 0, since we are working with $\mathbf{F}_p$ coefficients. So

$$H^*_G(GL(n)/S, \mathbf{F}_p(*)) = H^*(BG, \mathbf{F}_p(*))$$

$$\langle x_1, \ldots, x_n, y_1, \ldots, y_n\rangle/(e_i(y_1, \ldots, y_n) = c_i),$$

where $|x_i| = (1, 1)$. This is a free module over $H^{*,*}(BG)$ with highest-degree generator (say, $x_1 \cdots x_n y_1^{n-1} y_2^{n-2} \ldots y_{n-1}$) of degree n^2. □

To prove Theorem 6.10, it now suffices to show that $M_{l,G}(GL(n)/S)$ is generated in bounded degrees as an $\mathbf{F}_p[c_1, \ldots, c_n]$-module, that it has regularity at most $n^2 + l - n$ as an $\mathbf{F}_p[c_1, \ldots, c_n]$-module, and that it has regularity at most n^2 if k is algebraically closed. (We will apply this to $l = j + n$, in view of Lemma 6.12.) Note that $|c_i| = 2i$ in this context. We can view the groups $M_{l,G}(GL(n)/S)$ as $M_{l,S}(GL(n)/G)$. Since CH_S^* is finite over $CH_{GL(n)}^*$, it suffices to show that $M_{l,S}(GL(n)/G)$ is generated in bounded degrees as a CH_S^*-module and that it has regularity at most $n^2 + l - n$ as a CH_S^*-module, or at most n^2 if k is algebraically closed.

Let U be the group of strictly upper-triangular matrices in $GL(n)$, and let $M = U\backslash GL(n)/G$. As in the proof of Theorem 6.5, M is a smooth quasi-projective variety of dimension $n(n+1)/2$, and T acts on M with finite stabilizer groups. Since U is an iterated extension of additive groups, homotopy invariance of motivic cohomology shows that $M_{l,S}(GL(n)/S) = M_{l,S}(M)$. So it suffices to show that $M_{l,S}(M)$ is generated in bounded degrees and has regularity at most $n^2 + l - n$ as a CH_S^*-module, or at most n^2 if k is algebraically closed.

Let M_i be the smooth closed subscheme in M of points with isotropy group in S of rank at least i, and let $M_{(i)}$ be the open subset of M_i consisting of points with isotropy group of rank equal to i. For each $d \geq 0$, let $M_{(i),d}$ be the union of the connected components of $M_{(i)}$ with codimension d in M. The analog for motivic cohomology of Duflot's theorem is:

Theorem 6.13 *There is a short exact sequence of CH_S^*-modules*

$$0 \to \oplus_d H_S^{*-2d}(M_{(i),d}, \mathbf{F}_p(* - d)) \to H_S^*(M - M_{i+1}, \mathbf{F}_p(*))$$
$$\to H_S^*(M - M_i, \mathbf{F}_p(*)) \to 0.$$

Proof This is the usual long exact sequence, Lemma 6.8, for equivariant motivic cohomology. If we prove the injectivity of the first homomorphism in all degrees, the exact sequence gives the surjectivity of the last homomorphism. As in the proof of Theorem 6.7, it suffices to show that for each subgroup $V \subset S$ of p-rank i, the Euler class of the normal bundle to $X := (M - M_{i+1})^V$ in $M - M_{i+1}$ is not a zero divisor in motivic cohomology $H_S^*((M - M_{i+1})^V, \mathbf{F}_p(*))$.

Since $S \cong (\mu_p)^n$, we can choose a splitting $S = V \times W$ for the subgroup $V \subset S$. Since V acts trivially on X while W acts freely on it, we have an isomorphism

$$H_S^*(X, \mathbf{F}_p(*)) \cong H^*(X/W, \mathbf{F}_p(*))\langle x_1, \ldots, x_v, y_1, \ldots, y_v\rangle,$$

where $|x_i| = (1, 1)$, $|y_i| = (2, 1)$, and $y_1, \ldots, y_v$ are the first Chern classes of 1-dimensional representations of $V \cong (\mu_p)^v$. As in the proof of Theorem 6.13, the Euler class of the normal bundle is a monic polynomial in $y_1, \ldots, y_v$ over $CH^*(X/W)/p$, and so it is a non-zero-divisor on this motivic cohomology ring. □

By Theorem 6.13, we have a filtration of $M_{l,S}(M)$ by finitely many CH_S^*-submodules such that the subquotients are isomorphic to $M_{l,S}(M^{(V)})$ for the subgroups V of S, shifted up in degree by twice the codimension of $M^{(V)}$ in M. Choose a splitting $S = V \times W$ for the subgroup $V = (\mu_p)^v \subset S \cong (\mu_p)^n$. The group V acts trivially on $M^{(V)}$ while W acts freely, and so we have

$$H_S^*(M^{(V)}, \mathbf{F}_p(*)) \cong H^*(M^{(V)}/W, \mathbf{F}_p(*))\langle x_1, \ldots, x_v, y_1, \ldots, y_v\rangle,$$

by the Künneth formula for motivic cohomology, Lemma 6.11.

Lemma 6.14 *Let X be a smooth scheme of dimension n over a field k. Then $H^{2i-l}(X, \mathbf{F}_p(i))$ is 0 if $2i - l > 2n + l$. If k is algebraically closed and p is invertible in k, then $H^{2i-l}(X, \mathbf{F}_p(i))$ is 0 if $2i - l > 2n$.*

Proof The first bound is trivial from the identification of motivic cohomology with higher Chow groups (these groups would be represented by cycles of negative dimension). For k algebraically closed, Suslin showed that $H^{2i-l}(X, \mathbf{F}_p(i))$ maps isomorphically to etale cohomology $H_{\text{et}}^{2i-l}(X, \mathbf{F}_p(i))$ for $i \geq n$ [127, corollary 4.3]. We can assume that $l \geq 0$; otherwise motivic cohomology is zero. So our assumption $2i - l > 2n$ implies that $i \geq n$, and so the given motivic cohomology group is isomorphic to an etale cohomology group in degree $> 2n$, which is known to be zero since k is algebraically closed. □

By Lemma 6.14, $M_l(M^{(V)})$ is concentrated in a bounded set of degrees for each $l \geq 0$. Therefore, the formula before the lemma shows that $H_S^*(M^{(V)}, \mathbf{F}_p(*))$ is generated in a bounded set of degrees as a module over $CH_S^* = \mathbf{F}_p[y_1, \ldots, y_n]$. By Theorem 6.13, it follows that $M_{l,S}(M)$ is generated in a bounded set of degrees as a CH_S^*/p-module. By our earlier arguments, we have proved the first part of Theorem 6.10, on bounded generation.

Moreover, in the notation before Lemma 6.14, the regularity of $M_{l,S}(M^{(V)})$ as a CH_S^*-module is equal to the regularity of $M_l(M^{(V)}/W)$ as a CH_W^*-module. Since $M_l(M^{(V)}/W)$ is 0 in high degrees, its regularity as a CH_W^*/p-module is equal to the highest degree in which it is nonzero, by Lemma 3.9. If we can show that $M_l(M^{(V)}/W)/p$ is zero in degrees greater than $2\dim(M^{(V)}/W) - 2n + l$, or greater than $2\dim(M^{(V)}/W) - n$ for k algebraically closed, then our discussion before Lemma 6.14 shows that $M_{l,S}(M)$ has regularity at most $2\dim(M) - 2n + l = n^2 + l - n$ as a CH_S^*-module, or n^2 for k algebraically closed, which will finish the proof of Theorem 6.10.

The torus $T = (G_m)^n$ acts with finite stabilizers on $M = U\backslash GL(n)/G$. Because T commutes with the action of S, T also acts on the locally closed subscheme $M^{(V)}$ and hence on $M^{(V)}/W$, again with finite stabilizers. The following lemma proves the vanishing of $M_l(M^{(V)}/W)$ in degrees greater than $2\dim(M^{(V)}/W) - 2n + l$, or greater than $2\dim(M^{(V)}/W) - n$ for k algebraically closed, as we want. Theorem 6.10 is proved. □

Lemma 6.15 *Let X be a smooth variety over a field k. Let p be a prime number that is invertible in k. Let T be a split torus that acts on X with finite stabilizer groups. Then $H^{2i-j}(X, \mathbf{F}_p(i)) = 0$ for $2i - j > 2\dim(X) + j - 2\dim(T)$. If k is algebraically closed, then $H^{2i-j}(X, \mathbf{F}_p(i)) = 0$ for $2i - j > 2\dim(X) - \dim(T)$.*

Proof The T-action on X has only finitely many different stabilizer group schemes in T. Stratify X according to the possible stabilizer groups. The strata are smooth, because the fixed points of any finite subgroup scheme of T on X form a smooth subscheme. By the localization sequence for motivic cohomology (Lemma 6.8), the lemma reduces to the special case where T acts on X with the same finite stabilizer group A everywhere. Here T/A is a split torus, and so we have reduced to the case where a split torus T acts freely on a smooth variety X. That is, X is a principal $(G_m)^n$-bundle over a smooth variety B, for some n.

We can view a G_m-bundle E over a smooth variety M as the complement of the zero section in the total space of a line bundle over M. Using homotopy invariance for motivic cohomology, the localization sequence has the form:

$$H^{2i-j}(M, \mathbf{F}_p(i)) \to H^{2i-j}(E, \mathbf{F}_p(i)) \to H^{2i-j-1}(M, \mathbf{F}_p(i-1)).$$

Thus $H^{2i-j}(E, \mathbf{F}_p(i))$ vanishes if the two other groups vanish. Applying this repeatedly, we find (whether k is algebraically closed or not) that the lemma reduces to the case of the base variety B, with a trivial torus action.

We want to show that B smooth over an arbitrary field k, $H^{2i-j}(B, \mathbf{F}_p(i))$ vanishes for $2i - j > 2\dim(B) + j$, while for k algebraically closed, $H^{2i-j}(B, \mathbf{F}_p(i))$ vanishes for $2i - j > 2\dim(B)$. These statements are exactly Lemma 6.15. ☐

Corollary 6.16 *Let G be a finite group. Suppose that G has a faithful representation of dimension n over an algebraically closed field k in which the order of G and a fixed prime number p are invertible. Then the cycle map from $M_j(G)$ to etale cohomology in degrees $\equiv j \pmod 2$ is an isomorphism for all $j \geq n^2$.*

This means that the motivic cohomology of BG is closer to etale cohomology than is true for smooth varieties in general. Namely, for a smooth variety X over a field, the Beilinson-Lichtenbaum conjecture (Theorem 6.9) gives that the cycle map from $M_j(X)$ to etale cohomology is an isomorphism in degrees at most j. But Corollary 6.16 says that $M_j(G)$ (meaning $M_j(BG)$) is equal to etale cohomology in all degrees when j is large.

The hypothesis $j \geq n^2$ in Corollary 6.16 can be weakened to $j \geq \sigma(CH_G^*)$, with CH^i considered as having degree $2i$. Theorem 7.1 gives a good bound for $\sigma(CH_G^*)$ when G is a p-group.

Proof By the Beilinson-Lichtenbaum conjecture, $M_j(G)$ maps isomorphically to etale cohomology in degrees at most j, and injectively in degrees at most $j+2$. (Notice that $M_j(G)$ is concentrated in degrees congruent to j modulo 2, so we are only considering etale cohomology in degrees congruent to j modulo 2.)

By Theorem 6.10, $M_j(G)$ is generated by elements of degree at most n^2 as a module over the Chern classes $\mathbf{F}_p[c_1, \ldots, c_n]$. The same is true for the etale cohomology of BG (which is simply the cohomology of G) by Symonds's theorem, Corollary 4.3. So the map from $M_j(G)$ to etale cohomology is surjective in all degrees if $j \geq n^2$. Symonds's theorem also gives that all relations among a minimal set of generators of the etale cohomology of BG as an $\mathbf{F}_p[c_1, \ldots, c_n]$-module are in degrees at most $n^2 + 1$. Therefore $M_j(G)$ maps isomorphically to etale cohomology in all degrees if $j \geq n^2$, since that implies that $j + 2 \geq n^2 + 1$. □

7

Bounds for p-Groups

As we have seen in Section 5.2, the bounds in Theorem 5.2 and Symonds's Corollary 4.3 for the Chow and cohomology rings in terms of the dimension of a faithful representation are essentially optimal among arbitrary finite groups. In this chapter, we give much better bounds for the Chow and cohomology rings of a p-group. It is natural to concentrate on the case of p-groups, since the mod p cohomology of any finite group is a summand of the cohomology of a Sylow p-subgroup, and likewise for Chow rings. I don't know whether these improved bounds for p-groups are anywhere near optimal.

The bounds involve a constant that we now define. For each prime number p, let

$$\alpha_p = 1 + p + \sum_{d=2}^{p-1} \left(\left\lfloor \frac{p-2}{d-1} \right\rfloor - \left\lfloor \frac{p-2}{d} \right\rfloor \right) d.$$

For example, $\alpha_2 = 3, \alpha_3 = 6, \alpha_5 = 14$, and $\alpha_7 = 23$. For large primes p, comparing the sum with an integral shows that α_p is asymptotic to $p \log p$.

Theorem 7.1 *Let G be a p-group with a faithful complex representation V. Write V as a direct sum of irreducibles, $V = V_1 \oplus \ldots \oplus V_s$. The dimensions of the irreducible representations V_i are powers of p, say* $\dim(V_i) = p^{m_i}$*. Then the mod p Chow ring CH_G^* is generated as a module over certain transferred Euler classes of degree at most* $\max(p^{m_i})$ *by elements of degree at most* $\sum_i (\alpha_p^{m_i} - p^{m_i})$*. A fortiori, the ring CH_G^* is generated by elements of degree at most* $\max(p^{m_1}, \ldots, p^{m_s}, \sum_i (\alpha_p^{m_i} - p^{m_i}))$.

We also prove a strong bound for the cohomology of p-groups. This result for p-groups is significantly better than the bound n^2 that holds for arbitrary finite groups, as we will discuss.

Theorem 7.2 *Let G be a p-group with a faithful complex representation V. Write V as a direct sum of irreducibles, $V = V_1 \oplus \ldots \oplus V_s$. The dimensions* *of the irreducible representations* V_i *are powers of* p, *say* $\dim(V_i) = p^{m_i}$. *Then the cohomology ring* $H_G^* = H^*(BG, \mathbf{F}_p)$ *is generated as a module over certain transferred Euler classes of degree at most* $2\max(p^{m_i})$ *by elements of degree at most* $\sum_i (2\alpha_p^{m_i} - p^{m_i})$. *A fortiori, the ring* H_G^* *is generated by elements of degree at most* $\max(2p^{m_1}, \ldots, 2p^{m_s}, \sum_i (2\alpha_p^{m_i} - p^{m_i}))$.

For example, let G be a 2-group that has a faithful irreducible complex representation of dimension n, which must be of the form $n = 2^m$. Then Theorems 7.1 and 7.2 give that the Chow ring of BG is generated by elements of degree at most $3^m - 2^m$, and the cohomology ring is generated by elements of degree at most $2 \cdot 3^m - 2^m$, for $m \geq 2$. These bounds are on the order of $n^{\log 3/\log 2} \doteq n^{1.58}$ for $n = 2^m$ large, which significantly improves the bounds on the order of n^2 that hold for arbitrary finite groups (Theorem 5.2 and Corollary 4.3).

The bounds in Theorems 7.1 and 7.2 are even better for other classes of p-groups. For example, if G is a p-group with a faithful irreducible complex representation of dimension $n = p$, then the Chow ring of BG is generated in degree at most $\alpha_p - p$ and the cohomology ring is generated in degree at most $2\alpha_p - p$, for p odd. For large primes p, these bounds are on the order of $p \log p$, which is far better than the bounds on the order of $n^2 = p^2$ given by the results for general finite groups.

The bounds in Theorems 7.1 and 7.2 are also very good for a faithful representation that is a sum of low-dimensional irreducible representations. For example, let G be a 2-group with a faithful complex representation of dimension $n = 2s$ that is a sum of s 2-dimensional irreducible representations, say with $s \geq 2$. Then the Chow ring of BG is generated in degree at most s $(= n/2)$, and the cohomology ring is generated in degree at most $4s$ $(= 2n)$. Although these bounds are much smaller than n^2, they happen to coincide with the bounds given by Theorems 5.4 and 5.5 for an arbitrary finite group with a faithful representation that is a sum of 2-dimensional irreducibles.

7.1 Invariant theory of the group $\mathbf{Z}/p$

As a first step toward our bounds for the cohomology and Chow ring of p-groups, we need the following new bound in the invariant theory of the group $\mathbf{Z}/p$. Namely, we describe a system of parameters (Definition 3.4) for the ring of invariants.

Lemma 7.3 *Let* p *be a prime number. Let the group* $G = \mathbf{Z}/p$ *act on the polynomial ring* $R = \mathbf{F}_p[y_0, \ldots, y_{p-1}]$ *by cyclically permuting the variables. Then the ring of invariants* R^G *has a system of parameters that consists of one* *element* $y_0 + \cdots + y_{p-1}$ *of degree 1,* $\lfloor \frac{p-2}{d-1} \rfloor - \lfloor \frac{p-2}{d} \rfloor$ *elements of degree* d *for* $2 \leq d \leq p-1$, *and one element* $y_0 \cdots y_{p-1}$ *of degree* p.

The proof uses the following standard constructions from invariant theory.

Definition 7.4 For a finite group G acting on an abelian group M, the *trace* is the homomorphism $\text{tr} = \text{tr}_1^G : M \to M^G$ defined by $\text{tr}(x) = \sum_{g \in G} gx$. For G acting on a commutative ring R, the *norm* is the function $N = N_1^G : R \to R^G$ given by $N(x) = \prod_{g \in G} gx$.

The following simplification of my first proof of Lemma 7.3 is due to Jim Shank and David Wehlau.

Proof of Lemma 7.3 This is clear for $p = 2$, where R^G is the polynomial ring $\mathbf{F}_2[y_0 + y_1, y_0 y_1]$. For larger primes p, the ring of invariants for the regular representation of $\mathbf{Z}/p$ in characteristic p becomes more complicated; for example, it is not Cohen-Macaulay for $p \geq 5$, by Ellingsrud and Skjelbred [39].

Let g be the generator of $\mathbf{Z}/p$, and let $\Delta = g - 1$ in the group algebra $\mathbf{F}_p[\mathbf{Z}/p]$. Change the basis of the representation to $x_0 = y_0$ and $x_i = \Delta x_{i-1}$ for $1 \leq i \leq p-1$. (We have $\Delta x_{p-1} = 0$.) The action of $\mathbf{Z}/p$ is given in this basis by $g^j x_i = \sum_{l \geq 0} \binom{j}{l} x_{i+l}$, where we define $x_i = 0$ for $i \geq p$.

Consider the graded reverse lexicographic ordering on monomials in $x_0, \ldots, x_{p-1}$ with

$$x_0 > \cdots > x_{p-1}.$$

That is, we have $x^I < x^J$ if the degree $|I|$ is less than $|J|$, or if $|I| = |J|$ and $i_{p-1} > j_{p-1}$, or if $|I| = |J|$ and $i_{p-1} = j_{p-1}$ and $i_{p-2} > j_{p-2}$, and so on.

The norm $N(x_0) = \prod_{j=0}^{p-1} g^j x_0$ is an invariant whose leading term is x_0^p. We will show that for each $1 \leq j \leq p-1$, there is an invariant with leading term x_j^d, where d is the least integer such that $dj \geq p-1$. (Or, equivalently, $j > \lfloor (p-2)/d \rfloor$.) By the leading terms, these elements form a system of parameters for $k[x_0, \ldots, x_{p-1}]$ and hence for the subring of invariants. The number of these elements of degree d is 1 for $d = 1$, $\lfloor \frac{p-2}{d-1} \rfloor - \lfloor \frac{p-2}{d} \rfloor$ for $2 \leq d \leq p-1$, and 1 for $d = p$, as we want.

We define the invariant as the trace $\text{tr}(x_0^{d-1} x_m)$, where $m = dj - (p-1)$. By definition of d, we have $0 \leq m \leq j-1$. We have

$$\text{tr}(x_0^{d-1} x_m) = \sum_{l_1, \ldots, l_d \geq 0} \left[\sum_{j=0}^{p-1} \binom{j}{l_1} \cdots \binom{j}{l_d} \right] x_{l_1} \cdots x_{l_{d-1}} x_{m+l_d}.$$

We use the following simple identity, sometimes called Newton's formula. It follows, for example, from the fact that the sum of the elements of a nontrivial subgroup of $\mathbf{F}_p^*$ is zero.

Lemma 7.5 *In the integers modulo a prime number p,*

$$\sum_{j=0}^{p-1} j^e = \begin{cases} 0 & \text{if } 0 \le e \le p-2 \\ -1 & \text{if } e = p-1, \end{cases}$$

where we define $0^0 = 1$.

As a result, the expression in brackets in the formula above for $\text{tr}(x_0^{d-1} x_m)$ is zero unless $l_1 + \cdots + l_d \ge p - 1$. It follows that the leading term of the invariant $\text{tr}(x_0^{d-1} x_m)$ is x_j^d, with a nonzero coefficient in $\mathbf{F}_p$. □

7.2 Wreath products

Lemma 7.6 *Let p be a prime number and m a positive integer. Let G be the wreath product* $\mathbf{Z}/p \wr \cdots \wr \mathbf{Z}/p$ *with m copies of* $\mathbf{Z}/p$. *Let R be the polynomial ring over* $\mathbf{F}_p$ *with* p^m *variables, and let G act on R by the natural permutation action on the variables. Let the variables have degree* $b > 0$. *Then*

$$\sigma(R^G) \le b\alpha_p^m - p^m.$$

The constant α_p was defined before Theorem 7.1, and $\sigma(R)$ is defined in Definition 6.1.

Proof The lemma is proved by constructing a system of parameters with known degrees for the ring of invariants R^G. For example, when $p = 2$, our system of parameters for the invariants of the m-fold wreath product of $\mathbf{Z}/2$ has degrees $1, 2$ if $m = 1$; $1, 2, 2, 4$ if $m = 2$; $1, 2, 2, 4, 2, 4, 4, 8$ if $m = 3$; and so on. That is, for $p = 2$, each of these sequences of degrees starts with the previous sequence, and then multiplies the previous sequence by 2.

For $m = 1$, so that $G = \mathbf{Z}/p$, Lemma 7.3 gives a system of parameters $f_1(y_0, \ldots, y_{p-1}) = y_0 + \cdots + y_{p-1}, \ldots, f_p(y_0, \ldots, y_{p-1}) = y_0 \cdots y_{p-1}$ for R^G. We have $\sum_{i=1}^{p}(|f_i| - 1) = b\alpha_p - p$ by definition of α_p, since we define $|y_i| = b$ for all i. So we have $\sigma(R^G) \le b\alpha_p - p$ for $m = 1$.

Suppose by induction that we have constructed a system of parameters $u_1, \ldots, u_{p^{m-1}}$ for the invariants of the iterated wreath product H of $m - 1$ copies of $\mathbf{Z}/p$ acting on the polynomial ring in p^{m-1} variables. Then the product group H^p acts on the polynomial ring R in p^m variables by acting separately on p sets of p^{m-1} variables. There is an obvious system of parameters for R^{H^p} that we call u_{ij} for $0 \le i \le p - 1$ and $1 \le j \le p^{m-1}$, where the polynomial u_{ij} is the polynomial u_j in the ith set of p^{m-1} variables.

We have $G = \mathbf{Z}/p \ltimes H^p$, and so R^G is the ring of invariants of the cyclic group $\mathbf{Z}/p$ acting on R^{H^p}. In particular, a generator σ of $\mathbf{Z}/p$ acts on $u_{ij} \in R^{H^p}$ by $\sigma(u_{ij}) = u_{i+1,j}$, where i is understood modulo p. Therefore, for each $1 \leq j \leq p^{m-1}$, the polynomials $f_1(u_{0j}, \ldots, u_{p-1,j}) = u_{0j} + \cdots + u_{p-1,j}, \ldots, f_p(u_{0j}, \ldots, u_{p-1,j}) = u_{0j} \cdots u_{p-1,j}$ are G-invariant. The polynomials u_{ij} are all integral over this set of p^m elements $v_1, \ldots, v_{p^m}$ of R^G, and so the whole polynomial ring R is finite over $\mathbf{F}_p[v_1, \ldots, v_{p^m}]$. As a result, $v_1, \ldots, v_{p^m}$ is a system of parameters for R^G.

By induction, we have $\sum_{i=1}^{p^{m-1}} |u_i| = a\alpha_p^{m-1}$, and the construction shows that

$$\sum_{i=1}^{p^m} |v_i| = \alpha_p \sum_{i=1}^{p^{m-1}} |u_i|$$
$$= b\alpha_p^m.$$

So $\sigma(R^G) \leq \sum_{i=1}^{p^m}(|u_i| - 1) = b\alpha_p^m - p^m$. □

Corollary 7.7 *Let G be the wreath product of m copies of $\mathbf{Z}/p$ with the multiplicative group over a field k with p invertible in k, $\mathbf{Z}/p \wr \cdots \wr \mathbf{Z}/p \wr G_m$. Then $\sigma(CH_G^*) \leq \alpha_p^m - p^m$. Also, taking k to be the complex numbers, $H_G^* = H^*(BG, \mathbf{F}_p)$ has $\sigma(H_G^*) \leq 2\alpha_p^m - p^m$.*

Proof Let G be the wreath product $G = \mathbf{Z}/p \wr \cdots \wr \mathbf{Z}/p \wr G_m$. with $\mathbf{Z}/p$ occurring m times. By Lemma 2.21, the Chow ring CH_G^* consists of transferred Euler classes. For $m \geq 1$, write $G = \mathbf{Z}/p \wr H$. Then Lemma 2.21 and its proof show that

$$CH_G^* \to CH_{H^p}^* \times CH_{\mathbf{Z}/p \times H}^*$$

is injective, and that CH_G^* maps onto the invariants $(CH^* H^p)^{\mathbf{Z}/p}$. Here $\mathbf{Z}/p \times H$ is included in $G = \mathbf{Z}/p \wr H$ with H as the diagonal subgroup in H^p.

Suppose by induction that we have constructed a system of parameters $u_1, \ldots, u_{p^{m-1}}$ for CH_H^* such that $|u_1| = 1$ and $\sum |u_i| = \alpha_p^{m-1}$. Pulling these classes back by the p projections $H^p \to H$, we get a system of parameters u_{ij} for $CH_{H^p}^*$ with $0 \leq i \leq p-1$ and $1 \leq j \leq p^{m-1}$, where the classes u_{ij} are pulled back from the ith factor H. We know that $CH_{H^p}^* = (CH_H^*)^{\otimes p}$, by the assumption above on H (Lemma 2.12).

Since CH_G^* maps onto $(CH_{H^p}^*)^{\mathbf{Z}/p}$, there are elements $v_1, \ldots, v_{p^m}$ of CH_G^* that restrict to the $\mathbf{Z}/p$-invariants

$$f_1(u_{0j}, \ldots, u_{p-1,j}) = u_{0j} + \cdots + u_{p-1,j},$$
$$\cdots$$
$$f_p(u_{0j}, \ldots, u_{p-1,j}) = u_{0j} \cdots u_{p-1,j}$$

for $1 \leq j \leq p^{m-1}$. Here the polynomials $f_1, \ldots, f_p$ are the system of parameters constructed in Lemma 7.3 for the invariants of $\mathbf{Z}/p$ on its regular representation. Explicitly, the elements v_i that restrict to

$$f_1(u_{0j}, \ldots, u_{p-1,j}), \ldots, f_{p-1}(u_{0j}, \ldots, u_{p-1,j})$$

can be defined as transfers from $CH^*_{H^p}$ to CH^*_G, while the element v_i that restricts to $u_{0j}\cdots u_{p-1,j}$ can be defined as the multiplicative transfer (or Evens norm) of $u_{0j}\in CH^*_{H^p}$, which was defined on Chow rings by Fulton and MacPherson [45]. See the summary in Section 8.1.

We can say that v_1 is the element that restricts to $u_{01}+\cdots+u_{p-1,1}$, so that $|v_1|=1$. By adding an element pulled back from $CH^1_{\mathbf{Z}/p}\cong \mathbf{F}_p$ by the surjection $G=\mathbf{Z}/p\wr H\to \mathbf{Z}/p$ if necessary, we can assume that v_1 has nonzero restriction to the subgroup $\mathbf{Z}/p$ of G. This does not change the restriction of v_1 to H^p.

Clearly the elements u_{ij} in CH^*_H are all integral over $v_1,\ldots,v_{p^m}$, and so the whole ring CH^*_H is finite over $v_1,\ldots,v_{p_m}$ in CH^*_G. If we can show that $CH^*_{\mathbf{Z}/p\times H}$ is also finite over $v_1,\ldots,v_{p^m}$, then $v_1,\ldots,v_{p^m}$ form a system of parameters in CH^*_G as we want, using that CH^*_G injects into $CH^*_{H^p}\times CH^*_{\mathbf{Z}/p\times H}$.

We know that v_1 restricts to $u_{01}+\cdots+u_{p-1,1}$ in $CH^1_{H^p}$, and so it restricts to pu_1, which is zero, in CH^1_H under the diagonal inclusion $H\to H^p$. We have arranged that v_1 has nonzero restriction to $CH^1_{\mathbf{Z}/p}\cong \mathbf{F}_p$. So we know the restriction of v_1 to $CH^1_{\mathbf{Z}/p\times H}=CH^1_{\mathbf{Z}/p}\oplus CH^1_H$. It follows that the quotient ring of

$$CH^*_{\mathbf{Z}/p\times H}\cong (CH^*_H)[c_1]$$

by v_1 is isomorphic to CH^*_H, with c_1 being sent to zero. The remaining elements $v_2,\ldots,v_{p^m}$ include some that restrict in H^p to $u_{0j}\cdots u_{p-1,j}$ for each $1\le j\le p^{m-1}$. Those elements restrict in the diagonal subgroup $H\subset H^p$ to u_j^p, for $1\le j\le p^{m-1}$. We know that $u_1,\ldots,u_{p^{m-1}}$ form a system of parameters in CH^*_H, and so their pth powers do as well. Thus we have shown that $CH^*_{\mathbf{Z}/p\times H}$ is finite over $\mathbf{F}_p[v_1,\ldots,v_{p^m}]$. By the previous paragraph, this completes the proof that CH^*_G is finite over $\mathbf{F}_p[v_1,\ldots,v_{p^m}]$.

The construction shows that $\sum_{i=1}^{p^m}|v_i|=\alpha_p\sum_{i=1}^{p^{m-1}}|u_i|$, and so $\sum_{i=1}^{p^m}|v_i|=\alpha_p^m$ by induction. It follows that $\sigma(CH^*_G)\le\sum_{i=1}^{p^m}(|v_i|-1)=\alpha_p^m-p^m$, as we want.

We now prove the corresponding result for cohomology. It suffices to show that the images of the algebraic cycles $u_1,\ldots,u_{p^m}$ in H^*_G form a system of parameters in H^*_G. Again, we prove this by induction on m, where G is the m-fold wreath product of $\mathbf{Z}/p$ with the multiplicative group G_m over the complex numbers.

We have $G=\mathbf{Z}/p\wr H$, where H is the $(m-1)$-fold wreath product of $\mathbf{Z}/p$ with G_m. We use Quillen's theorem that the restriction map $H^*_{\mathbf{Z}/p\wr H}\to H^*_{H^p}\times H^*_{\mathbf{Z}/p\times H}$ is injective (Theorem 2.19). So to show that $H^*_{\mathbf{Z}/p\wr H}$ is finite over the polynomial ring $\mathbf{F}_p[v_1,\ldots,v_{p^m}]$, it suffices to show that $H^*_{H^p}$ and $H^*_{\mathbf{Z}/p\times H}$ are finite over $\mathbf{F}_p[v_1,\ldots,v_{p^m}]$.

By induction on m, the ring H_H^* is finite over the analogous polynomial ring $\mathbf{F}_p[u_1, \ldots, u_{p^{m-1}}]$. Therefore $H_{H^p}^* = (H_H^*)^{\otimes p}$ is finite over the elements u_{ij} for $0 \leq i \leq p-1$ and $1 \leq j \leq p^{m-1}$, in the notation used above. The definition of the elements v_i shows that the elements u_{ij} are all integral over $\mathbf{F}_p[v_1, \ldots, v_{p^m}]$. Therefore $H_{H^p}^*$ is finite over this polynomial ring.

It remains to show that $H_{\mathbf{Z}/p \times H}^*$ is finite over $\mathbf{F}_p[v_1, \ldots, v_{p^m}]$. We know that v_1 restricts to c_1 in $H_{\mathbf{Z}/p \times H}^* \cong H_H^*\langle x_1, c_1 \rangle$, by the corresponding statement in the Chow ring. Here $|x_1| = 1$ and $|c_1| = 2$ in $H_{\mathbf{Z}/p}^*$; we have $x_1^2 = 0$ if p is odd and $x_1^2 = c_1$ if $p = 2$. Therefore the quotient ring of $H_{\mathbf{Z}/p \times H}^*$ by v_1 is isomorphic to $H_H^*[x_1]/(x_1^2)$. Some of the remaining elements $v_2, \ldots, v_{p^m}$ restrict to the pth powers of the elements $u_1, \ldots, u_{p^{m-1}}$ in H_H^*. Since H_H^* is finite over $u_1, \ldots, u_{p^{m-1}}$, it is also finite over their pth powers. Therefore $H_{\mathbf{Z}/p \times H}^*$ is finite over $\mathbf{F}_p[v_1, \ldots, v_{p^m}]$, as we wanted.

That completes the proof that H_G^* is finite over $\mathbf{F}_p[v_1, \ldots, v_{p^m}]$. Here $\sum |v_i| = 2\alpha_p^m$ (the factor of 2 is because CH^i maps to H^{2i}). Therefore $\sigma(H_G^*) \leq \sum_{i=1}^{p^m}(|v_i| - 1) = 2\alpha_p^m - p^m$. □

7.3 Bounds for the Chow ring and cohomology of a p-group

Proof (Theorem 7.1) Let G be a p-group with a faithful complex representation V. Write V as a direct sum of irreducibles, $V = V_1 \oplus \ldots \oplus V_s$. The dimensions of the irreducible representations V_i are powers of p, say $\dim(V_i) = p^{m_i}$. More precisely, Blichfeldt showed that each irreducible complex representation V_i of a p-group G is induced from a 1-dimensional representation of some subgroup H [124, theorem 16]. Moreover, for any subgroup H in a p-group G, there is a chain of subgroups $H = H_0 \subset H_1 \subset \cdots \subset H_{m_i} = G$ such that each subgroup has index p in the next one. It follows that the homomorphism from G into $GL(p^{m_i})$ corresponding to V_i factors through the m_i-fold wreath product of $\mathbf{Z}/p$ with the multiplicative group G_m, $\mathbf{Z}/p \wr \cdots \wr \mathbf{Z}/p \wr G_m \subset GL(p^{m_i})$.

Since the whole representation V of G is faithful, G is a subgroup of the product K over $i = 1, \ldots, s$ of the m_i-fold wreath products of $\mathbf{Z}/p$ with G_m. By Lemma 6.2, $\sigma(CH_G^*) \leq \sigma(CH_K^*)$. By Lemma 2.12, CH_K^* is the tensor product of the Chow rings of the wreath products we mentioned, and so $\sigma(CH_K^*)$ is at most the sum of σ of these wreath products. Combining this with Lemma 7.6, we have

$$\sigma(CH_G^*) \leq \sum_{i=1}^{s}(\alpha_p^{m_i} - p^{m_i}).$$

By Theorem 6.5, it follows that CH_G^* is generated as a module over our system of parameters by elements of degree at most $\sum_{i=1}^{s}(\alpha_p^{m_i} - p^{m_i})$. This system of parameters consists of transferred Euler classes, since they are pulled back from the wreath products $\mathbf{Z}/p \wr \cdots \wr \mathbf{Z}/p \wr G_m$, whose Chow ring consists entirely of transferred Euler classes (Lemma 2.21). (Lemma 2.16 gives that transfer commutes with pullback.) It follows from the bound on CH_G^* as a module that CH_G^* is generated as an algebra by elements of degree at most $\max(p^{m_1}, \ldots, p^{m_s}, \sum_i(\alpha_p^{m_i} - p^{m_i}))$. □

Proof of Theorem 7.2 As in the previous proof, G is a subgroup of the product from $i = 1$ to s of the m_i-fold wreath products of $\mathbf{Z}/p$ with the multiplicative group G_m (or, if one prefers compact Lie groups, the circle group S^1). By Lemma 6.2, $\sigma(H_G^*)$ is at most the sum from $i = 1$ to s of σ of the cohomology of these wreath products. By Lemma 7.6, we have

$$\sigma(H_G^*) \leq \sum_{i=1}^{s}(2\alpha_p^{m_i} - p^{m_i}).$$

By Symonds (Theorem 4.1), it follows that H_G^* is generated as a module over our system of parameters by elements of degree at most $\sum_{i=1}^{s}(\alpha_p^{m_i} - p^{m_i})$. This system of parameters consists of transferred Euler classes, by the corresponding statement in the Chow ring. It follows that from this bound on H_G^* as a module that H_G^* is generated as an algebra by elements of degree at most $\max(2p^{m_1}, \ldots, 2p^{m_s}, \sum_i(2\alpha_p^{m_i} - p^{m_i}))$. □

8

The Structure of Group Cohomology and the Chow Ring

In this chapter we prove some of the main results on the cohomology of a finite group, and generalize them to the Chow ring. First, there is Quillen's theorem that, up to F-isomorphism (loosely, "up to pth powers"), the cohomology ring of a finite group is determined in a simple way by its elementary abelian subgroups. We prove Yagita's theorem that the Chow ring of a finite group, up to F-isomorphism, has the same description in terms of the elementary abelian subgroups. In fact, we extend Yagita's theorem from finite groups viewed as algebraic groups over the complex numbers to finite groups over any field containing the pth roots of unity (Theorem 8.10). It follows that the cycle map from the Chow ring of a finite group to the cohomology ring is an F-isomorphism.

The basic tool for the proof of Quillen's and Yagita's theorems is the norm map, described in Section 8.1. This is an operation in cohomology or the Chow ring which has had several important applications: a construction of Steenrod operations, a formula for the Chern classes of an induced representation, as well as the general properties of group cohomology and the Chow ring proved by Quillen and Yagita.

We also prove Carlson's theorem that, up to pth powers, describes the sum of the images of transfer from all proper subgroups in the cohomology ring of a p-group. Given Yagita's theorem, Carlson's theorem for cohomology immediately implies the corresponding statement for the Chow ring. These results are applied in Chapters 9 and 12 to show that the cohomology ring modulo transfers from proper subgroups is significantly simpler than the whole cohomology ring, and likewise for Chow rings.

Quillen proved his theorem relating group cohomology to elementary abelian subgroups for the classifying space of any compact Lie group G, not only for finite groups G. For a compact *connected* Lie group G and any prime number p, Adams conjectured that the $\mathbf{F}_p$-cohomology ring H^*_G is detected on elementary abelian subgroups [141, conjecture 1.1]. For finite groups, this holds only in some special cases. See Chapter 12 on more general detection theorems for finite groups.

8.1 The norm map

Given a covering map $f: X \to Y$ with degree n of topological spaces, there is a "transfer" or pushforward map $f_*: H^i(X, \mathbf{Z}) \to H^i(Y, \mathbf{Z})$, but also a "multiplicative transfer" or "norm" map $N: H^i(X, \mathbf{Z}) \to H^{ni}(Y, \mathbf{Z})$ for i even. (For i odd, the norm sends $H^i(X, \mathbf{Z})$ to $H^{ni}(Y, \widetilde{\mathbf{Z}})$, where $\widetilde{\mathbf{Z}}$ is a locally constant sheaf associated to the covering $X \to Y$.) The norm map was first defined by Evens in the context of group cohomology [12, vol. 2, section 4.1]. The geometric intuition is simple. If X is a manifold, we can represent an element u of $H^i(X, \mathbf{Z})$ by a submanifold S of codimension i (possibly with singularities). Move S into general position. Then the norm of u is represented by the submanifold of Y that, on a contractible open subset U of Y, is the intersection of the n submanifolds $S \cap U_i$, where $f^{-1}(U)$ is the disjoint union of copies $U_1, \ldots, U_n$ of U.

Evens defined the norm map for group cohomology, and applied it to give an algebraic proof that the cohomology ring of a finite group is finitely generated. The Chern classes of an induced representation can be expressed using the norm (together with the usual additive transfer), by Evens and Fulton-MacPherson [41, 45]. Finally, the norm map can be used to give an algebraic definition of Steenrod operations on group cohomology [12, vol. 2, section 4.4].

Fulton and MacPherson defined the norm map on Chow groups for a finite etale morphism of smooth k-schemes $X \to Y$ of degree n, $N_X^Y: CH^i X \to CH^{ni} Y$. (There is no need to restrict to even degrees in the Chow ring.) The norm has a useful extension to non-homogeneous elements, $N_X^Y: CH^* X \to CH^* Y$. We record the formal properties of the norm map in this section. In the rest of this chapter, we use the norm map to prove some fundamental properties of the Chow ring of a classifying space.

As in Lemma 2.15, write res^G_H for the restriction map $CH^*BG \to CH^*BH$ and $x \mapsto gx$ for the conjugation isomorphism $CH^*BH \to CH^*B(gHg^{-1})$.

Lemma 8.1

(i) *Let $X_1 \to Y$ and $X_2 \to Y$ be finite etale morphisms of smooth k-schemes. Let $x_1 \in CH^*X_1$, $x_2 \in CH^*X_2$, and let x be the element of $CH^*(X_1 \coprod X_2)$ that restricts to x_1 and x_2. Then*

$$N^Y_{X_1 \coprod X_2}(x) = N^Y_{X_1}(x_1) N^Y_{X_2}(x_2).$$

(ii) *Let H be a subgroup of a finite group G, considered as an algebraic group over a field k. For $x, y \in CH^*BH$, $N^G_H(xy) = N^G_H(x) N^G_H(y)$.*

(iii) *For H a subgroup of a finite group G and $x, y \in CH^*BH$,*

$$N_H^G(x+y) = N_H^G(x) + N_H^G(y)$$

plus a sum of transfers from proper subgroups of G. Explicitly, let $W \to BG$ denote the principal S_n-bundle corresponding to the action $f: G \to S_n$ of G on the set G/H, so that W is the disjoint union of copies of $B\ker(f)$ indexed by the set $S_n/f(G)$. Let $Z_r = W/(S_r \times S_{n-r})$, $X_r = W/(S_{r-1} \times S_{n-r})$, and $Y_r = W/(S_r \times S_{n-r-1})$, so that we have finite etale morphisms $X_r \to Z_r$ of degree r and $Y_r \to Z_r$ of degree $n-r$. We also have natural maps $X_r \to BH$ and $Y_r \to BH$, since $BH = W/S_{n-1}$. Then

$$N_H^G(x+y) = N_H^G(x) + N_H^G(y) + \sum_{r=1}^{n-1} \operatorname{tr}_{Z_r}^{BG}(N_{X_r}^{Z_r}(x) N_{Y_r}^{Z_r}(y)).$$

(iv) *(Double coset formula) Let K and H be subgroups of a finite group G, viewed as an algebraic group over a field. Then*

$$\operatorname{res}_K^G N_H^G x = \prod_{g \in K\backslash G/H} N_{K\cap gHg^{-1}}^K \operatorname{res}_{K\cap gHg^{-1}}^{gHg^{-1}} gx$$

for x in CH^*BH.

(v) *For H a normal subgroup of a finite group G,*

$$\operatorname{res}_H^G N_H^G x = \prod_{g \in G/H} gx$$

for x in CH^*BH.

A more complete list of properties of norm and transfer maps says that for a finite group G viewed as an algebraic group over a field k, the assignment $H \mapsto CH^*BH$ for subgroups H of G is a *Tambara functor* [133]. (The main property not mentioned in Lemmas 2.15 and 8.1 is the formula for the norm of a transfer, generalizing Lemma 8.1(iii).) A more classical example of a Tambara functor is the assignment $H \mapsto H^{\mathrm{ev}}(H, R)$ for subgroups H of G, when the finite group G acts on a commutative ring R [133, section 3.4].

Proof Fulton and MacPherson proved the corresponding properties for the norm map associated to a finite etale morphism of smooth varieties: (ii) is [45, property 7.1], (iii) is [45, theorem 8.1], and (iv) is [45, property 7.3]. These imply the corresponding properties for the Chow groups of a classifying space BG, since each Chow group CH^iBG is defined as $CH^i(V-S)/G$ for a suitable smooth variety $(V-S)/G$. (v) is a special case of (iv). □

Note that $N_H^G(a) = a^n$ for an integer $a \in CH^0BH$ and a subgroup $H \subset G$ of index n. Using formula (3) for the norm of a sum, it follows that the norm $CH^*BH \to CH^*BG$ passes to a well-defined norm map on Chow groups modulo a prime number p, $N_H^G: CH_H^* \to CH_G^*$.

8.2 Quillen's theorem and Yagita's theorem

In this section we prove Quillen's theorem, probably the most important result on the cohomology of finite groups. The theorem says that the cohomology ring of a finite group G has a simple description up to F-isomorphism: it is determined by the elementary abelian subgroups of G. By definition, an *F-isomorphism* $f\colon A \to B$ is a homomorphism of $\mathbf{F}_p$-algebras such that every element of the kernel of f is nilpotent, and for every element b in B, there is a natural number r such that b^{p^r} is in the image of f.

We also prove Yagita's theorem that the Chow ring of a finite group has exactly the same description, up to F-isomorphism. It follows that the map from the Chow ring to the cohomology ring is an F-isomorphism. Thus the Chow ring of a finite group is qualitatively similar to the cohomology ring.

Yagita's theorem was originally proved for finite groups viewed as algebraic groups over the complex numbers [153, theorem 3.1]. In this section, we prove it over any base field that contains the p^2 roots of unity, which requires some new arguments using motivic cohomology. Finally, in Theorem 8.10, we prove Yagita's theorem over any base field that contains the pth roots of unity, which examples show is an optimal assumption.

For a fixed prime number p, write $CH_G^* = CH^*(BG)/p$ and $H_G^* = H^*(BG, \mathbf{F}_p)$. Write k_s for the separable closure of a field k.

Theorem 8.2 *Let G be a finite group and p a prime number. View G as an algebraic group over a field k of characteristic not p that contains the p^2 roots of unity. Then the cycle map $CH_G^* \to \oplus_i H_{\text{et}}^{2i}(BG_{k_s}, \mathbf{F}_p(i))$ is an F-isomorphism. For $k \subset \mathbf{C}$, it is equivalent to say that $CH_G^* \to H_G^*$ is an F-isomorphism.*

That is: every element of the kernel of $CH_G^* \to H_G^*$ is nilpotent, and for every element x of H_G^*, there is an $r \geq 0$ such that x^{p^r} is in the image of the cycle map. Later we will extend the theorem to the case where k contains only the pth roots of unity (Theorem 8.10).

Let us begin the proof of Theorem 8.2.

Lemma 8.3 *Let G be a finite group, viewed as an algebraic group over a field k. Let M be a torsion* $\mathrm{Gal}(k_s/k)$*-module such that all elements have order invertible in k. Then the etale cohomology $H_{\text{et}}^*(BG_k, M)$ is isomorphic to the continuous cohomology $H^*(G \times \mathrm{Gal}(k_s/k), M)$. In particular, for k separably closed, the etale cohomology of BG_k is simply the cohomology of the group G.*

Grothendieck studied the continuous cohomology of the profinite group $G \times \mathrm{Gal}(k_s/k)$ (viewed as the G-equivariant etale cohomology of Spec k) before our algebro-geometric model for BG was defined [58].

Proof Under our assumption on M, etale cohomology is homotopy invariant, and so $H^*_{\text{et}}(EG_k, M)$ is isomorphic to $H^*_{\text{et}}(k, M) = H^*(\text{Gal}(k_s/k), M)$ [104, corollary VI.4.20]. There is a natural map from the continuous cohomology $H^*(G \times \text{Gal}(k_s/k), M)$ to $H^*_{\text{et}}(BG_k, M)$. We have a Künneth spectral sequence

$$E_2^{**} = H^*(G, H^*(\text{Gal}(k_s/k), M)) \Rightarrow H^*(G \times \text{Gal}(k_s/k), M),$$

and this maps to the Hochschild-Serre spectral sequence for etale cohomology,

$$E_2^{**} = H^*(G, H^*(EG_k, M)) \Rightarrow H^*(BG_k, M)$$

[104, theorem 2.20]. Since this homomorphism of spectral sequences is an isomorphism on E_2 terms, it gives an isomorphism $H^*(G \times \text{Gal}(k_s/k), M) \to H^*(BG_k, M)$. $\square$

For the proof of Theorem 8.2, let us write H^*_G for either the ring $H^*(BG, \mathbf{F}_p)$ or $\oplus_i H^{2i}_{\text{et}}(BG_{k_s}, \mathbf{F}_p(i))$ (isomorphic to the even-degree subring of the former ring). The arguments work the same way in both cases.

Following Quillen, consider the category of elementary abelian p-subgroups of G with morphisms being the homomorphisms given by conjugation by elements of G together with inclusions. Then restriction gives a ring homomorphism

$$CH^*_G \to \varprojlim CH^*_A.$$

(Explicitly, an element of $\varprojlim CH^*_A$ is an element x_A of CH^*_A for every elementary abelian p-subgroup A such that $x_B = x_A|_B$ whenever $B \subset A$ and $x_{gAg^{-1}} = gx_A$ for every $g \in G$.) We will show that this map is an F-isomorphism. The same arguments prove Quillen's theorem [114], [12, vol. 2, corollary 5.6.4]:

Theorem 8.4 *Let G be a finite group, p a prime number. Then*

$$H^*_G \to \varprojlim H^*_A$$

is an F-isomorphism.

A morphism of schemes over a field is called a universal homeomorphism if it is a homeomorphism (for the Zariski topology), and remains so after any extension of the base field. For example, the morphism $x \mapsto x^p$ is a universal homeomorphism from the affine line over $\mathbf{F}_p$ to itself. Quillen observed that Theorem 8.4 gives a simple description of the "variety" Spec H^{ev}_G, considered up to universal homeomorphisms: it is the union of affine spaces over $\mathbf{F}_p$ corresponding to the elementary abelian subgroups of G, glued together by inclusion and conjugation of such subgroups. Here "glued" includes the possibility of taking the quotient of the affine space Spec H^{ev}_E by the finite group $N_G(E)/C_G(E)$. In particular, all irreducible components of Spec H^{ev}_G

are unirational. Thus the cohomology rings of finite groups are very special among finitely generated $\mathbf{F}_p$-algebras.

In particular, Quillen's theorem implies that the dimension of the ring H_G^* is equal to the *p-rank* of G (the maximal rank of the elementary abelian p-subgroups of G). (For a ring R that is finite over a central subring A, we define the *dimension* of R to be the Krull dimension of A; this is independent of the choice of A. The ring H_G^* is finite over its commutative subring H_G^{ev}, and so this definition applies.) The irreducible components of Spec H_G^{ev} are in one-to-one correspondence with the conjugacy classes of maximal elementary abelian p-subgroups of G.

Since k contains the pth roots of unity, we have $CH_A^* = \mathbf{F}_p[y_1, \ldots, y_n]$. So $CH_A^* \to H_A^*$ is an F-isomorphism, and one deduces easily that $\varprojlim CH_A^* \to \varprojlim H_A^*$ is an F-isomorphism. So proving that $CH_G^* \to \varprojlim CH_A^*$ is an F-isomorphism will prove Theorem 8.2.

The proof of F-surjectivity follows Quillen's and Yagita's arguments [114, 153]. Here we only need k to contain the pth roots of unity.

Lemma 8.5 *Let A be an elementary abelian subgroup of G. Write $[N_G(A)\colon A] = qb$, where q is a power of p and b is prime to p. Let u be a positive-degree element of $(CH_A^*)^{N_G(A)}$ that restricts to zero on all proper subgroups of A. Then there is an element v in CH_G^* that restricts to u^q on A and to zero on all elementary abelian subgroups A' of G that are not conjugate to a subgroup containing A.*

Proof Let $w = N_A^G(1+u)$. By the double coset formula for the norm (Lemma 8.1),

$$w|_A = \prod_{g \in A\backslash G/A} N^A_{A\cap gAg^{-1}} \operatorname{res}^{gAg^{-1}}_{A\cap gAg^{-1}} g(1+u).$$

For $g \notin N_G(A)$, the factor shown is equal to 1, since u restricts to zero on proper subgroups of A. So

$$\begin{aligned} w|_A &= \prod_{g \in N_G(A)/A} g(1+u) \\ &= (1+u)^{qb} \\ &= 1 + bu^q + \text{terms of higher degree,} \end{aligned}$$

using that u is fixed by $N_G(A)$. Let v be $1/b \in \mathbf{F}_p$ times the term of degree $q|u|$ in w. Then v restricts to u^q on A, as we want. For an elementary abelian subgroup A' of G that is not conjugate to a subgroup containing A, we have

$$\begin{aligned} w|_{A'} &= \prod_{g \in A'\backslash G/A} N^{A'}_{A'\cap gAg^{-1}} \operatorname{res}^{gAg^{-1}}_{A'\cap gAg^{-1}} g(1+u) \\ &= 1, \end{aligned}$$

since $A' \cap gAg^{-1}$ is always a proper subgroup of gAg^{-1}. So $v|_{A'} = 0$, as we want. □

We can now prove the F-surjectivity of $CH^*_G \to \varprojlim CH^*_A$. Let u be an element of $\varprojlim CH^*_A$. We want to show that some p-power of u is the restriction of an element of CH^*_G. We can assume that u has positive degree. Let A be an elementary abelian subgroup of smallest order among those such that $u_A \neq 0$. We know that u_A is in $(CH^*_A)^{N_G(A)}$. By assumption, u restricts to zero on all proper subgroups of A. Let q be the maximum power of p dividing $[N_G(A) \colon A]$. By Lemma 8.5, there is an element v of CH^*_G that restricts to u^q_A on A and to zero on all elementary abelian subgroups not conjugate to a subgroup containing A.

Let $u_2 = u^q - v$ in $\varprojlim CH^*_A$. We know that u_2 vanishes on all elementary abelian subgroups of order smaller than A, and also on A, while it is equal to u^q on the elementary abelian subgroups that have the same order as A but are not conjugate to it. By induction on the order of A and on the set of conjugacy classes of elementary abelian subgroups of a given order, some p-power of u_2 is a restriction from G. Therefore, some p-power of u is a restriction from G. We have proved the F-surjectivity of $CH^*_G \to \varprojlim CH^*_A$.

Lemma 8.6 *Let G be a finite group and p a prime number. View G as an algebraic group over a field k of characteristic not p that contains the p^2 roots of unity. Then any element of CH^*_G that restricts to zero on all elementary abelian subgroups of G is nilpotent.*

We follow Yagita's arguments on the case $k = \mathbf{C}$, inspired by Minh's proof of the corresponding result in cohomology, originally due to Quillen. There is some extra work for k not algebraically closed.

Proof We consider all Chow rings modulo p. The Chow ring of G injects into the Chow ring of a Sylow p-subgroup, and so we can assume that G is a p-group. If G is elementary abelian, then the result is clear. Let G be a p-group that is not elementary abelian. Let $V = G/[G, G]G^p$ be the maximal elementary abelian quotient group of G. Let $x_1, \ldots, x_n$ be a basis for H^1_V and $y_i = \beta x_i$; then H^*_V is a free module over $\mathbf{F}_p[y_1, \ldots, y_n]$ with basis the monomials $\prod_{i \in I} x_i$ for $I \subset \{1, \ldots, n\}$. Also, since k contains the pth roots of unity,

$$CH^*_V = \mathbf{F}_p[y_1, \ldots, y_n].$$

For cohomology, the following result is due to Serre [123].

Lemma 8.7 *Let e_V be the product of one nonzero element from each line in the $\mathbf{F}_p$-vector space $CH^1_V = \mathbf{F}_p\{y_1, \ldots, y_n\}$. Then e_V pulls back to zero in CH^*_G.*

Proof Since the p-group G is nilpotent but not elementary abelian, it maps onto a nontrivial central extension H of V by $\mathbf{Z}/p$. Such an extension is classified by an element of H^2_V that pulls back to zero in H^2_H. Therefore we have a relation

$$f = \sum_{i<j} c_{ij} x_i x_j + \sum_k d_k y_j = 0$$

in H^*_G for some $c_{ij}, d_k \in \mathbf{F}_p$, not all zero. The lemma is trivially true if some nontrivial linear combination of $y_1, \ldots, y_n$ is zero in H^*_G. (That is, the lemma is easy if the abelianization of G is not elementary abelian.) So we can assume that some c_{ij} is not zero.

We want to deduce an analogous relation in motivic cohomology $H^2_M(BG, \mathbf{Z}/p(2))$. By the Beilinson-Lichtenbaum conjecture (Theorem 6.9), the latter group is isomorphic to $H^2_{\text{et}}(BG, \mathbf{Z}/p(2))$. For k separably closed, that would be isomorphic to $H^2(BG, \mathbf{Z}/p)$, but we need more care since we are assuming only that k contains the p^2 roots of unity. By Lemma 8.3, $H^*_{\text{et}}(BG_k, M) \cong H^*(G \times \text{Gal}(k_s/k), M)$ for every finite $\text{Gal}(k_s/k)$-module M of order invertible in k. In particular, we have a splitting

$$\begin{aligned} H^1_M(BV, \mathbf{Z}/p(1)) &\cong H^1_{\text{et}}(BV, \mathbf{Z}/p(1)) \\ &\cong H^1(V, \mu_p(k)) \oplus H^1_{\text{et}}(k, \mu_p). \end{aligned}$$

We can view $x_1, \ldots, x_n$ in H^1_V as a basis for the summand of $H^1_M(BV, \mathbf{Z}/p(1))$ that restricts to zero in $H^1(k, \mu_p)$. Let $y_i = \beta x_i$ in $H^2_M(BV, \mathbf{Z}/p(1)) = CH^1_V$. Then y_i has a class in

$$H^2_{\text{et}}(BV, \mathbf{Z}/p(1)) = H^2(V, \mu_p(k)) \oplus H^1(V, H^1_{\text{et}}(k, \mu_p)) \oplus H^2_{\text{et}}(k, \mu_p),$$

which I claim lies in the summand $H^2(V, \mu_p(k))$. Clearly y_i restricts to zero in $H^2_{\text{et}}(k, \mu_p)$, because x_i restricts to zero in $H^1_{\text{et}}(k, \mu_p)$. If we assumed only that k contained the pth roots of unity, then y_i could have nonzero component in $H^1(V, H^1_{\text{et}}(k, \mu_p))$; a similar observation was made by Grothendieck [58, equation 5.6]. But since k contains the p^2 roots of unity, we have an exact sequence

$$0 \to \mu_p(k) \to \mu_{p^2}(k) \to \mu_p(k) \to 0$$

of abelian groups. Viewing these as G-modules with G acting trivially, we get a Bockstein map on $H^*(G, H^0(k, \mu_p))$, and this is compatible with the Bockstein map on $H^*_{\text{et}}(BG_k, \mu_p) = H^*(BG \times BGal(k_s/k), \mu_p)$ via the homomorphism $G \times Gal(k_s/k) \to G$ of profinite groups. We conclude that the class of y_i in etale cohomology lies in the summand $H^2(V, \mu_p(k))$.

Write τ for a generator of $H^0_M(k, \mathbf{Z}/p(1)) = \mu_p(k)$. Then the elements $x_i x_j$ for $i < j$ and τy_i in $H^2_M(BV, \mathbf{Z}/p(2))$ both map into the summand

$H^2(V, H^0_{\text{et}}(k, \mathbf{Z}/p(2))) \cong H^2_V$ of $H^2_{\text{et}}(BV, \mathbf{Z}/p(2))$, by what we have shown. Since the element

$$f = \sum_{i<j} c_{ij}x_i x_j + \tau \sum_k d_k y_j$$

of $H^2_M(BV, \mathbf{Z}/p(2))$ pulls back to zero in the ordinary cohomology of G, it pulls back to zero in $H^2_{\text{et}}(BG, \mathbf{Z}/p(2))$, and so it pulls back to zero in $H^2_M(BG, \mathbf{Z}/p(2))$ by the Beilinson-Lichtenbaum conjecture (Theorem 6.9).

At this point, we can return to Yagita's arguments. Voevodsky defined the Milnor operation

$$Q_r \colon H^i_M(X, \mathbf{Z}/p(j)) \to H^{i+2p^r-1}_M(X, \mathbf{Z}/p(j + p^r - 1))$$

on motivic cohomology inductively by writing Q_0 for the Bockstein β and $Q_{i+1} = [P^{p^i}, Q_i]$ [150, section 2.2]. (For $p = 2$, this formula uses our assumption that k contains the 4th roots of unity; otherwise Q_i is defined differently [144, example 13.7].) The operation Q_i is a derivation, and we have $Q_r x_i = y_i^{p^r}$ and $Q_r y_i = 0$, using $y_i = \beta x_i$ together with the unstable properties of Steenrod operations (Section 6.3). Therefore

$$Q_r Q_s f = \sum_{i<j} c_{ij}(y_i^{p^r} y_j^{p^s} - y_i^{p^s} y_j^{p^r}) = 0$$

in CH^*_G, for any r and s. Note the remarkable feature of the Milnor operation in this argument, emphasized by Yagita: it produces a relation among algebraic cycles from purely topological input.

Let $L = (CH^*_G)^{\oplus n}$, viewed as a free module over CH^*_G, and let $e_i \in L$ be the vector $(y_i, y_i^p, \ldots, y_i^{p^{n-1}})$. The relation above in CH^*_G (applied for $r, s \leq n-1$) implies that $\sum_{i<j} c_{ij} e_i \wedge e_j = 0$ in $\Lambda^2 L$. We know that c_{ij} is not zero in $\mathbf{F}_p$ for some $i < j$; after changing the numbering, we can assume that $c_{12} \neq 0$. Multiplying by $e_3 \wedge \cdots \wedge e_n$, we find that $c_{12} e_1 \wedge \cdots \wedge e_n = 0$. So the determinant

$$\det(y_i^{p^{j-1}})_{1 \leq i,j \leq n} = e_1 \wedge \cdots \wedge e_n$$

is zero in $\Lambda^n L \cong CH^*_G$.

The determinant of this matrix

$$\begin{pmatrix} y_1 & y_1^p & \cdots & y_1^{p^{n-1}} \\ y_2 & y_2^p & \cdots & y_2^{p^{n-1}} \\ \vdots & \vdots & \ddots & \vdots \\ y_n & y_n^p & \cdots & y_n^{p^{n-1}} \end{pmatrix}$$

is known as the *Moore determinant*, an analog in characteristic p of the Vandermonde determinant. The Moore determinant is the product of one nonzero vector in each line in $\mathbf{F}_p\{y_1, \ldots, y_n\}$. (Indeed, the Moore determinant vanishes whenever there is an $\mathbf{F}_p$-linear relation between $y_1, \ldots, y_n$, since that gives a linear relation among the rows. So the Moore determinant, as a polynomial in $y_1, \ldots, y_n$, is a multiple of the product mentioned here. Since both have degree $1 + p + \cdots + p^{n-1}$, they are equal, up to a scalar.) We have therefore shown that the product of one nonzero vector in each line in $\mathbf{F}_p\{y_1, \ldots, y_n\}$ is equal to zero in CH_G^*. □

The following statement is an analog of a theorem on cohomology by Minh. For Chow rings with $k = \mathbf{C}$, it was proved by Yagita [153, theorem 3.4].

Lemma 8.8 *Let G be a p-group that is not elementary abelian. Consider G as an algebraic group over a field k of characteristic not p that contains the p^2 roots of unity. If $u \in CH_G^*$ restricts to zero on all proper subgroups, then $u^p = 0$.*

Proof Let $V = G/[G, G]G^p$ be the maximal elementary abelian quotient group of G. Let x be a nonzero element of $H_G^1 = H_V^1 \cong \mathrm{Hom}(V, \mathbf{Z}/p)$. Then $\ker(x)$ is a subgroup of index p in G. We will show that if u restricts to zero in the Chow ring of $\ker(x)$ for all such x, then u^p is a product of some element of CH_G^* with the product e_V of one nonzero element from each $\mathbf{F}_p$-line in CH_V^1 restricted to G. That will imply that $u^p = 0$, by Lemma 8.7 (using that k contains the p^2 roots of unity). Identifying H_V^1 with a summand of $H_M^1(V, \mathbf{Z}/p(1))$, we can say that the Bockstein gives an isomorphism from H_V^1 to CH_V^1. Write $y = \beta x$.

Consider the norm $v := N_G^{G\times\mathbf{Z}/p}(u)$ in $CH^*_{G\times Z/p}$. By the Chow Künneth formula (Lemma 2.12), we can write v as a sum $\sum_{i\geq 0} f_i(u)t^i$, where t is a generator of $CH^1_{\mathbf{Z}/p} \cong \mathbf{Z}/p$ and $f_i(u)$ is in CH_G^*. Restricting v to the normal subgroup $G \subset G \times \mathbf{Z}/p$ gives u^p by Lemma 8.1 (since $\mathbf{Z}/p$ acts trivially by conjugation on G), and so we have $f_0(u) = u^p$. (The elements $f_i(u)$ are in fact the Steenrod operations of u, suitably renumbered, but we do not use that [12, vol. 2, definition 4.4.1].)

Intuitively, intersecting two cycles cannot be made a strictly commutative operation, because of the need to perturb cycles when they do not intersect transversely, although in a sense it is commutative up to all higher homotopies. As Benson observed, this failure is manifested by the norm of the restriction of an element (for example from $B(G \times \mathbf{Z}/p)$ to BG) not always being a power of that element, in other words by the non-vanishing of Steenrod operations. By contrast, the transfer of a restriction is a multiple of the original element [12, vol. 2, section 4.4]. With rational coefficients, there is no obstruction to making the product commutative at the level of cycles, as shown in topology by de Rham cohomology.

Let $x = x_1, \ldots, x_n$ be a basis for $H_V^1 = \mathrm{Hom}(G, \mathbf{Z}/p)$, and let $a_1, \ldots, a_n$ be elements of G such that $x_i(a_j) = \delta_{ij} \in \mathbf{Z}/p$. Then $a_1, \ldots, a_n$ generate the group G. Let G_x be the subgroup $\langle a_1 b, a_2, \ldots, a_n \rangle \subset G \times \mathbf{Z}/p$, where we write $\mathbf{Z}/p = \langle b \rangle$, and let $f_x \colon G \to G_x$ be the isomorphism given by

$f_x(a_1) = a_1 b$ and $f_x(a_i) = a_i$ for $i > 1$. Then G and G_x are both normal subgroups of index p in $G \times \mathbf{Z}/p$, and their intersection is $\ker(x) \subset G$. We have

$$f_x^*(v|_{G_x}) = u^p + \sum_{i>0} f_i(u) y^i.$$

But the element on the left is zero, by the double coset formula for the norm (Lemma 8.1), since u restricts to zero on $\ker(x)$. Adding up these formulas for all $0 \neq x$ in H^1_V, we find that $u^p = \sum_{i>0}(f_i(u) \sum_{y \neq 0 \in V^*} y^i)$. For each $i > 0$, $\sum_{y \neq 0 \in V^*} y^i$ is a polynomial function on $V \otimes_{\mathbf{F}_p} \overline{\mathbf{F}_p}$ that vanishes on all codimension-1 linear subspaces defined over $\mathbf{F}_p$. So it is a multiple of the product e_V of one nonzero element y from each $\mathbf{F}_p$-line in $CH^1_V = V^*$. Thus u^p is a multiple of e_V, pulled back to CH^*_G, and hence is zero. □

We can now finish the proof of Lemma 8.6. As discussed earlier, this complete the proof of Theorem 8.2. Let u be an element of CH^*_G that restricts to zero on all elementary abelian subgroups of G. If G is elementary abelian, then $u = 0$ and we are done. Otherwise, by induction on the order of G, the restriction of u to each proper subgroup of G is nilpotent. So there is an $r \geq 0$ such that u^{p^r} restricts to zero on all proper subgroups of G. Then $u^{p^{r+1}} = 0$ by Lemma 8.8. □

8.3 Yagita's theorem over any field containing the pth roots of unity

We now strengthen Lemma 8.6 by removing the assumption that the base field k contains the p^2 roots of unity.

Lemma 8.9 *Let G be a finite group and p a prime number. View G as an algebraic group over a field k of characteristic not p. Then any element of $CH^*_G = CH^*(BG)/p$ that restricts to zero on all elementary abelian subgroups of G is nilpotent.*

Proof Let E be the extension field of k obtained by adjoining the pth roots of unity. The degree of E over k divides $p - 1$ and hence is prime to p. Also, E is a separable extension of k. Using transfer for the finite etale morphism $BG_E \to BG_k$, it follows that the pullback map $CH^*_{G_k} \to CH^*_{G_E}$ is injective. Thus it suffices to prove the lemma when k contains the pth roots of unity. We can also assume that k is infinite, since we can replace a finite field k by the direct limit over an infinite sequence of extensions of degree prime to p.

We have proved the lemma when k contains the p^2 roots of unity (Lemma 8.9). Suppose that k does not contain the p^2 roots of unity. Let F be the extension field of k obtained by adjoining the p^2 roots of unity; then F is a cyclic extension of k of degree p. It follows that there is a morphism $\alpha\colon \operatorname{Spec}(k) \to B(\mathbf{Z}/p)_k$ of k-schemes that classifies the principal $\mathbf{Z}/p$-bundle $\operatorname{Spec}(F) \to \operatorname{Spec}(k)$. (To prove this, consider a finite-dimensional approximation $(V - S)/(\mathbf{Z}/p)$ of $B(\mathbf{Z}/p)_k$, where V is a representation of $\mathbf{Z}/p$ over k and $S \subsetneq V$ is a closed G-invariant subset such that G acts freely on $V - S$. Then $(\operatorname{Spec}(F) \times_k V)/(\mathbf{Z}/p)$ is a vector bundle over $\operatorname{Spec}(F)/(\mathbf{Z}/p) = \operatorname{Spec}(k)$, by faithfully flat descent as described in section 2.2. Since k is infinite, this k-vector space has a k-rational point in the open subset $(\operatorname{Spec}(F) \times_k (V - S))/(\mathbf{Z}/p)$. The image of this point in $(V - S)/(\mathbf{Z}/p)$ is a morphism $\alpha\colon \operatorname{Spec}(k) \to (V - S)/(\mathbf{Z}/p)$ of k-schemes that classifies the principal $\mathbf{Z}/p$-bundle $\operatorname{Spec}(F) \to \operatorname{Spec}(k)$, as we want.)

It follows that we have a pullback diagram

$$\begin{array}{ccc} BG_F & \longrightarrow & BG_k \\ \downarrow & & \downarrow \gamma \\ BG_k & \underset{\beta}{\longrightarrow} & B(G \times \mathbf{Z}/p)_k, \end{array}$$

where the right vertical map γ is given by the inclusion $G \subset G \times \mathbf{Z}/p$, whereas the bottom map β is given by the identity on BG_k together with the morphism $BG_k \to \operatorname{Spec}(k) \underset{\alpha}{\to} B(\mathbf{Z}/p)_k$. (To be precise, on finite-dimensional approximations to the classifying spaces, this is an actual pullback diagram of finite etale morphisms.)

Since k contains the pth roots of unity, we have $CH^*_{(G\times\mathbf{Z}/p)_k} = CH^*_{G_k}[t]$ with $|t| = 1$. Therefore, for any element y in $CH^i_{G_k}$, its norm via γ has the form $N_G^{G\times\mathbf{Z}/p}(y) = \sum_{j=0}^{pi} f_j(y)t^j$ for some $f_j(y) \in CH_G^{pi-j}$. Restricting the principal $\mathbf{Z}/p$-bundle γ to BG_k by the morphism $\gamma\colon BG_k \to B(G \times \mathbf{Z}/p)_k$ gives the trivial $\mathbf{Z}/p$-bundle over BG_k, from which we read off that $f_0(y) = y^p$. (The other elements $f_j(y)$ are certain Steenrod operations of y, as mentioned in the proof of Lemma 8.8, but we do not need that.)

The norm is compatible with pullback [45, property 7.3], and so our diagram implies that

$$\begin{aligned} N_{G_F}^{G_k}\operatorname{res}_{G_F}^{G_k} y &= \beta^* N_G^{G\times\mathbf{Z}/p} y \\ &= \beta^* \sum_{j=0}^{pi} f_j(y)t^i \end{aligned}$$

in $CH^*_{G_k}$. The important point is that $CH^1 \operatorname{Spec}(k) = 0$. So $\beta^* t = 0$ in $CH^1 BG_k$, because β factors through $\operatorname{Spec}(k)$. As a result,

$$N_{G_F}^{G_k}\operatorname{res}_{G_F}^{G_k} y = y^p.$$

Therefore, if $y \in CH^i_{G_k}$ restricts to zero in $CH^i_{G_F}$, then $y^p = 0$. Thus, since we know the lemma for G_F, it holds for G_k. □

We can now prove Yagita's theorem assuming only that the base field contains the pth roots of unity.

Theorem 8.10 *Let G be a finite group and p a prime number. View G as an algebraic group over a field k of characteristic not p. Then the cycle map $CH^*_G \to \oplus_i H^{2i}_{\text{et}}(BG_{k_s}, \mathbf{F}_p(i))$ has nilpotent kernel. If k contains the pth roots of unity, then the cycle map is an F-isomorphism. For $k \subset \mathbf{C}$ containing the pth roots of unity, it is equivalent to say that $CH^*_G \to H^*_G$ is an F-isomorphism.*

We do need k to contain the pth roots of unity, as shown by the example of $G = \mathbf{Z}/p$ over $\mathbf{Q}$ for p odd. In that case, $CH^*_G \to H^*_G$ is not F-surjective, since the image is the subring $\mathbf{F}_p[y^{p-1}]$ of the free graded-commutative algebra $H^*_G = \mathbf{F}_p\langle x, y\rangle$ [138, example 13.1].

Proof We already showed after Lemma 8.5 that the cycle map is F-surjective when k contains the pth roots of unity. It remains to show that the cycle map has nilpotent kernel, for any field k of characteristic not p.

Let E be the extension field of k obtained by adjoining the pth roots of unity. Then E has degree prime to p over k, and so $CH^*_{G_k} \to CH^*_{G_E}$ is injective. So it suffices to show that the kernel of the cycle map is nilpotent when k contains the pth roots of unity. In that case, for every elementary abelian p-subgroup A of G, CH^*_A is a polynomial ring over $\mathbf{F}_p$ and the cycle map $CH^*_A \to H^*_A$ is injective. So an element y of the kernel of $CH^*_G \to H^*_G$ restricts to zero in CH^*_A for all elementary abelian p-subgroups A. By Lemma 8.9, y is nilpotent. (In fact, the proof gives an explicit bound N such that all elements y of the kernel have $y^{p^N} = 0$.) □

8.4 Carlson's theorem on transfer

Carlson's theorem on transfer in group cohomology is important for the rest of the book. Here we prove it and generalize it to the Chow ring.

Lemma 8.11 *Let H be a subgroup of a finite group G. Then the transfer maps $\text{tr}^G_H \colon CH^*_H \to CH^*_G$ and $\text{tr}^G_H \colon H^*_H \to H^*_G$ commute with pth powers.*

Proof For each $i \geq 0$, the pth power map $CH^i_G \to CH^{pi}_G$ is equal to the Steenrod operation P^i defined by Voevodsky (Section 6.3). Moreover, Steenrod operations commute with pushforward maps for finite etale morphisms, hence with the transfer, for example by [21, definition 8.13 and proposition 9.11]. The same argument works for cohomology. □

We now prove Carlson's theorem [26, corollary 12.4.6]. We follow an argument by Benson, but with explicit estimates of the p-powers needed for the proof [11]. We will apply the theorem when G is a p-group. Readers may wish to concentrate on that case, where the statement is a little simpler.

Theorem 8.12 *Let G be a finite group and p a prime number. Let P be a Sylow p-subgroup of G, and let C be the p-torsion subgroup of the center of P. Then the kernel of $H^*_G \to H^*_C$ has the same radical as the sum of the images of all transfers to G from proper subgroups of P. Moreover, the sum of the images of all transfers from centralizers $C_G(E)$ of elementary abelian p-subgroups E such that $C_G(E)$ does not contain a Sylow p-subgroup of G has the same radical in H^*_G.*

Example Let G be the dihedral group of order 8. Then

$$CH^*_G = \mathbf{F}_2[c_1V, c_2V, a]/(a^2 = ac_1V)$$

by Lemma 13.2, and the kernel of restriction to $C = \mathbf{Z}/2$ is the ideal (a, c_1V) in CH^*_G. Thus the ring CH^*_G has Krull dimension 2 (and $\mathrm{Spec}(CH^*_G)$ is the union of two affine planes over $\mathbf{F}_2$ that meet in a line), while the image of CH^*_G in CH^*_C is the 1-dimensional ring $\mathbf{F}_2[c_2V]$. We use the notation from the proof of Lemma 13.2; so G has two non-central elementary abelian subgroups of G called A_1 and A_2, both isomorphic to $(\mathbf{Z}/2)^2$. Here A_1 and A_2 are their own centralizers in G. We compute that $\mathrm{tr}^G_{A_1}(t_1) = a + c_1V$ in CH^*_G, while $\mathrm{tr}^G_{A_2}(t_2) = a$. (These computations can be done by observing that CH^*_G is detected on A_1 and A_2 in this case. So it suffices to compute the restriction of these transferred elements to A_1 and A_2 using the double coset formula, Lemma 2.15.) Thus the sum of the images of transfer from $CH^*_{A_1}$ and $CH^*_{A_2}$ is the ideal $\ker(CH^*_G \to CH^*_C) = (a, c_1V)$, which checks Corollary 8.14 in this case.

Likewise, the cohomology ring of the dihedral group G is $H^*_G = \mathbf{F}_2[x_1, x_2, y]/(x_1^2 = x_1x_2)$, where $|x_1| = |x_2| = 1$ and $|y| = 2$. For a suitable choice of these generators, the Chow ring CH^*_G maps to cohomology by $a \mapsto x_1^2$, $c_1V \mapsto x_2^2$, and $c_2V \mapsto y^2$. The restriction $H^*_G \to H^*_C$ has image $\mathbf{F}_2[y]$ and kernel the ideal (x_1, x_2). We compute (directly, or using the calculation for Chow rings) that the images of the transfer maps $H^1_{A_1} \to H^1_G$ and $H^1_{A_2} \to H^1_G$ are spanned by $x_1 + x_2$ and x_1, respectively. So the sum of the images of transfer from $H^*_{A_1}$ and $H^*_{A_2}$ is the ideal $\ker(H^*_G \to H^*_C) = (x_1, x_2)$, which checks Theorem 8.12 in this case.

Proof (Theorem 8.12) One direction is easy. Let $H \subset P$ be a proper subgroup. Then the image of the transfer $\mathrm{tr}^G_H \colon H^*_H \to H^*_G$ is contained in the kernel of $H^*_G \to H^*_C$. Indeed,

$$\mathrm{res}^G_C \mathrm{tr}^G_H x = \sum_{g \in C\backslash G/H} \mathrm{tr}^C_{C \cap gHg-1} \mathrm{res}^{gHg^{-1}}_{C \cap gHg^{-1}} gx$$

for x in H^*_H (Lemma 2.15). The transfer from the cohomology of any proper subgroup of C to C is zero, and so the expression is zero except for terms corresponding to C-orbits on G/H of size 1. Since P is a p-group and H is a proper subgroup of P, all P-orbits on G/H have order a multiple of p. Finally, the terms corresponding to C-orbits in a single P-orbit are all equal, since P centralizes C. So the sum is a multiple of p and hence is zero in H^*_C.

For the reverse direction, we use the following result on invariant theory. The statement uses the trace for a finite group action, defined after Lemma 7.3. For a vector space V over $\mathbf{F}_p$, let $\sigma_V \in \mathbf{F}_p[V^*]$ be the product of all nonzero elements of V^*. (Thus $\sigma_V = e_V^{p-1}$, in the notation of Lemma 8.7.)

Lemma 8.13 *Let G be a finite group, and let V be a faithful representation of G of dimension n over $\mathbf{F}_p$. Then there is a polynomial $\alpha \in \mathbf{F}_p[V^*]$ such that $\mathrm{tr}_1^G(\alpha) = \sigma_V^{n-1}$.*

Proof It suffices to prove this when $G = GL(n, \mathbf{F}_p)$. In that case, this was shown by Campbell, Hughes, Shank, and Wehlau, with $\alpha = \prod_{i=1}^n x_i^{p^n-1-(p-1)p^{i-1}}$ [25, corollary 9.14]. (There is a related calculation by Priddy and Wilkerson, but as far as I can see, it shows only that σ_V^n is in the image of the trace [110, p. 784].) Shank and Wehlau gave a simpler proof, which also shows the optimality of this statement. In fact, the image of the trace for $GL(n, \mathbf{F}_p)$ on $\mathbf{F}_p[V^*]$ is the ideal generated by σ_V^{n-1} in the invariant ring [125, theorem 5.5]. □

We now prove Theorem 8.12. Let $y \in H^*_G$ be an element that restricts to zero on C. We will show that there is an $r \geq 0$ such that y^{p^r} is a sum of transfers from centralizers $C_G(E)$ of elementary abelian subgroups E with $C_G(E)$ not containing a Sylow p-subgroup of G.

It suffices to show: (*) let E be an elementary abelian subgroup of a p-group G. Let $y \in H^*_G$ be an element that restricts to zero on all proper subgroups of E. Then there is an $r \geq 0$ and an element $x \in H^*_{C_G(E)}$ such that (i) $\mathrm{res}^G_E(y^{p^r} - \mathrm{tr}^G_{C_G(E)}x) = 0$ and (ii) $\mathrm{res}^G_{E'}\mathrm{tr}^G_{C_G(E)}x = 0$ for all elementary abelian subgroups $E' \subset G$ that are not conjugate to a subgroup containing E. Indeed, suppose that we know (*), and let y be an element of H^*_G that restricts to zero on C. Then y restricts to zero on all elementary abelian subgroups of G whose centralizer contains a Sylow p-subgroup of G. Then, after raising y to a suitable p-power, (*) implies that we can subtract transfers from centralizers of elementary abelian subgroups E such that $C_G(E)$ does not contain a Sylow p-subgroup of G, and get an element of H^*_G that restricts to zero on all elementary abelian subgroups. (Here we use that transfers commute with p-powers, Lemma 8.11. Also, every element of H^*_G transferred from $C_G(E)$ is also transferred from a Sylow p-subgroup of $C_G(E)$, which is an element of $\mathcal{S}$.) But then raising to a further p-power gives zero, by Quillen (Theorem 8.4), and the theorem is proved.

To analyze property (ii) in (*), note that

$$\mathrm{res}^G_{E'}\mathrm{tr}^G_{C_G(E)}x = \sum_{E'\backslash G/C_G(E)} \mathrm{tr}^{E'}_{E'\cap gC_G(E)g^{-1}}\mathrm{res}^{gC_G(E)g^{-1}}_{E'\cap gC_G(E)g^{-1}}gx.$$

On the right side, we are restricting x to $g^{-1}E'g \cap C_G(E)$, up to conjugation. So (ii) holds if x restricts to zero on all elementary abelian subgroups $A \subset C_G(E)$ that do not contain E, as we will arrange.

To prepare to prove (*), note that the group $W_G(E) := N_G(E)/C_G(E)$ acts faithfully on the given elementary abelian subgroup E. Let $\alpha \in CH_E^* \subset H_E^*$ be the element given by Lemma 8.13, so that $\mathrm{tr}_1^{W_G(E)}(\alpha) = \sigma_E^{n-1}$, where E has rank n. Let $\beta = \sigma_E\alpha$; then $\mathrm{tr}_1^{W_G(E)}(\beta) = \sigma_E^n$ and β restricts to zero on all proper subgroups of E. By Lemma 8.5, there is an element of $H_{C_G(E)}^*$ (constructed using the norm) that restricts to β^{p^s} on E, where $[C_G(E) : E] = p^s$, and restricts to zero on all elementary abelian subgroups of $C_G(E)$ that do not contain E. Let $[G : E] = p^b$, where it is clear that $b \geq s$. For our purpose, raise the element just produced to the p^{b-s} power, so as to produce an element η of $H_{C_G(E)}^*$ that restricts to β^{p^b} on E and restricts to zero on all elementary abelian subgroups of $C_G(E)$ that do not contain E.

We will take $x \in H_{C_G(E)}^*$ to be of the form $x = \eta\, \mathrm{res}_{C_G(E)}^G z$ for a suitable $z \in H_G^*$. Then x restricts to zero on all elementary abelian subgroups of $C_G(E)$ that do not contain E, and so property (ii) holds. It remains to choose $z \in H_G^*$ in order to make x satisfy property (i).

Clearly $\mathrm{tr}_{C_G(E)}^G(x) = (\mathrm{tr}_{C_G(E)}^G \eta)z$ in H_G^*. The restriction of this to E is

$$(\mathrm{res}_E^G \mathrm{tr}_{C_G(E)}^G \eta)\mathrm{res}_E^G z.$$

We want to choose z to make this equal to some p-power of y restricted to E.

Here

$$\mathrm{res}_E^G \mathrm{tr}_{C_G(E)}^G \eta = \sum_{g \in E\backslash G/C_G(E)} \mathrm{tr}_{E\cap gC_G(E)g^{-1}}^E \mathrm{res}_{E\cap gC_G(E)g^{-1}}^{gC_G(E)g^{-1}} g\eta.$$

Up to conjugation, the expression on the right involves the restriction of η to $g^{-1}Eg \cap C_G(E)$. We know that η restricts to zero on all elementary abelian subgroups of $C_G(E)$ that do not contain E. So the only nonzero terms are those with $g \in N_G(E)$. That is,

$$\begin{aligned}
\mathrm{res}_E^G \mathrm{tr}_{C_G(E)}^G \eta &= \sum_{g \in W_G(E)} \mathrm{res}_E^{C_G(E)} g\eta \\
&= \sum_{g \in W_G(E)} g(\mathrm{res}_E^{C_G(E)} \eta) \\
&= \mathrm{tr}_1^{W_G(E)} \beta^{p^b} \\
&= (\mathrm{tr}_1^{W_G(E)} \beta)^{p^b} \\
&= \sigma_E^{np^b}.
\end{aligned}$$

(Here we write tr_1^W for the trace of W acting on H_E^*, and we use that the trace commutes with pth powers.) So we need to find an element $z \in H_G^*$ such that $\sigma_E^{np^b} \mathrm{res}_E^G z$ is equal to some p-power of y restricted to E.

In the assumption of (*), we are given that $y \in H_G^*$ restricts to zero on all proper subgroups of E. So there is a positive integer m such that the restriction of y^m to H_E^* is a multiple of σ_E. (We can take $m = 2$ if $p = 2$, and $m = p - 1$ if p is odd.) So a high enough power of y, which we can take to be a p-power, restricts on E to a multiple of σ_E^{n+1}; write $\mathrm{res}_E^G y^{p^t} = \sigma_E^{n+1} u$ for some $t \geq 0$ and some $u \in H_E^*$. So $\mathrm{res}_E^G y^{p^t} = \sigma_E^n v$ where $v := \sigma_E u$ in H_E^* restricts to zero on all proper subgroups of E. Since σ_E is not a zero divisor in H_E^* and $\mathrm{res}_E^G y^{p^t}$ is $W_G(E)$-invariant, v is also $W_G(E)$-invariant. By Lemma 8.5, there is an element $z \in H_G^*$ with $\mathrm{res}_E^G z = v^{p^b}$, where $[G : E] = p^b$. So $\mathrm{res}_E^G y^{p^{t+b}} = \sigma_E^{np^b} \mathrm{res}_E^G z$. By the previous paragraph, Theorem 8.12 is proved. □

Corollary 8.14 *Let G be a finite group and p a prime number. Let P be a Sylow p-subgroup of G, and let C be the p-torsion subgroup of the center of P. Consider G as an algebraic group over a field k of characteristic not p that contains the pth roots of unity. Then the kernel of $CH_G^* \to CH_C^*$ is the radical of the sum of the images of all transfers to G from proper subgroups of P. Moreover, the sum of the images of all transfers to G from centralizers in P of non-central elementary abelian subgroups of P has the same radical.*

Proof (Corollary 8.14) Given our assumption on k, the cycle map $CH_G^* \to H_G^*$ is an F-isomorphism for all finite groups G, by Theorem 8.10. So Theorem 8.12 on cohomology implies the result for Chow rings, using that transfers commute with pth powers (Lemma 8.11). One could also repeat the argument essentially verbatim for Chow groups. □

9

Group Cohomology and the Chow Ring Modulo Transfers Are Cohen-Macaulay

In this chapter, we show that the structure of group cohomology and Chow rings is significantly simplified by working modulo transfers from proper subgroups. Namely, the cohomology ring modulo transfers and the Chow ring modulo transfers are Cohen-Macaulay rings. In this respect, they are much simpler than the whole cohomology or Chow ring. Nonetheless, if we can compute the cohomology modulo transfers for a group and its subgroups, then we can read off additive generators for the whole cohomology.

More precisely, these results work for cohomology or Chow rings modulo transfers from a smaller class of subgroups. For a p-group, the relevant subgroups are the centralizers of non-central elementary abelian subgroups.

Theorem 9.3 proves an analogous statement in invariant theory: for any linear representation of a finite group G over any field, the quotient of the invariant ring by traces from a certain class of subgroups to G is a Cohen-Macaulay ring. This generalizes a result of Fleischmann's [42].

9.1 The Cohen-Macaulay property

Fix a prime number p. For a finite group G, we continue to write $H^*_G = H^*(BG, \mathbf{F}_p)$ and $CH^*_G = CH^*(BG)/p$. Let P be a Sylow p-subgroup. Let $\mathcal{S}$ be the set of centralizers $C_G(E)$ of elementary abelian p-subgroups E such that $C_G(E)$ contains no Sylow p-subgroup of G. Let

$$T(G) = H^*_G \Big/ \sum_{H \in \mathcal{S}} \operatorname{tr}_H^G H^*_H$$

and

$$A(G) = CH^*_G \Big/ \sum_{H \in \mathcal{S}} \operatorname{tr}_H^G CH^*_H.$$

Note that the rings CH^*BG and $A(G)$ depend on a choice of field k, with G being viewed as an algebraic group over k. We usually do not indicate k in the notation. Our results also hold for the quotient rings of H^*_G and CH^*_G by transfers from all proper subgroups of P, but it should be more useful to understand the richer quotient rings $T(G)$ and $A(G)$.

For any subgroup H of a finite group G, the transfer map $H^*_H \to H^*_G$ is H^*_G-linear (Lemma 2.15(i)), and so its image is an ideal in H^*_G. It follows that $T(G)$ is a graded-commutative $\mathbf{F}_p$-algebra. Likewise, $A(G)$ is a commutative graded $\mathbf{F}_p$-algebra.

Example Let G be the dihedral group of order 8. Then $H^*_G = \mathbf{F}_2[x_1, x_2, y]/(x_1^2 = x_1x_2)$, where $|x_1| = |x_2| = 1$ and $|y| = 2$. By the example after Theorem 8.12, the two elementary abelian subgroups of rank 2 in G are their own centralizers, and the images of transfer from those two subgroups span H^1_G. So the quotient ring $T(G)$ of H^*_G is the polynomial ring $\mathbf{F}_2[y]$. The 2-dimensional irreducible complex representation V of G is faithful, and its Euler class $\chi(V)$ (meaning c_2V) is equal to y^2 in $T(G)$. Thus $T(G)$ is a free $\mathbf{F}_2[\chi(V)]$-module, which agrees with Theorem 9.1.

In this section we show that the rings $T(G)$ and $A(G)$ are always Cohen-Macaulay. An earlier result with a similar flavor is Green's theorem that the essential ideal in H^*_G is a Cohen-Macaulay module [50]; the method goes back to Duflot's lower bound for the depth of the cohomology ring (Corollary 3.19). It is striking that working modulo transfers simplifies the cohomology ring in this way.

For comparison, the whole cohomology ring of a finite group is Cohen-Macaulay in some examples, but in most cases it is not; this seems to be a large part of what makes the cohomology ring hard to compute. Among the main successes in the cohomology of finite groups are Quillen's calculation of the cohomology of $GL(n, \mathbf{F}_q)$ with mod l coefficients when q and l are relatively prime (Theorem 2.23), and Quillen's calculation of the cohomology of an extraspecial 2-group [12, vol. 2, section 5.5], [112]. (By definition, a p-group G is *extraspecial* if it is a central extension of an elementary abelian group by $\mathbf{Z}/p$ and the center of G is equal to $\mathbf{Z}/p$.) In those cases, the cohomology rings are Cohen-Macaulay.

The cohomology of $GL(n, \mathbf{F}_q)$ with mod p coefficients where q is a power of p, and the cohomology of an extraspecial p-group with p odd, remain unknown in general; these are non-Cohen-Macaulay examples. More broadly, for any finite group G that has maximal elementary abelian p-subgroups of different ranks, the scheme Spec H^{ev}_G is not equidimensional by Quillen's theorem (Theorem 8.4), and so the ring H^*_G is not Cohen-Macaulay. That argument shows that the cohomology of the symmetric group is not Cohen-Macaulay in most cases.

For some purposes, it is enough to consider the ring $T(G)$ for a p-group G, in which case the definition is a little simpler: $T(G)$ is the quotient of H_G^* by transfers from centralizers of non-central elementary abelian subgroups of G. But I have made an effort to formulate statements that are nontrivial for arbitrary finite groups. One reason is that the mod p cohomology of a finite group G can be simpler than the cohomology of a Sylow p-subgroup P, and so it may be reasonable to study the cohomology of G without complete knowledge of the cohomology of P.

For example, for l an odd prime number and q a prime power congruent to 1 modulo l, the mod l cohomology ring of the general linear group $GL(l, \mathbf{F}_q)$ is Cohen-Macaulay by Quillen (Theorem 2.23). But the cohomology of an l-Sylow subgroup P of G, which is a wreath product $\mathbf{Z}/l \wr \mathbf{Z}/l^r$ for some r, is not Cohen-Macaulay, since P has maximal elementary abelian l-subgroups of different ranks (namely, 2 and l). In another direction, for a prime power $q \equiv 3 \pmod 4$, Quillen showed that the mod 2 cohomology of $G = GL(2, \mathbf{F}_q)$ is detected on the diagonal subgroup $\{\pm 1\}^2$ [115, corollary to theorem 3]. (In the terminology of Section 12.1, the statement that H_G^* is detected on elementary abelian subgroups means that the topological nilpotence degree $d_0(H_G^*)$ is equal to zero.) But a Sylow 2-subgroup of G is a semidihedral group P, and the cohomology of P (e.g., H_P^1) is not detected on elementary abelian 2-subgroups (cf. Lemma 13.4).

Theorem 9.1 *Let G be a finite group, p a prime number, c the p-rank of the center of a Sylow p-subgroup of G. Then the ring $T(G)$ is Cohen-Macaulay of dimension c.*

Suppose in addition that G is a p-group. Let $V = V_1 \oplus \cdots \oplus V_c$ be a faithful complex representation of dimension n with c irreducible summands (the smallest possible number). Then the ring $T(G)$ is a finitely generated free $\mathbf{F}_p[\chi(V_1), \ldots, \chi(V_c)]$-module.

The same result holds if we replace $T(G)$ by a cruder ring, the quotient of H_G^* by transfers to G from all proper subgroups of a p-Sylow subgroup of G. The same proof works, and some steps are easier.

Let us explain why the smallest number of irreducible summands for a faithful complex representation $V = V_1 \oplus \cdots \oplus V_s$ of a p-group G is equal to the p-rank c of the center. Let $C = Z(G)[p] \cong (\mathbf{Z}/p)^c$ be the p-torsion subgroup of the center of G. By Schur's lemma, C acts by scalars, through some 1-dimensional representation of C, on each of the irreducible representations $V_1, \ldots, V_s$ of G. Since V is a faithful representation of G, it is faithful on C, which means that these 1-dimensional representations of C span $\mathrm{Hom}(C, \mathbf{C}^*) \cong (\mathbf{Z}/p)^c$. So $s \geq c$. Conversely, a representation of G is faithful if and only if its restriction to C is faithful, because every nontrivial normal subgroup of G has nontrivial intersection with the center [4, section 5.15] and hence with C. So G has a faithful representation with exactly c irreducible summands.

Corollary 10.3 will show that, for a p-group G, $T(G)$ is generated as an $\mathbf{F}_p[\chi(V_1), \ldots, \chi(V_c)]$-module in degrees at most $2n - c$. Since $T(G)$ is a free $\mathbf{F}_p[\chi(V_1), \ldots, \chi(V_c)]$-module, it is equivalent to say that the ring $T(G)$ has regularity at most zero.

Proof (Theorem 9.1) Let G be a finite group. Let P be a Sylow p-subgroup of G and $C = Z(P)[p] \cong (\mathbf{Z}/p)^c$. By Carlson's theorem, the kernel of the restriction map $T(G) \to T(C) = H_C^*$ is nilpotent (Theorem 8.12). By Venkov's theorem (Theorem 1.1), H_C^* is finite over H_G^*. So H_C^* is finite over the quotient ring $T(G)$. We conclude that the dimension of $T(G)$ is equal to the dimension of H_C^*, which is the p-rank c of C. It remains to show that the graded-commutative ring $T(G)$ has depth at least c.

Let $T_G(P)$ be the quotient ring of H_P^* by transfers from all intersections $P \cap C_G(E)$ such that E is an elementary abelian subgroup of G with $C_G(E)$ containing no Sylow p-subgroup of G. This is not a very natural ring to consider, but it serves our purpose. As the notation indicates, $T_G(P)$ depends on G as well as on P.

I claim that the restriction and transfer maps

$$H_G^* \xrightarrow[\mathrm{res}_P^G]{} H_P^* \xrightarrow[\mathrm{tr}_P^G]{} H_G^*$$

pass to well-defined homomorphisms on the quotient rings,

$$T(G) \to T_G(P) \to T(G).$$

Indeed, it is clear that the second map is well defined, and the double coset formula (as in Lemma 2.15(iii)) gives that the first map is well defined. The first map is a ring homomorphism, the second is $T(G)$-linear, and the composition is multiplication by $[G : P] \neq 0 \in \mathbf{F}_p$, by the corresponding properties of cohomology rings. Therefore, $T(G)$ is a summand of $T_G(P)$ as a $T(G)$-module. Also, by Venkov's theorem (Theorem 1.1), H_P^* is finite over H_G^*, and so $T_G(P)$ is finite over $T(G)$. Since $T(G)$ has dimension c, it follows that $T_G(P)$ also has dimension c. Suppose we can show that $T_G(P)$ is Cohen-Macaulay; then the summand statement implies that $T(G)$ is Cohen-Macaulay, as we want.

Let $V = V_1 \oplus \cdots \oplus V_c$ be a faithful complex representation of P of dimension n with c irreducible summands (the smallest possible number). As discussed after the statement of Theorem 9.1, C acts by scalars, through some 1-dimensional representation of C, on each of the irreducible representations $V_1, \ldots, V_c$ of P, and these 1-dimensional representations of C form a basis for $\mathrm{Hom}(C, \mathbf{C}^*) \cong (\mathbf{Z}/p)^c$. Write $y_1, \ldots, y_c$ for the Euler classes of these representations in H_C^2, and let ζ_i be the Euler class $\chi(V_i)$ in H_P^*. Then $\zeta_1, \ldots, \zeta_c$ restrict to $y_1^{p^{a_1}}, \ldots, y_c^{p^{a_c}}$ in H_C^*, where $\dim_{\mathbf{C}} V_i = p^{a_i}$ for $i = 1, \ldots, c$.

Recall that H_P^* is a comodule over the Hopf algebra H_C^*, via the product homomorphism $C \times P \to P$ (Lemma 3.18). The crucial point is that the quotient ring $T_G(P)$ of H_P^* inherits the structure of a H_C^*-comodule. To prove this, we first note that $T_G(P)$ can be defined as the quotient ring of H_P^* by transfers from those subgroups $H = P \cap C_G(E)$ such that E is an elementary abelian subgroup of G with $C_G(E)$ containing no Sylow p-subgroup of G, *and* such that C is contained in H. Indeed, if C is not contained in H, then the subgroup $CH \subset P$ is the product of H with a nontrivial elementary abelian subgroup. In that case, the transfer map from H to CH is zero, and so the transfer from H to P is zero. So we can omit a subgroup H from the definition of $T_G(P)$ except when C is contained in H.

For a subgroup H of P that contains C, the transfer map $\mathrm{tr}_H^P \colon H_H^* \to H_P^*$ is a map of H_C^*-comodules, by the pullback diagram

$$\begin{array}{ccc} C \times H & \longrightarrow & H \\ \downarrow & & \downarrow \\ C \times P & \longrightarrow & P. \end{array}$$

Therefore, the quotient ring $T_G(P)$ of H_P^* is an H_C^*-comodule.

Then the proof of Duflot's theorem on depth (Theorem 3.17) applies verbatim to show that $T_G(P)$ is a free graded module over the polynomial ring $\mathbf{F}_p[\zeta_1, \ldots, \zeta_c]$. Since the ring $T_G(P)$ has dimension c, it is Cohen-Macaulay. As mentioned earlier, this completes the proof that $T(G)$ is Cohen-Macaulay for every finite group G. When G is a p-group (so $T_G(P) = T(G)$), we have proved the more precise statement that $T(G)$ is a finitely generated free module over the polynomial ring $\mathbf{F}_p[\zeta_1, \ldots, \zeta_c]$ on Euler classes $\zeta_i = \chi(V_i)$. □

Theorem 9.2 *Let G be a finite group, p a prime number, k a field of characteristic not p that contains the pth roots of unity. Let c be the p-rank of the center of a Sylow p-subgroup of G. Then the ring $A(G)$ is Cohen-Macaulay of dimension c.*

Suppose in addition that G is a p-group. Let $V = V_1 \oplus \cdots \oplus V_c$ be a faithful k-representation of dimension n with c irreducible summands (the smallest possible number). Then the ring $A(G)$ is a free $\mathbf{F}_p[\chi(V_1), \ldots, \chi(V_c)]$-module with generators in bounded degrees.

Note that the ring $A(G)$ is not known to be noetherian. We use the definition of Cohen-Macaulayness in Definition 3.15.

Proof Essentially the same arguments work for $A(G)$ as in Theorem 9.1 for $T(G)$. In particular, Carlson's theorem on transfer works for the Chow ring (Corollary 8.14), using that k contains the pth roots of unity. Also, with that assumption, we know that CH_P^* is a CH_C^*-comodule, since the Chow Künneth formula gives that $CH_{C\times P}^* = CH_C^* \otimes_{\mathbf{F}_p} CH_P^*$ (Lemma 2.12). Here $CH_C^* = \mathbf{F}_p[y_1, \ldots, y_c]$ is slightly simpler than H_C^*, but the arguments work the same way. □

9.2 The ring of invariants modulo traces

There is a surprisingly strong analogy between the $\mathbf{F}_p$-cohomology ring of a finite group and the ring of invariants for a representation of a finite group over any field. There are obvious differences. Invariant rings are always normal, whereas cohomology rings are usually not even domains. But, for example, Symonds showed that the ring of invariants for a representation of a finite group over any field has regularity at most zero, just like the cohomology ring of a finite group [131, 132].

In this section, following Symonds's suggestion, we prove an analog for invariant rings of the theorem that cohomology rings modulo transfers are Cohen-Macaulay. The proof turns out to be closely analogous to that of Theorem 9.1. Fleischmann proved Theorem 9.3 in the special case that the representation V is a summand of a permutation representation [42, proposition 12.4 and theorem 12.7]. Theorem 9.3 also generalizes Hochster-Eagon's theorem that rings of invariants in characteristic zero are Cohen-Macaulay [36]. (As mentioned in Section 7.1, a ring of invariants even for the cyclic group $\mathbf{Z}/p$ in characteristic p need not be Cohen-Macaulay when p is at least 5.)

This section is not used in the rest of the book.

To begin, we define the trace map in invariant theory [10, section 1.5]. Let G be a finite group acting on an abelian group M, and let $H \subset G$ be a subgroup. Then we define the trace $\mathrm{tr}_H^G \colon M^H \to M^G$ by $\mathrm{tr}_H^G(x) = \sum_{g\in G/H} gx$. This is a special case of the transfer in group cohomology, since $M^H = H^0(H, M)$. As a result, the assignment $H \mapsto M^H$ for subgroups H of G is a cohomological Mackey functor (as described in Section 2.5). For R a commutative ring with G-action, we can also define the norm $N_H^G \colon R^H \to R^G$ by $N_H^G(x) = \prod_{g\in G/H} gx$, which makes $H \mapsto R^H = H^0(H, R)$ into a Tambara functor (as described in Section 8.1). These statements are elementary to check by hand. In particular, for R a commutative ring with G-action and a subgroup $H \subset G$, write res_H^G for the inclusion $R^G \to R^H$, which is a ring homomorphism. Then the trace $\mathrm{tr}_H^G \colon R^H \to R^G$ is R^G-linear. That is,

$$\mathrm{tr}_H^G(\mathrm{res}_H^G(x)y) = x\ \mathrm{tr}_H^G(y).$$

Theorem 9.3 *Let G be a finite group, V a representation of G over a field k of characteristic p, P a Sylow p-subgroup of G. (For k of characteristic zero, let P be the trivial subgroup of G.) Let $\mathcal{S}$ be the class of all stabilizer subgroups*

G_u in G of $\overline{k}$-points u of V such that G_u contains no Sylow p-subgroup of G. Let $O(V)$ be the polynomial ring of regular functions on V. Then the quotient of the ring of invariants $O(V)^G$ by the sum of the traces $\mathrm{tr}_H^G \colon O(V)^H \to O(V)^G$ from all subgroups H in $\mathcal{S}$ is a Cohen-Macaulay ring, of dimension equal to $\dim(V^P)$*. The same goes for the quotient of $O(V)^G$ by traces from all proper subgroups of P.*

For applications, it should be useful that we can say something about the ring of invariants modulo traces from a restricted class $\mathcal{S}$ of subgroups, not just modulo traces from all proper subgroups. At least for G a p-group, the class $\mathcal{S}$ is an analogue in invariant theory of the class of centralizers of non-central elementary abelian subgroups in group cohomology; compare Theorem 9.1.

Proof Let $R = O(V)$. Let $I \subset R^G$ be the sum of all traces to G from all stabilizers G_u of $\overline{k}$-points u of V such that G_u contains no Sylow p-subgroup of G. Let $t(G, V)$ be the quotient ring R^G/I.

Lemma 9.4 *The closed subset of the quotient variety $V/G =$ Spec R^G associated to the ideal I in R^G is the image of the linear subspace $V^P \subset V$.*

This result can be viewed as the analog for invariant rings of Carlson's theorem in group cohomology, Theorem 8.12. We only need Lemma 9.4 to deduce that the quotient ring $t(G, V)$ has dimension equal to the dimension of V^P. Fleischmann proved a version of Lemma 9.4 for G a p-group [42, proposition 12.5(ii)]. The proof of Lemma 9.4 shows that the image in R^G of traces from all proper subgroups of P defines the same closed subset. So the latter ideal and the one defined using stabilizer subgroups have the same radical, although they may not be equal.

Proof (Lemma 9.4) Let T be the closed subset of $V/G =$ Spec R^G associated to the ideal in the lemma. Then a $\overline{k}$-point v of V maps into T if and only if every element of the given ideal vanishes at v, which in turn means that for every $\overline{k}$-point u of V with stabilizer G_u containing no p-Sylow subgroup of G and every $f \in R^{G_u} = O(V)^{G_u}$, the trace $\mathrm{tr}_{G_u}^G(f)$ vanishes at v. This is equivalent to the same statement for every f in $O(V_{\overline{k}})^{G_u}$. So we can assume from now on that k is algebraically closed and work only with k-points. The G-orbit of v has the form G/G_v, and there is a G_u-invariant polynomial on V that takes any values we like in $\overline{k}$ on the G_u-orbits in G/G_v. So $\mathrm{tr}_{G_u}^G(f)$ vanishes at v for every $f \in O(V)^{G_u}$ if and only if all G_v-orbits on G/G_u have order a multiple of p. These G_v-orbits have the form $G_v/(G_v \cap gG_ug^{-1}) = G_v/(G_v \cap G_{gu})$ for $g \in G$. We conclude that a point v in V maps into T if and only if for every point u in V with G_u containing no Sylow p-subgroup of G, $G_v/(G_v \cap G_u)$ has order a multiple of p.

If G_v contains a Sylow p-subgroup of G, then $G_v/(G_v \cap G_u)$ has order a multiple of p for all points u as above, and so v maps into T. Conversely, if G_v contains no Sylow p-subgroup of G, then we can take $u = v$, and then $G_v/(G_v \cap G_u)$ has order 1, which is not a multiple of p. So v does not map into T. Note that $V \to V/G$ is surjective (as it is a finite dominant morphism [10, theorem 1.4.4]). So we have identified the closed subset $T \subset V/G$. $\square$

Let $t_G(P, V)$ be the quotient ring of R^P by the sum of all traces to P from all subgroups $P \cap G_u$ with u a $\overline{k}$-point of V such that G_u contains no Sylow p-subgroup of G. The point of defining $t_G(P, V)$ is that the restriction and trace maps

$$R^G \to R^P \to R^G$$

pass to well-defined maps on the quotient rings

$$t(G, V) \to t_G(P, V) \to t(G, V).$$

Indeed, the restriction map is well-defined by the double coset formula, which applies because the assignment $H \mapsto R^H$ for subgroups H of G is a Mackey functor. The trace map is well-defined simply because $P \cap G_u$ is contained in G_u. The composition of the two maps is multiplication by $[G : P]$, which is nonzero in k. We deduce that the ring $t(G, V)$ is a summand of $t_G(P, V)$ as a $t(G, V)$-module.

Also, R^P is finite over R^G (because R is finite over R^G), and so $t_G(P, V)$ is finite over its subring $t(G, V)$. It follows that the two rings have the same dimension, which is equal to $\dim(V^P)$ as we have shown. Therefore, the ring $t(G, V)$ is Cohen-Macaulay if we can show that $t_G(P, V)$ is Cohen-Macaulay, or equivalently that $t_G(P, V)$ has depth at least $\dim(V^P)$.

Let $S = O(V^P)$, the ring of regular functions on the linear subspace $V^P \subset V$. Addition $V^P \times V \to V$ is a P-equivariant morphism of k-schemes, and so it gives a P-equivariant homomorphism

$$\varphi\colon R \to S \otimes_k R$$

of k-algebras. Since this homomorphism comes from an action of the commutative algebraic group V^P on V, it makes R into a comodule over the Hopf algebra S. Also, since φ is P-equivariant, it commutes with trace maps. Explicitly, for any subgroup H of P, φ maps $\mathrm{tr}_H^P R^H$ into

$$\begin{aligned}\mathrm{tr}_H^P((S \otimes_k R)^H) &= \mathrm{tr}_H^P(S \otimes_k R^H)\\ &= S \otimes_k \mathrm{tr}_H^P R^H.\end{aligned}$$

Thus φ makes the quotient ring $t_G(P, V)$ of R^P into a comodule over S.

Let $x_1, \ldots, x_r$ be linear functions on V that restrict to a basis for the vector space $(V^G)^*$. Then the norms $\zeta_i = N_1^P(x_i)$ (Definition 7.4) are P-invariant regular functions that restrict to $x_1^{|P|}, \dots, x_r^{|P|}$. In particular, they restrict to a regular sequence in $S = k[x_1, \dots, x_r]$. Then the same argument as in the proof of Duflot's theorem in group cohomology (Theorem 3.17) shows that $t_G(P, V)$ is a free module over the polynomial ring $k[\zeta_1, \dots, \zeta_r]$. By the earlier part of this proof, it follows that the ring $t(G, V)$ is Cohen-Macaulay. □

10

Bounds for Group Cohomology and the Chow Ring Modulo Transfers

Symonds's regularity theorem reduces the problem of computing the cohomology ring of a finite group G to calculations in degree at most n^2 (Corollary 4.3), when G has a faithful complex representation of dimension n. Can the calculations be reduced further? One answer is that the bounds can be improved when G is a p-group (Theorem 7.2). In this chapter, we prove even stronger bounds for the cohomology ring or Chow ring modulo transfers.

These results come from conversations I had with Symonds. First I showed that the Chow ring of a finite group modulo transfers from proper subgroups has regularity at most zero. Then Symonds showed that the cohomology ring modulo transfers has regularity at most zero. We present a version of Symonds's argument here, and also extend it to Chow rings and motivic cohomology.

These results give very strong bounds for the degrees of generators of the Chow ring or cohomology ring of a finite group modulo transfers from proper subgroups. In Corollary 10.5, we show that for a finite p-group G with a faithful representation of dimension n and p-rank of the center equal to c, the Chow ring modulo transfers of G is generated over the Euler classes of certain representations by elements of degree at most $n - c$. This is an optimal bound, much better than the bounds for the whole Chow ring. For cohomology modulo transfers, Symonds proved the optimal degree bound $2n - c$ (Corollary 10.3). Compare Theorem 12.4, where we prove that a different measure of the complexity of the cohomology ring, Henn-Lannes-Schwartz's topological nilpotence degree, is at most $2n - c$.

Theorem 10.1 *Let p be a prime number. Let M be a smooth manifold with finite-dimensional $\mathbf{F}_p$-cohomology. Let G be a finite group acting on M. Let $\mathcal{S}$ be a collection of subgroups of G. Then the quotient of $H_G^*(M, \mathbf{F}_p)$ by the sum over all subgroups $H \in \mathcal{S}$ of transfers from $H_H^*(M, \mathbf{F}_p)$ has regularity at most* $\dim(M)$.

If every elementary abelian p-subgroup of G is contained in some element of $\mathcal{S}$, then this quotient ring has regularity at most $\dim(M) - 1$.

Here the regularity is defined by viewing the given quotient ring as a module over H_G^*. The regularity is the same with respect to any ring over which the given ring is a finitely generated module, by Lemma 3.10.

Corollary 10.2 *For a finite group G and a prime number p, and any collection $\mathcal{S}$ of subgroups H of G, the quotient of the graded ring $H_G^* = H^*(BG, \mathbf{F}_p)$ by transfers from $\mathcal{S}$ has regularity at most zero. If every elementary abelian subgroup of G is contained in some element of $\mathcal{S}$, then this quotient ring has regularity at most -1.*

Corollary 10.3 *Let p be a prime number. Let G be a p-group, c the p-rank of the center of G. Let $V = V_1 \oplus \cdots \oplus V_c$ be a faithful complex representation of G with c irreducible summands (the smallest possible number). Let n be the dimension of V. Let $\mathcal{S}$ be the set of centralizers of non-central elementary abelian subgroups of G. Then the ring*

$$T(G) = H_G^* \Big/ \sum_{H \in \mathcal{S}} \operatorname{tr}_H^G H_H^*$$

is a free $\mathbf{F}_p[\chi(V_1), \ldots, \chi(V_c)]$-module, with generators in degrees at most $2n - c$. Moreover, the same is true for the quotient ring of H_G^* by the sum of all transfers from all proper subgroups.

For any p-group G that is not p-central, $T(G)$ is in fact a free module over $\mathbf{F}_p[\chi(V_1), \ldots, \chi(V_c)]$ with generators in degrees at most $2n - c - 1$. For any p-group G that is not elementary abelian, the quotient of H_G^* by transfers from all proper subgroups of G is a free $\mathbf{F}_p[\chi(V_1), \ldots, \chi(V_c)]$-module with generators in degrees at most $2n - c - 1$.

By definition, a p-group G is *p-central* if every element of order p of G belongs to the center of G. Equivalently, every elementary abelian p-subgroup of G is contained in the center. We have to exclude p-central groups G from the last statement of Corollary 10.3 about $T(G)$, because for G p-central, $T(G)$ is equal to H_G^*, which has regularity equal to zero by Benson-Carlson (Corollary 4.2). That is, for a p-central group G, H_G^* is a free $\mathbf{F}_p[\chi(V_1), \ldots, \chi(V_c)]$-module with top generator in degree $2n - c$.

Proof of Corollary 10.3 By Theorem 9.1, the ring $T(G)$ is Cohen-Macaulay. More precisely, it is a finitely generated free module over $\mathbf{F}_p[\chi(V_1), \ldots, \chi(V_c)]$. Also, $T(G)$ has regularity at most zero by Theorem 10.1. So this free module is generated in degrees at most $\sum_{i=1}^{c}(|\chi(V_i)| - 1) = 2n - c$. The same arguments apply to the quotient ring of H_G^* by transfers from centralizers of non-central elementary abelian groups, rather than from all proper subgroups. Corollary 10.2 also explains when this degree bound can be improved by 1. □

Proof of Theorem 10.1 We follow the proof of Symonds's Theorem 4.1 as far as possible. Choose a faithful complex representation $G \subset U(n)$. Let $S = (\mathbf{Z}/p)^n \subset (S^1)^n \subset U(n)$. Let G act on $M \times U(n)$ by the given action on M and by left multiplication on $U(n)$, and let S act trivially on M and by right multiplication on $U(n)$ (so the two actions commute).

By Lemma 4.5, for every subgroup H of G, $H^*_H(M \times U(n)/S)$ is a free $H^*_H M$-module with top generator in degree n^2. Moreover, these free modules can be taken to have the same set of generators for each subgroup H, pulled back from H^*_S. Therefore, the quotient ring of $H^*_G(M \times U(n)/S)$ by the sum over $H \in \mathcal{S}$ of transfers from $H^*_H(M \times U(n)/S)$ is a free module over $H^*_G M / \sum_{H \in \mathcal{S}} \operatorname{tr}^G_H(H^*_H M)$, with top generator in degree n^2. So it suffices to show that

$$H^*_G(M \times U(n)/S) \Big/ \sum_{H \in \mathcal{S}} \operatorname{tr}^G_H H^*_H(M \times U(n)/S)$$

has regularity at most $\dim(M) + n^2$.

At first we view this quotient as a module over H^*_G or $H^*_{U(n)}$. But it has the same regularity as a module over H^*_S by Lemma 3.10, and from now on, we view it as a module over H^*_S. Let N be the manifold $M \times U(n)$; then N has dimension $\dim(M) + n^2$, and has commuting actions of G and S, with G acting freely. We want to show that the quotient ring

$$H^*_S(N/G) \Big/ \sum_{H \in \mathcal{S}} \operatorname{tr}^G_H H^*_S(N/H)$$

has regularity at most $\dim(N)$ as a module over H^*_S. For each subgroup H of G, let $N_{H,i}$ be the closed subset of N of points whose image in N/H has stabilizer in S of rank at least i, $N_{H,(i)}$ the submanifold of N where this rank is equal to i, and $N_{H,(i),d}$ the union of the connected components of $N_{H,(i)}$ that have codimension d in N. Define a decreasing filtration of $H^*_S(N/H)$ as an H^*_S-module by $F^i_H = \ker(H^*_S(N/H) \to H^*_S((N - N_{H,i})/H))$. By Theorem 4.6, the ith graded piece F^i_H/F^{i+1}_H is isomorphic to $\oplus_d H^{*-d}_S(N_{H,(i),d}/H)$. In particular, that result used the observation that the normal bundle of the submanifold $N_{H,(i),d}/H$ in N/H is orientable when p is odd. (For $p = 2$, there is no issue about orientability.) That holds because the normal bundle to a connected component of $(N/H)^V$, for a subgroup $V \cong (\mathbf{Z}/p)^i$ of S, is a real representation of V with no trivial summands, and such a representation can be given a complex structure in a canonical way, when p is odd.

Thus the bottom piece (a submodule) of $H^*_S(N/H)$ corresponds to the points of N/H with stabilizer in S of the highest rank i, and hence is a free module over a big polynomial quotient ring of H^*_S, whereas the higher graded pieces of $H^*_S(N/H)$ are direct sums of free modules over successively smaller polynomial quotient rings of H^*_S. The generators for these free modules over free graded-commutative quotient rings H^*_V are in degrees at most dim(N). (This implies that the whole H^*_S-module $H^*_S(N/H)$ has regularity at most dim(N).)

For a subgroup H of G, the stabilizer in S of a point in N/H is at most the stabilizer in S of the image point in N/G, although it is important to realize that these stabilizers can be different. So the filtrations of N associated to different subgroups of G can be different.

We will show by decreasing induction on j (starting with $j = \text{rank}(S) + 1$) that the quotient of $H^*_S(N/G)$ by the sum over all H in $\mathcal{S}$ of the image of the transfer map from $F^j_H H^*_S(N/H)$ to $H^*_S(N/G)$ is filtered (by the image of the given filtration of $H^*_S(N/G)$) with ith graded piece a direct sum over the subgroups $V \subset S$ of rank i of free modules over H^*_V with generators in degrees $\leq \dim(N)$. We know this for $j = \text{rank}(S) + 1$, and the theorem follows if we can prove it for $j = 0$.

Suppose we know this statement for $j + 1$, and we want to prove it for j. Let Q be the quotient of $H^*_S(N/G)$ by the images of $F^{j+1}_H H^*_S(N/H)$ for all H in $\mathcal{S}$. The filtration F^*_G of $H^*_S(N/G)$ induces a filtration of Q, which we also call F^*_G. We know that the filtration F^*_G of Q has the property we want (previous paragraph). Clearly we have a natural map from $\text{gr}^j_H H^*_S(N/H)$ to the quotient Q for all H in $\mathcal{S}$, and we want to show that the cokernel of this map has the property we want. Let V be any subgroup of rank j in V, and fix a splitting $S \cong V \times W$. For each subgroup H in $\mathcal{S}$, let $N^{(V)}/H$ be a connected component of the locally closed submanifold of N/H on which S has stabilizer equal to V, and let d be the codimension of N_V/H in N/H. Then we know that $\text{gr}^j H^*_S(N/H)$ is the direct sum of the groups $H^{*-d}_S(N_V/H)$ over all subgroups V in S of rank j and all connected components N_V/H.

Here $H^{*-d}_S(N_V/H)$ is a free module over H^*_V with generators in degree $\leq \dim(N)$. In geometric terms, its support as an H^*_S-module has dimension equal to j. Therefore, the intersection of the sum of the images of all the modules $H^{*-d}_S(N_V/H)$ in Q with $F^{j+1}_G Q$ is zero. Indeed, by our inductive hypothesis, the latter submodule of Q is filtered with quotients that are free modules on polynomial quotient rings of H^*_S of rank greater than j.

So the quotient of Q by the sum over $H \in \mathcal{S}$ of the images of $H^{*-d}_S(N_V/H)$ is filtered with all quotients of the form we want, except possibly for the jth graded piece. This graded piece is a direct sum over the subgroups $V \subset S$ of rank j, and the summand corresponding to the group V we are considering is the quotient of $H^{*-e}_S(N^{(V)}_G/G)$ by the images of $H^{*-d}_S(N^{(V)}_H/H)$. (Here $N^{(V)}_G/G$ means the locally closed submanifold of N/G on which S has stabilizer equal to V, and e means the codimension of $N^{(V)}_G/G$ in N/G.) This map can be described as the composite of the restriction to an open subset, $H_S^{*-d}(N_H^{(V)}/H) \to H_S^{*-d}(N_H^{(V)} \cap N_G^{(V)}/H)$, and proper pushforward maps $H_S^{*-d}((N_H^{(V)} \cap H_G^{(V)})/H) \to H_S^{*-e}(N_G^{(V)}/H) \to H_S^{*-e}(N_G^{(V)}/G)$. In particular, V acts trivially on all the spaces involved here. So this map can be viewed as a map of modules concentrated in degrees at most $\dim(N)$, $H_W^{*-d}(N_H^{(V)}/H) \to H_W^{*-e}(N_G^{(V)}/G)$, tensored over $\mathbf{F}_p$ with H_V^*. Therefore the cokernel of this map has the form we want: a module in degrees $\leq \dim(N)$ tensored over $\mathbf{F}_p$ with H_V^*.

This completes the induction. The first part of the theorem is proved.

Now suppose that every elementary abelian p-subgroup of G is contained in some element of $\mathcal{S}$. We have to show that the quotient of $H_G^* M$ by the sum over all subgroups $H \in \mathcal{S}$ of transfers from $H_H^* M$ has regularity at most $\dim(M) - 1$. As above, let $N = M \times U(n)$, which has commuting actions of G and $S = (\mathbf{Z}/p)^n$, with G acting freely. It suffices to show that the quotient ring

$$H_S^*(N/G) \Big/ \sum_{H \in \mathcal{S}} \operatorname{tr}_H^G H_S^*(N/H)$$

has regularity at most $\dim(N) - 1$.

Our argument shows that this ring is filtered as an H_S^*-module with subquotients that are modules in degrees $\leq \dim(N)$ tensored over $\mathbf{F}_p$ with H_V^*, for subgroups V of S. We have to show that those modules are in fact in degrees $\leq \dim(N) - 1$. In the notation above, we need to show that the transfer map

$$\sum_{H \in \mathcal{S}} H_W^{\text{top}}(N_H^{(V)}/H) \to H_W^{\text{top}}(N_G^{(V)}/G)$$

is surjective. Here S is isomorphic to $V \times W$, and the elementary abelian group W acts freely on the manifolds $N_H^{(V)}/H$ and $N_G^{(V)}/G$. We write $H^{\text{top}} X$ for the cohomology $H^n X$ of an n-manifold X; more precisely, if X has connected components of different dimensions, we mean the product of the top-degree cohomology groups of the components.

For a connected manifold X, $H^{\text{top}} X = 0$ if X is noncompact or nonorientable, whereas $H^{\text{top}} X$ is generated by the class of a point if X is compact and oriented. So it suffices to show that every point in $N_G^{(V)}/G$ is in the image of $N_H^{(V)}/H$ for some subgroup $H \in \mathcal{S}$. For that, it suffices to show that $N_G^{(V)}$ is the union of the subsets $N_H^{(V)}$.

Thus, let x be a point in $N_G^{(V)} \subset N$. This means that

$$V = \{s \in S : xs = gx \text{ for some } g \in G\},$$

where we write the action of G on the left and S on the right. Thus for any subgroup H of G, the subgroup

$$S_H := \{s \in S : xs = hx \text{ for some } h \in H\}$$

is contained in V, and we want to show that there is a subgroup H in $\mathcal{S}$ such that S_H is equal to V.

Let A be the stabilizer subgroup in $G \times S$ of the point x in N. We know that for every $s \in V$, there is an element $(g, s) \in A$. Since G acts freely on N, A has trivial intersection with $G \times 1$. So A projects isomorphically to some subgroup of S, and hence A is an elementary abelian p-group. So the projection of A to G is also an elementary abelian p-group. By our assumption on $\mathcal{S}$, there is a subgroup H in $\mathcal{S}$ that contains that subgroup of G. Then, for every $s \in V$, there is an element $(h, s) \in A$ with $h \in H$. That is, in the previous paragraph's notation, S_H is equal to V, as we want. □

Theorem 10.4 *Let G be a finite group scheme over a field k. Suppose that the order of G is invertible in k. Let p be a prime number, and let $\mathcal{S}$ be any collection of subgroup schemes H of G. Consider $CH_G^* = CH^*(BG)/p$ as a graded ring with CH^i in degree i. Then the quotient ring of CH_G^* by transfers from $\mathcal{S}$ has regularity at most zero.*

If k is algebraically closed and every abelian p-subgroup of G is contained in some element of $\mathcal{S}$, then this quotient ring has regularity at most -1.

The following corollary gives a very strong bound on the degrees of generators for $A(G)$, as defined in Section 9.1. For an r-dimensional representation V of a finite group G over a field k, the Euler class $\chi(V)$ denotes the top Chern class $c_r(V)$ in CH^*BG or in the quotient ring $A(G)$.

Corollary 10.5 *Let p be a prime number, and let k be a field of characteristic not p that contains the pth roots of unity. Let G be a p-group, and let c be the p-rank of the center of G. Let $V = V_1 \oplus \cdots \oplus V_c$ be a faithful representation of G over k with c irreducible summands (the smallest possible number). Let n be the dimension of V. Let $\mathcal{S}$ be the set of centralizers of non-central elementary abelian subgroups of G. Then the ring*

$$A(G) = CH_G^* \Big/ \sum_{H \in \mathcal{S}} \operatorname{tr}_H^G CH_H^*$$

is a free $\mathbf{F}_p[\chi(V_1), \ldots, \chi(V_c)]$-module, with generators in degrees at most $n - c$. Moreover, the same is true for the quotient ring of CH_G^* by the sum of all transfers from proper subgroups of G.

Suppose in addition that k is algebraically closed. If G is a p-group such that every abelian subgroup centralizes some non-central elementary abelian subgroup, then $A(G)$ is in fact a free $\mathbf{F}_p[\chi(V_1), \ldots, \chi(V_c)]$-module with generators in degrees at most $n - c - 1$. If k is algebraically closed and G is a p-group that is not abelian, then the quotient of CH_G^* by transfers from all proper subgroups is again a free $\mathbf{F}_p[\chi(V_1), \ldots, \chi(V_c)]$-module with generators in degrees at most $n - c - 1$.

The bounds in Corollary 10.5 are optimal for some p-groups. For example, let G be any finite abelian p-group, viewed as an algebraic group over $\mathbf{C}$. The group G has a faithful complex representation $V = V_1 \oplus \cdots \oplus V_n$ with each V_i of dimension 1. Then $A(G) = CH_G^*$ is equal to $\mathbf{F}_p[\chi(V_1), \ldots, \chi(V_n)]$ and hence is generated in degree zero as a free module over that ring. That agrees with the upper bound $n - c = n - n = 0$ in this case.

For a nonabelian example, let G be the modular p-group $\mathbf{Z}/p \ltimes \mathbf{Z}/p^2$ for an odd prime number p. In this case, G is not abelian, but it does not satisfy the other condition in Corollary 10.5 (the abelian subgroup $H = \mathbf{Z}/p^2$ does not centralize any non-central elementary abelian subgroup). Here G has a faithful irreducible complex representation V of dimension p, and so $n - c = p - 1$. Let A be an elementary abelian subgroup of rank 2 in G, which is unique up to conjugacy. Then $A(G)$ is the quotient ring $CH_G^*/\mathrm{tr}_A^G CH_A^*$, which is isomorphic to $\mathbf{F}_p[b, \chi(V)]/(b^p)$ with $|b| = 1$ and $|\chi(V)| = p$, by the computation of CH_G^* in Lemmas 13.8 and 13.9. So $A(G)$ is generated in degrees up to $p - 1$ as a free $\mathbf{F}_p[\chi(V)]$-module, showing that the upper bound $n - c$ is optimal in this case. On the other hand, the quotient ring of CH_G^* by transfers from all proper subgroups is $CH_G^*/(\mathrm{tr}_A^G CH_A^* + \mathrm{tr}_H^G CH_H^*)$, which we compute by the same lemmas to be $\mathbf{F}_p[b, \chi(V)]/(b^{p-1})$. So this quotient ring is generated in degrees up to $p - 2$ as a free $\mathbf{F}_p[\chi(V)]$-module, which shows the optimality of the upper bound $n - c - 1$ for G nonabelian in Corollary 10.5.

Theorem 10.4 The proof follows that of Theorem 6.5, modified to take account of transfers as in the proof of Theorem 10.1. To summarize: we choose a faithful representation $G \subset GL(n)$ over k, and let S be the subgroup scheme $(\mu_p)^n \subset (G_m)^n \subset GL(n)$. Then $CH_H^* GL(n)/S$ is a free module over CH_H^* with top generator in degree $n(n-1)/2$, for every subgroup H of G. So we reduce to showing that the CH_S^*-module $(CH_S^* GL(n)/G)/\sum_{H \in S} \mathrm{tr}_H^G CH_S^* GL(n)/H$ has regularity at most $n(n-1)/2$. Let U be the group of strictly upper-triangular matrices in $GL(n)$, and let $N = U\backslash GL(n)$. Taking the quotient by a free action of U does not change Chow groups. So it suffices to show that the CH_S^*-module $(CH_S^* N/G)/\sum_{H \in S} \mathrm{tr}_H^G CH_S^* N/H$ has regularity at most $n(n-1)/2$, which is $\dim(N) - n$. In short, the $-n$ here comes from the action of the diagonal torus $T = (G_m)^n$ on N/H for all subgroups H. This action has finite stabilizers and commutes with the action of S (because S is contained in T).

It suffices to show that the CH_S^*-module above is filtered, with quotients that are direct sums of free modules over polynomial quotient rings of CH_S^* with generators in degree at most $\dim(N) - n$. We showed this without taking quotients by transfers in the proof of Theorem 6.5, and the same holds after taking quotients by transfers by the proof of Theorem 10.1. Let $N = U\backslash GL(n)$. For subgroups H of G and V of S, let $N_H^{(V)}/H$ be the locally closed smooth subscheme of N/H on which S has stabilizer equal to V, and let e_H be the codimension of $N_H^{(V)}/G$ in N/H. Choose a splitting $S = V \times W$. The last step is to prove that the quotient of $CH^*(W\backslash N_G^{(V)}/G)$ by the images of $CH^{*+e_G-e_H}(W\backslash N_H^{(V)}/H)$ for H in $\mathcal{S}$ is concentrated in degrees at most $\dim(N_G^{(V)}/G) - n$. This follows from the same statement without taking the quotient, which holds by Lemma 5.3 because the torus T acts on the smooth k-scheme $W\backslash N^{(V)}/G$ with finite stabilizers. This completes the proof that the quotient ring of CH_G^* by transfers from $\mathcal{S}$ has regularity at most zero.

Now suppose that k is algebraically closed and every abelian p-subgroup of G is contained in some element of $\mathcal{S}$. In this case, we have to improve our regularity bound for the quotient ring above by 1. By the same argument as for cohomology in the proof of Theorem 10.1, it suffices to show that the transfer map

$$\sum_{H \in \mathcal{S}} CH_n(W\backslash N_H^{(V)}/H) \to CH_n(W\backslash N_G^{(V)}/G)$$

is surjective, where S splits as $V \times W$ and the elementary abelian group W acts freely on the smooth schemes $N_H^{(V)}/H$ and $N_G^{(V)}/G$. Note that n is the lowest dimension in which these Chow groups could be nonzero, because the torus $T = (G_m)^n$ acts with finite stabilizers on the varieties $W\backslash N_H^{(V)}/H$.

By the proof of Lemma 5.3, the Chow group of $W\backslash N_G^{(V)}/G$ in degree n is generated by T-orbits of k-points, using that k is algebraically closed. The surjectivity on Chow groups above holds if, for every k-point y of $N_G^{(V)}/G$, there is a subgroup $H \in \mathcal{S}$ and a point z in $(N_G^{(V)} \cap N_H^{(V)})/H$ such that z maps to y and the stabilizer in T of z in $W\backslash N_H^{(V)}/H$ has index prime to p in the stabilizer in T of y in $W\backslash N_G^{(V)}/G$.

This is a formal consequence of having commuting actions of T and G on N $(= U\backslash GL(n))$ such that each of T and G acts freely on N. Indeed, that implies that the stabilizer subgroup in $T \times G$ of any k-point of N projects isomorphically to both T and G. So, for any point y in N, the stabilizer subgroup in T of y in N/G can be identified with the stabilizer subgroup in G of y in $T\backslash N$. Since T is abelian, the latter subgroup $G_y \subset G$ is abelian. So there is a subgroup $H \in \mathcal{S}$ that contains the Sylow p-subgroup of G_y. Then, using again the identification we mentioned, the stabilizer subgroup in T of y in N/H contains a p-Sylow subgroup of the stabilizer subgroup in T of y in N/G.

This implies the facts we wanted, two paragraphs back. Namely, if we write V for the stabilizer in $S = (\mu_p)^n \subset T$ of y in N/G, then the stabilizer in S of y in N/H is also equal to V, because S is a p-subgroup of the torus T. Also, choose a splitting $S = V \times W$, and let K_L be the stabilizer in T of y in N/L, for any subgroup L of G. Then the stabilizer in T of the point y in $W\backslash N/L$ is equal to $K_L W$, for any subgroup L in G. For the subgroup $H \in \mathcal{S}$ constructed in the previous paragraph, K_H has index prime to p in K_G, and so the stabilizer in T of y in $W\backslash N/H$ (namely, $K_H W$) has index prime to p in the stabilizer of y in $W\backslash N/G$, as we wanted. □

Finally, we state the analogous regularity bound for motivic cohomology modulo transfers. Fix a prime number p. We recall the notation from Theorem 6.10: for an affine group scheme G and any integer j, define

$$M_j(G) = \oplus_i H_M^{2i-j}(BG, \mathbf{F}_p(i)).$$

We view $M_j(G)$ as a graded abelian group, graded by the degree of H^*; thus $M_j(G)$ is concentrated in degrees $\equiv j \pmod 2$. In particular, $M_0(G)$ is the Chow ring $CH_G^* = CH^*(BG)/p$. For each j, $M_j(G)$ is a module over the Chow ring CH_G^*.

Theorem 10.6 *Let G be a finite group scheme over a field k, p a prime number, and j a natural number. Suppose that the order of G is invertible in k. Then, for any faithful representation $G \to GL(n)$, $M_j(G)$ is generated by elements of degree at most $n^2 + j$ as a module over $\mathbf{F}_p[c_1, \ldots, c_n]$, and at most n^2 if k is algebraically closed. Moreover, let $\mathcal{S}$ be any collection of subgroup schemes of G. Then $M_j(G)/\sum_{H\in\mathcal{S}} \operatorname{tr}_H^G M_j(H)$ has regularity at most j as an $\mathbf{F}_p[c_1, \ldots, c_n]$-module, and at most 0 if k is algebraically closed.*

The proof is the same as for Theorem 6.10, modified to take account of transfers as in the proof of Theorem 10.1.

11

Transferred Euler Classes

For many complex algebraic groups G, the Chow ring of the classifying space BG is generated by Chern classes of complex representations of G. This fails in general for the symmetric groups, but at least the Chow ring of any symmetric group consists of transferred Euler classes, meaning **Z**-linear combinations of transfers to G of Euler classes (top Chern classes) of representations of subgroups (Corollary 2.22). We will see that transferred Euler classes form a subring of the Chow ring (Lemma 11.2), and that they include all transferred Chern classes (Lemma 11.3).

Schuster and Yagita gave an example of a finite group, the extraspecial 2-group 2_+^{1+6} of order 2^7 contained in Spin(7), whose complex cobordism does not consist of transferred Euler classes [121]. Guillot showed that the Chow ring of the same group (specifically, CH^3BG) also does not consist of transferred Euler classes [62], thus answering negatively a question in [138]. Presumably there are similar examples at odd primes. There is no conjecture now about what sort of elements suffice to generate the Chow ring of an arbitrary finite group.

Nonetheless, the following theorem gives a fairly large class of finite groups for which the Chow ring is generated by transferred Euler classes. Whenever that holds, it follows that CH^iBG is finite for all $i > 0$, or equivalently (by Theorem 5.2) that CH^*BG is a finitely generated **Z**-algebra. That remains an open question for arbitrary finite groups.

By definition, the *exponent* of a finite group G is the least common multiple of the orders of all elements of G.

Theorem 11.1 *Let G be a finite group, and let p be a prime number. Suppose that some Sylow p-subgroup P of G has a faithful complex representation of dimension at most $p+2$. Then the mod p Chow ring of $BG_{\mathbf{C}}$ consists of transferred Euler classes.*

More generally, consider a finite group G as an algebraic group over a field k of characteristic not p that contains the pth roots of unity for p odd, or the 4th roots of unity for $p = 2$. Suppose that P has a faithful representation of dimension n over k with c irreducible summands such that $n - c \leq p$. Then the mod p Chow ring of BG_k consists of transferred Euler classes.

The theorem applies to infinitely many groups, including some with fairly complicated p-local structure. An example is $G = GL(4, \mathbf{F}_2) \cong A_8$ with $p = 2$, since the 2-Sylow subgroup S of G (the group of 4×4 strictly upper-triangular matrices over $\mathbf{F}_2$, of order 64) has a faithful irreducible complex representation of dimension 4. It follows that the mod 2 Chow rings of S and G (as groups over $\mathbf{C}$) consist of transferred Euler classes. Another example is the Mathieu group M_{12}. Its mod 2 Chow ring is generated by transferred Euler classes, since the Sylow 2-subgroup of M_{12} (another group of order 64) has a faithful irreducible complex representation of dimension 4. The cohomology of M_{12} and other sporadic simple groups is discussed in Adem-Milgram [1, chapter VIII].

Theorem 11.1 can also be used to show that the Chow ring of any 2-group of order at most 32 consists of transferred Euler classes. Indeed, the only 2-group of order at most 32 that does not have a faithful complex representation of dimension 4 is $(\mathbf{Z}/2)^5$, for which the Chow ring certainly consists of transferred Euler classes. The statement about faithful representations can be checked using the free group-theory program GAP [46], or by the methods of Cernele-Kamgarpour-Reichstein [28, proof of Lemma 13].

The result that the Chow rings of the 51 groups of order $2^5 = 32$ consist of transferred Euler classes is nearly optimal, in view of Guillot's counterexample of order 2^7. I would guess that the Morava K-theories of groups of order 32 also consist of transferred Euler classes; this is true at least for $K(2)$, by Schuster [120].

For an odd prime p, Theorem 11.1 implies that the Chow ring of every p-group of order at most p^4 consists of transferred Euler classes. Indeed, every such group has a faithful complex representation of dimension at most $p + 1$ [28, theorem 1].

11.1 Basic properties of transferred Euler classes

In this section we define transferred Euler classes, which form a subring of the Chow ring of a finite group. This subring contains all Chern classes. We prove a theorem of Green and Leary: the homomorphism from transferred Euler classes to the cohomology of a finite group is an F-isomorphism. Finally, we note that the analog of Green and Leary's result holds for the Chow ring.

Let G be a finite group, viewed as an algebraic group over a field k. We define the subgroup of *transferred Euler classes* $\text{tr}\,\chi(G)$ in CH^*BG to be the subgroup generated by elements $\text{tr}_H^G \chi(V)$ for all subgroups H in G and all representations V of H over k. We can also talk about transferred Euler classes in $CH_G^* = CH^*(BG)/p$ or (for $k = \mathbf{C}$) in $H_G^* = H^*(BG, \mathbf{F}_p)$. The transferred Euler classes in these other rings are, by definition, the images of $\text{tr}\,\chi(G) \subset CH^*BG$.

Lemma 11.2 *Let G be a finite group, viewed as an algebraic group over a field k. Then the transferred Euler classes form a subring of CH^*BG.*

Proof We follow an argument by Hopkins-Kuhn-Ravenel [70, proposition 7.2]. Let H_1 and H_2 be subgroups of G, with representations V_1 of H_1 and V_2 of H_2 over k. We want to show that $(\text{tr}_{H_1}^G \chi(V_1))(\text{tr}_{H_2}^G \chi(V_2))$ is a transferred Euler class. Rewrite this product as

$$\begin{aligned} &= \text{res}_G^{G\times G} \text{tr}_{H_1\times H_2}^{G\times G} \pi_1^* \chi(V_1) \otimes \pi_2^* \chi(V_2) \\ &= \text{res}_G^{G\times G} \text{tr}_{H_1\times H_2}^{G\times G} \chi(V_1 \oplus V_2), \end{aligned}$$

where π_1 and π_2 are the two projections on $H_1 \times H_2$. By the double coset formula (Lemma 2.15), that element is a transferred Euler class in CH^*BG. □

The following lemma shows that transferred Chern classes are no more general than transferred Euler classes.

Lemma 11.3 *Let G be a finite group, p a prime number, k a field such that p is invertible in k. Suppose that k contains the pth roots of unity for p odd, or the 4th roots of unity for $p = 2$. Consider G as an algebraic group over k. Then all Chern classes in $CH_G^* = CH^*(BG)/p$ of representations of G over k are transferred Euler classes.*

Proof Using transfer, it suffices to prove this for G a p-group. Blichfeldt showed that every irreducible k-representation of G over an algebraically closed field is monomial (induced from a 1-dimensional representation of a subgroup). Kahn generalized Blichfeldt's theorem to an arbitrary field k satisfying our assumptions: again, every irreducible k-representation of G is induced from a 1-dimensional k-representation [75, propositions 7.4, 9.1, and 9.2].

A monomial representation $G \to GL(p^r)$ factors through the wreath product $S_{p^r} \wr G_m$. Lemma 2.21 gives that that all elements of the Chow ring of $S_{p^r} \wr G_m$ are transferred Euler classes, using again that k contains the pth roots of unity. That completes the proof. □

Evens and Fulton-MacPherson gave formulas for the Chern classes of an induced representation [41, 45], but they seem not to imply Lemma 11.3

directly, because the formulas involve the Chern classes of a permutation representation. In the most important case $k = \mathbf{C}$, Lemma 11.3 can be deduced from Symonds's formula for the Chern classes of any complex representation in terms of transfers, norms, and products starting from first Chern classes of 1-dimensional representations [129]. Using that formula, one can write Chern classes as transferred Euler classes.

We now prove a result of Green and Leary for group cohomology [53, corollary 8.3], and its analog for Chow rings. Recall that a homomorphism $f: A \to B$ of graded-commutative $\mathbf{F}_p$-algebras is an *F-isomorphism* if every element of the kernel is nilpotent, and for every element $b \in B$ there is a natural number r and an element $a \in A$ such that $f(a) = b^{p^r}$.

Theorem 11.4

(i) *Let G be a finite group and p a prime number. Then the inclusion from the subring of transferred Euler classes to $H^*_G = H^*(BG, \mathbf{F}_p)$ is an F-isomorphism. In fact, it suffices to consider transferred Euler classes of representations defined over $\mathbf{Q}(\mu_p) \subset \mathbf{C}$.*

(ii) *Let G be a finite group and p a prime number. Let k be a field of characteristic not p that contains the pth roots of unity. View the finite group G as an algebraic group over k. Then the inclusion from the subring of transferred Euler classes to CH^*_G is an F-isomorphism.*

Green and Leary observed that transfers are essential here: the subring of Chern classes need not be F-isomorphic to the whole cohomology ring of a finite group [53].

Also, it would not be enough to use representations defined over $\mathbf{Q}$ in part (i), by the example of $G = \mathbf{Z}/p$ for p odd. The subring of transferred Euler classes of representations over $\mathbf{Q}$ in $H^*_{\mathbf{Z}/p} = \mathbf{F}_p\langle x, y\rangle$ is $\mathbf{F}_p[y^{p-1}]$, and the inclusion from that subring into $H^*_{\mathbf{Z}/p}$ is not an F-isomorphism.

Proof We first prove (ii). Let k contain the pth roots of unity. By Theorem 8.10, the restriction map $CH^*_G \to \varprojlim CH^*_A$ is an F-isomorphism. For any element y of CH^*_G, consider the image of y in $\varprojlim CH^*_A$. We showed after Lemma 8.5 that any element of $\varprojlim CH^*_A$, raised to some p-power, is the restriction of an $\mathbf{F}_p$-linear combination of graded pieces of elements $N^G_A(1 + u) \in CH^*_G$ for A an elementary abelian subgroup of G and $u \in CH^*_A$ of positive degree. Therefore, y raised to a possibly higher power of p is equal to an $\mathbf{F}_p$-linear combination of graded pieces of elements $N^G_A(1 + u) \in CH^*_G$ for A an elementary abelian subgroup of G and $u \in CH^*_A$ of positive degree. So it suffices to show that $N^G_A(1 + u)$ is a transferred Euler class.

By the formulas for the norm of a sum or product (Lemma 8.1), together with the fact that CH^*_A is generated by elements of degree 1, it suffices to show that $N^G_H(u)$ is a transferred Euler class for every subgroup H of a finite group G and every $u \in CH^1_H$. The element u is the first Chern class of a 1-dimensional representation L of H over k. Then $N^G_H(u)$ is the Euler class of the representation of G induced from L, by Fulton-MacPherson, generalizing Evens's theorem in cohomology [45, corollary 5.5], [41, theorem 4]. (ii) is proved.

We now prove (i). We will use what we have shown about Chow rings to prove this result on cohomology, although one could also give a more direct argument.

Consider the finite group G as an algebraic group over $k = \mathbf{Q}(\mu_p)$. We know by Theorem 8.10 that the cycle map $CH^*_G \to H^*_G$ is an F-isomorphism. Then part (1) implies that the subring of transferred Euler classes in H^*_G of representations defined over $\mathbf{Q}(\mu_p)$ is F-isomorphic to the whole ring H^*_G. □

11.2 Generating the Chow ring

Proof of Theorem 11.1 We can assume that G is a p-group. Consider G as an algebraic group over the field k. We are given a faithful representation V of dimension n over k. We have two different assumptions: for $k = \mathbf{C}$, we assume that $n \leq p + 2$; or, for any field k, we assume that n minus the number of irreducible summands of V is at most p. By omitting some irreducible summands of V, we can assume that the number of irreducible summands of V is equal to the p-rank c of the center of G.

Either assumption passes from G to its subgroups, and so it suffices to show that the quotient ring $A'(G)$ of CH^*_G by transfers from all proper subgroups is generated by Euler classes of k-representations of G. When $k = \mathbf{C}$ and G is abelian, the ring $A'(G)$ is generated by Euler classes, because the whole Chow ring is generated by Euler classes. So we can assume that G is not abelian if $k = \mathbf{C}$.

By Corollary 10.5, the ring $A'(G)$ is generated as a module over Euler classes by elements of degree at most $n - c$, and at most $n - c - 1$ if $k = \mathbf{C}$, since G is not abelian. We have $c \geq 1$ if $k = \mathbf{C}$ since G is nontrivial, and so $A'(G)$ is generated by elements of degree at most p under either of our assumptions. For any affine group scheme G of finite type over a field, the mod p Chow group CH^i_G is spanned by Chern classes of representations of G for $i \leq p$, by Theorem 2.25. Thus the ring $A'(G)$ is generated by Chern classes of representations. By Lemma 11.3, using our assumption on the roots of unity in k, it follows that the Chow ring modulo transfers, $A'(G)$, is generated by Euler classes of representations. □

12

Detection Theorems for Cohomology and Chow Rings

In 1991, Henn, Lannes, and Schwartz gave a powerful approach to computing the cohomology of finite groups [69]. Using the work of Miller and Lannes on unstable modules over the Steenrod algebra, they showed that the cohomology of any finite group G is determined in all degrees by the cohomology of certain subgroups (centralizers of elementary abelian subgroups) up to some finite degree, which we denote $d_0(H_G^*)$, the "topological nilpotence degree" of H_G^*. It is a fundamental problem to estimate the number $d_0(H_G^*)$.

Henn, Lannes, and Schwartz showed that if G has a faithful complex representation of dimension n, then $d_0(H_G^*)$ is at most n^2. Kuhn improved their bound, and conjectured that in fact $d_0(H_G^*)$ is at most $2n - c$, for c the p-rank of the center of a Sylow p-subgroup [86]. We prove Kuhn's conjecture in Theorem 12.4. This should be valuable for computations. There is a close connection between the ideas here and Symonds's result on the cohomology ring modulo transfers, Corollary 10.3.

In Theorem 12.7, we prove an analogous detection theorem for Chow rings: if a finite group G has a faithful complex representation of dimension n, then CH_G^* is determined by the Chow rings of centralizers of elementary abelian subgroups in degrees at most $n - c$. Compare the regularity theorem 6.5, which says only that the ring CH_G^* has generators and relations in degrees at most about n^2. Nonetheless, the proof of Theorem 12.7 uses the regularity theorem.

The proof of this detection theorem for the Chow ring is very different from Henn-Lannes-Schwartz's arguments for cohomology. Henn-Lannes-Schwartz's argument uses some of the deepest results in homotopy theory, including Lannes's computation of the cohomology of certain mapping spaces using his T-functor. Those results are not available for Chow rings, where we do not even have the spectral sequence of a group extension. Our approach is strongly influenced by Kuhn's work on group cohomology [86].

12.1 Nilpotence in group cohomology

In this section, we prove Kuhn's conjectured bound, Theorem 12.4, on the detection of group cohomology using centralizers of elementary abelian subgroups. The proof turns out to be an easy extension of Kuhn's results [86].

Let p be a prime number. For a finite group G, we write H_G^* for the ring $H^*(BG, \mathbf{F}_p)$. Let $c = c(G)$ be the p-rank of the center of a Sylow p-subgroup of G, and $r = r(G)$ the p-rank of G (i.e., the maximal rank of an elementary abelian p-subgroup of G).

We start by defining the "topological nilpotence degree" of Henn-Lannes-Schwartz. Since H_G^* is the mod p cohomology of a topological space, it belongs to the category $\mathcal{U}$ of *unstable* modules over the mod p Steenrod algebra. We recall the definition [122, definition 1.3.1]:

Definition 12.1 A graded module M over the mod p Steenrod algebra is *unstable* if

(i) for $p = 2$, $\mathrm{Sq}^i x = 0$ for all $x \in M$ and all $i > |x|$;
(ii) for p odd, $\beta^e P^i x = 0$ for all $x \in M$ and all $e + 2i > |x|$ with $e = 0$ or 1.

Let $\Sigma^d M$ denote the dth suspension (upward shift) of a graded module M. For an unstable module $N \in \mathcal{U}$, define the *topological nilpotence degree* $d_0(N)$ to be the supremum of the natural numbers d such that N contains a nonzero submodule of the form $\Sigma^d M$ with $M \in \mathcal{U}$.

For example, H_G^* is called *reduced* (as a module over the Steenrod algebra $\mathcal{A}$) if $d_0(H_G^*) = 0$. For $p = 2$, this is equivalent to H_G^* being reduced as a commutative ring; that is, every nilpotent element is zero. For p odd, $d_0(H_G^*)$ is zero if and only if every element x of H_G^* such that θx is nilpotent for every $\theta \in \mathcal{A}$ is zero [122, lemma 2.6.4]. For example, $d_0(H_{\mathbf{Z}/p}^*) = 0$ for an odd prime p, even though the degree-1 generator x in $H_{\mathbf{Z}/p}^* = \mathbf{F}_p\langle x, y\rangle$ is nilpotent, because $\beta x = y$ is not nilpotent.

The invariant $d_0(H_G^*)$ clearly only depends on H_G^* as a module over the Steenrod algebra. Henn-Lannes-Schwartz gave a remarkable interpretation of $d_0(H_G^*)$ that does not mention the Steenrod algebra [69, theorem 0.2].

Theorem 12.2 *Let G be a finite group. Then $d_0(H_G^*)$ is the least natural number d such that the algebra homomorphism*

$$H_G^* \to \prod_V H_V^* \otimes_{\mathbf{F}_p} H_{C_G(V)}^{\leq d}$$

is injective.

Here the product is over the elementary abelian p-subgroups V of G, the component maps are induced by the group homomorphisms $V \times C_G(V) \to G$, and $M^{\leq d}$ denotes the quotient of a graded module M by all elements of degree greater than d. For example, $d_0(H_G^*)$ is zero if and only if H_G^* is detected on elementary abelian subgroups; that is a very special situation, but it holds in some important examples such as the symmetric groups at any prime p, by Corollary 2.20.

One reason to be interested in the topological nilpotence degree of H_G^*, perhaps not the most important, is that it gives a bound for the nilpotence of H_G^* in the following algebraic sense [86, corollary 2.6]. The proof is straightforward.

Theorem 12.3 *Let G be a finite group and p a prime number. Let e be $d_0(H_G^*)$ for $p = 2$, or $d_0(H_G^*) + r(G)$ for p odd. Then $\mathrm{rad}(H_G^*)^e = 0$. That is, the product of any e nilpotent elements in H_G^* is zero.*

We now prove the bound for $d_0(H_G^*)$ conjectured by Kuhn [86, section 1].

Theorem 12.4 *Let G be a finite group with a faithful complex representation of dimension n. Let c be the p-rank of the center of a Sylow p-subgroup of G. Then $d_0(H_G^*) \leq 2n - c$.*

By Henn-Lannes-Schwartz's interpretation of $d_0(H_G^*)$ (Theorem 12.2), Theorem 12.4 means that computing the cohomology of a finite group G essentially reduces to a computation in degrees at most $2n - c$, for G and certain subgroups. Henn-Lannes-Schwartz's original bound for their invariant was that $d_0(H_G^*) \leq n^2$ [69, theorem 0.5].

Proof Let P be a Sylow p-subgroup of G. Then H_G^* is a summand of H_P^* as a module over the Steenrod algebra, using transfer. It follows that $d_0(H_G^*) \leq d_0(H_P^*)$. So we can assume from now on that G is a p-group.

Let $C = C(G)$ be the p-torsion subgroup of the center of G. The ring H_C^* is a commutative cocommutative Hopf algebra, since C is an abelian group; namely, the group homomorphism $C \times C \to C$, $(x, y) \mapsto xy$, gives the coproduct $H_C^* \to H_C^* \otimes_{\mathbf{F}_p} H_C^*$. Next, H_G^* is a comodule over H_C^*, using the group homomorphism $C \times G \to G$. Although H_G^* is not a Hopf algebra, the image of the restriction map $H_G^* \to H_C^*$ is a Hopf subalgebra of H_C^*, as one shows using the commutative diagram

$$\begin{array}{ccc} C \times C & \longrightarrow & C \\ \downarrow & & \downarrow \\ C \times G & \longrightarrow & G \end{array}$$

and the analogous diagram for $G \times C \to G$. Aguadé-Smith [2] and Broto-Henn [22, remark 1.3] deduced (using the Borel structure theorem on the structure of Hopf algebras [105, theorem 7.11]) that there is a basis $x_1, \ldots, x_c$

for H_C^1 such that, writing $y_i = \beta x_i$,

$$\operatorname{im}(H_G^* \to H_C^*) = \begin{cases} \mathbf{F}_2[x_1^{2^{j_1}}, \ldots, x_c^{2^{j_c}}] & \text{if } p = 2 \\ \mathbf{F}_p\langle y_1^{p^{j_1}}, \ldots, y_b^{p^{j_b}}, y_{b+1} \ldots, y_c, x_{b+1}, \ldots, x_c\rangle & \text{if } p \text{ is odd,} \end{cases}$$

for some natural numbers $j_1 \geq j_2 \geq \cdots$. For p odd, $c - b$ is the largest number of copies of $\mathbf{Z}/p$ that factor off G as a product group.

Definition 12.5 In this notation, we say that G has *type* $[a_1, \ldots, a_c]$, where

$$[a_1, \ldots, a_c] = \begin{cases} [2^{j_1}, \ldots, 2^{j_c}] & \text{if } p = 2 \\ [2p^{j_1}, \ldots, 2p^{j_b}, 1, \ldots, 1] & \text{if } p \text{ is odd,} \end{cases}$$

following Kuhn [86, section 2.6].

Define $e(G)$ to be the maximum degree of a generator for H_C^* as a module over H_G^*. By our description of the image of restriction, we have

$$e(G) = \sum_{i=1}^{c} (a_i - 1).$$

Because the image of $H_G^* \to H_C^*$ is so special, $e(G)$ is often easy to compute.

Using Symonds's theorem, Kuhn proved the following strong bound for $d_0(H_G^*)$ [86, theorem 1.5].

Theorem 12.6 *Let G be a p-group. Then*

$$d_0(H_G^*) \leq e(G).$$

If G is p-central, then equality holds.

It remains to prove Kuhn's conjecture that $e(G)$ is at most $2n - c$.

Let V be a faithful complex representation of G of dimension n. We can assume (by omitting some irreducible summands, if necessary) that V has exactly c irreducible summands, $V = V_1 \oplus \cdots \oplus V_c$. The dimension of an irreducible representation of G is a power of p, and so we can write $\dim(V_i) = p^{b_i}$. By Schur's lemma, C acts by scalars on each V_i, through some 1-dimensional complex representation L_i of C. Since V is faithful, the 1-dimensional representations $L_1, \ldots, L_c$ form a basis for $\operatorname{Hom}(C, \mathbf{C}^*) \cong (\mathbf{Z}/p)^c$. Let $y_i = c_1(L_i)$ in H_C^2. We can write $y_i = \beta x_i$ for elements x_i in H_C^1. In terms of these elements, the cohomology ring of C is $\mathbf{F}_2[x_1, \ldots, x_c]$ for $p = 2$, or $\mathbf{F}_p\langle x_1, \ldots, x_c, y_1, \ldots, y_c\rangle$ for p odd.

Because the restriction of the irreducible representation V_i to the subgroup C is a sum of p^{b_i} copies of L_i, the Euler class $\chi(V_i)$ in H_G^* restricts to $y_i^{p^{b_i}}$ in H_C^*, for $i = 1, \ldots, n$. As a result, we can write down generators for H_C^* as a module over $\mathbf{F}_p[\chi(V_1), \ldots, \chi(V_c)]$, with the highest-degree generator being $x_1 \ldots x_c y_1^{p^{b_1}-1} \cdots y_c^{p^{b_c}-1}$. That element has degree $c + \sum_i (2(p^{b_i} - 1)) =$

$2n - c$. A fortiori, H_C^* is generated as an H_G^*-module by elements of degree at most $2n - c$. □

12.2 The detection theorem for Chow rings

We now define the topological nilpotence degree of the Chow ring, and prove an upper bound for it. The bound is analogous to the bound for the topological nilpotence degree of the cohomology ring proved in Theorem 12.4. It is a strong computational tool, as we will see repeatedly in Chapter 13.

Fix a prime number p, and write CH_G^* for Chow groups with $\mathbf{F}_p$ coefficients. Let G be a finite group, which we view as an algebraic group over a field k. Define the *topological nilpotence degree* $d_0(CH_G^*)$ to be the least natural number d such that the $\mathbf{F}_p$-algebra homomorphism

$$CH_G^* \to \prod_V CH_V^* \otimes_{\mathbf{F}_p} CH_{C_G(V)}^{\leq d}$$

is injective. As in Theorem 12.2, the product is over the elementary abelian p-subgroups V of G, the component maps are induced by the group homomorphisms $V \times C_G(V) \to G$, and $M^{\leq d}$ denotes the quotient of a graded module M by all elements of degree greater than d. For example, $d_0(CH_G^*)$ is zero if and only if CH_G^* is detected on elementary abelian subgroups; that is a very special situation, but it holds in some important examples such as the symmetric groups at any prime p (Corollary 2.22).

Theorem 12.7 *Let G be a finite group, p a prime number. Let k be a field of characteristic not p that contains the pth roots of unity. Suppose that G has a faithful representation of dimension n over k, and let c be the p-rank of the center of a Sylow p-subgroup of G. Then*

$$d_0(CH_G^*) \leq n - c,$$

where G is viewed as an algebraic group over k.

Theorem 12.7 reduces the problem of checking relations in the Chow ring to computations in degrees at most the dimension n of a faithful representation, in fact a bit better than that.

Conjecture 12.8 Let p and the field k be as in Theorem 12.7. Then $d_0(CH_G^*)$ is equal to the supremum of the natural numbers d such that CH_G^*, as a module over the Steenrod algebra $\mathcal{A}$, contains a nonzero submodule of the form $\Sigma^d M$, with M an unstable $\mathcal{A}$-module. (For this purpose, "unstable" means that $P^i x = 0$ for all $x \in M$ and all $i > |x|$. Following section 6.3, remember that P^i sends CH_G^j to $CH_G^{j+i(p-1)}$.)

One direction of Conjecture 12.8 is easy: $d_0(CH_G^*)$ (as we have defined it) is *at least* the number $s_0(CH_G^*)$ defined using the Steenrod algebra. To prove that, use that the ring CH_V^* is reduced for an elementary abelian p-group V. Conversely, if $s_0(CH_G^*) = 0$, then CH_G^* is reduced as a ring (using the identity $P(x^p) = (Px)^p$ for the total Steenrod operation $P = P^0 + P^1 + \cdots$). Then Yagita's theorem (Lemma 8.9) gives that CH_G^* is detected on elementary abelian subgroups, that is, $d_0(CH_G^*) = 0$. So Conjecture 12.8 seems to be a plausible extension of Yagita's theorem.

Conjecture 12.8 would say that $d_0(CH_G^*)$ is determined by CH_G^* as a module over the Steenrod algebra. That would be interesting to know, by analogy with the Henn-Lannes-Schwartz theorem, Theorem 12.2. But it does not matter for the applications we have in mind. Since we defined $d_0(CH_G^*)$ in terms of injectivity of restriction maps, an upper bound for $d_0(CH_G^*)$ (such as Theorem 12.7) is immediately useful for computing CH_G^*.

Proof (Theorem 12.7) We can assume that G is a p-group, because the Chow ring of any finite group injects into that of a Sylow p-subgroup.

Let $C = Z(G)[p]$. We will prove a stronger version of the theorem (although it is easily seen to be equivalent): it suffices to consider elementary abelian subgroups that contain C. That is, we will show that

$$CH_G^* \to \prod_{\substack{C \subset V \\ V \text{ elem ab}}} CH_V^* \otimes CH_{C_G(V)}^{\leq e}$$

is injective, for $e = n - c$.

The ring CH_C^* is a commutative cocommutative Hopf algebra, since C is an abelian group; namely, the group homomorphism $C \times C \to C$, $(x, y) \mapsto xy$, gives the coproduct $CH_C^* \to CH_C^* \otimes_{\mathbf{F}_p} CH_C^*$. Next, CH_G^* is a comodule over CH_C^*, using the group homomorphism $C \times G \to G$. Although CH_G^* is not a Hopf algebra, the image of the restriction map $CH_G^* \to CH_C^*$ is a Hopf subalgebra of CH_C^*, by the same argument as in group cohomology (proof of Theorem 12.4). Therefore, there is a basis $y_1, \ldots, y_c$ for CH_C^1 such that $\mathrm{im}(CH_G^* \to CH_C^*)$ is equal to the subring

$$\mathbf{F}_p[y_1^{p^{a_1}}, \ldots, y_c^{p^{a_c}}]$$

of $CH_C^* = \mathbf{F}_p[y_1, \ldots, y_c]$ for some natural numbers $a_1 \geq \cdots \geq a_c$. This follows from the Borel structure theorem [105, theorem 7.11] applied to the Hopf algebra quotient $CH_C^*//\mathrm{im}(CH_G^* \to CH_C^*)$, which is finite over $\mathbf{F}_p$.

In this notation, we say that G has *Chow type* $[p^{a_1}, \ldots, p^{a_c}]$, by analogy with Definition 12.5 of the type for group cohomology. Define $e^{CH}(G)$ to be the maximum degree of a generator for CH_C^* as a module over CH_G^*. By our description of the image of the restriction homomorphism $CH_G^* \to CH_C^*$, we have

$$e^{CH}(G) = \sum_{i=1}^{c}(p^{a_i} - 1).$$

(After Definition 12.5, we defined $e(G)$ to be the analogous invariant using group cohomology rather than the Chow ring.)

Since $\mathrm{im}(CH_G^* \to CH_C^*)$ is a polynomial ring, we can fix a graded subring $B \subset CH_G^*$ that maps isomorphically to $\mathrm{im}(CH_G^* \to CH_C^*)$. We call such a subring a *Duflot algebra*, following Kuhn's definition in group cohomology [86, definition 2.7]. Beware that a Duflot algebra need not be a CH_C^*-submodule of CH_G^*, and it need not be closed under the Steenrod operations on CH_G^*.

For a graded module M over a graded ring B, the space of *indecomposables* for M is

$$Q_B M := M/B^{>0}M.$$

Define an ideal $\mathrm{CEss}(CH_G^*)$ in CH_G^*, the *central essential ideal*, as the elements that restrict to 0 on $C_G(V)$ for all elementary abelian subgroups $C \subsetneq V$ of G. Kuhn considered the analogous ideal in group cohomology [86, definition 2.7]. Define $e_{\mathrm{indec}}^{CH}(G)$ to be the supremum of the degrees of generators for $\mathrm{CEss}(CH_G^*)$ as a module over the Duflot algebra $B \subset CH_G^*$. (We will show that $e_{\mathrm{indec}}^{CH}(G) < \infty$ as well as more precise results.) Equivalently, $e_{\mathrm{indec}}^{CH}(G)$ is the maximum degree of the space of indecomposables $Q_B\mathrm{CEss}(CH_G^*)$.

Since all the groups $C_G(V)$ contain C, $\mathrm{CEss}(CH_G^*)$ is a sub-CH_C^*-comodule of CH_G^*. For a CH_C^*-comodule M, write $P_C M$ for the *primitive* subspace in M,

$$P_C M = \{x \in M : \Delta(x) = 1 \otimes x\},$$

where $\Delta\colon M \to CH_C^* \otimes M$ is the coproduct. Let $e_{\mathrm{prim}}^{CH}(G)$ be the supremum of the degrees in which the graded vector space $P_C\mathrm{CEss}(CH_G^*)$ is not zero. The main step toward the theorem is to prove the inequalities

$$e_{\mathrm{prim}}^{CH}(G) \le e_{\mathrm{indec}}^{CH}(G) \le e^{CH}(G) \le n - c.$$

The inequality $e^{CH}(G) \le n - c$ follows by the arguments used for Theorem 12.4. Let V be a faithful representation of G of dimension n over k. We can assume (by omitting some irreducible summands, if necessary) that V has exactly c irreducible summands, $V = V_1 \oplus \cdots \oplus V_c$. Since k contains the pth roots of unity, the dimension of an irreducible representation of G over k is a power of p [29, theorem 70.12], and so we can write $\dim(V_i) = p^{b_i}$. Since C is central in G, G preserves the isotypic decomposition of any kG-module W as a kC-module; so if W is irreducible for G, then it is isotypic for C (a direct sum of copies of one irreducible k-representation). Since k contains the pth roots of unity, all irreducible k-representations of C are 1-dimensional, and so C acts by scalars on each V_i, through some 1-dimensional k-representation L_i of C. Since V is faithful, the 1-dimensional representations $L_1, \ldots, L_c$ form a basis for $CH^1_C = \text{Hom}(C, k^*)/p$. So the Chow ring of C can be written as $\mathbf{F}_p[y_1, \ldots, y_c]$ with $y_i = c_1(L_i)$.

Because the restriction of the irreducible representation V_i to the subgroup C is a sum of p^{b_i} copies of L_i, the Euler class $\chi(V_i)$ in CH^*_G restricts to $y_i^{p^{b_i}}$ in H^*_C, for $i = 1, \ldots, n$. Thus $\text{im}(CH^*_G \to CH^*_C)$ contains $y_1^{p^{b_1}}, \ldots, y_c^{p^{b_c}}$ in this basis. So CH^*_C is generated as a module over $\text{im}(CH^*_G \to CH^*_C)$ by elements of degree at most

$$\sum_{i=1}^{c}(p^{b_i} - 1) = n - c.$$

That is, $e^{CH}(G) \leq n - c$, as we want.

To prove that $e^{CH}_{\text{prim}}(G) \leq e^{CH}_{\text{indec}}(G)$, we start with the following algebraic statement from Kuhn [85, lemma 5.2].

Lemma 12.9 *Let K be a Hopf subalgebra of a graded connected Hopf algebra H over a field F. Let M be a graded K-module that is also an H-comodule in a compatible way, meaning that the multiplication $M \otimes K \to M$ is a map of H-comodules. Then*

(a) M is a free K-module, and

(b) the composite $P_H M \hookrightarrow M \twoheadrightarrow Q_K M$ is injective.

The following lemma applies Lemma 12.9 to our situation, although some care is required.

Lemma 12.10 *Let G be a p-group and let $C = Z(G)[p]$. Let M be a non-negatively graded CH^*_G-module that is also a CH^*_C-comodule such that the map $CH^*_G \otimes M \to M$ is a map of CH^*_C-comodules. Let $B \subset CH^*_G$ be a Duflot algebra. Then*

(a) M is a free A-module, and

(b) the composite $P_C M \hookrightarrow M \twoheadrightarrow Q_B M$ is injective.

Proof The difficulty is that B need not be a sub-CH^*_C-comodule of CH^*_G. To deal with that, we filter M into simpler pieces. Let $L = \ker(CH^*_G \to CH^*_C)$, and let $M_i = L^i M \subset M$ for $i \geq 0$, where we write L^i for the $\mathbf{F}_p$-linear span of all products of i elements of the ideal L. Clearly $M = M_0 \supset M_1 \supset M_2 \supset \cdots$ are CH^*_G-modules.

Next, $L = \ker(CH_G^* \to CH_C^*)$ is a sub-CH_C^*-comodule of CH_G^*, by the commutative diagram

$$\begin{array}{ccc} CH_G^* & \longrightarrow & CH_C^* \\ {\scriptstyle\Delta}\downarrow & & {\scriptstyle\Delta}\downarrow \\ CH_G^* \otimes CH_C^* & \longrightarrow & CH_C^* \otimes CH_C^*. \end{array}$$

That follows from the commutative diagram

$$\begin{array}{ccc} G & \longleftarrow & C \\ \uparrow & & \uparrow \\ G \times C & \longleftarrow & C \times C. \end{array}$$

We now show that $LM \subset M$ is a sub-CH_C^*-comodule, which implies by induction that $M_i = L^i M$ is a sub-CH_C^*-comodule for all $i \geq 0$. We are assuming that M is a CH_G^*-module and a CH_C^*-comodule in a compatible way, as expressed by the commutative diagram:

$$\begin{array}{ccc} CH_G^* \otimes M & \longrightarrow & M \\ {\scriptstyle\Delta\otimes\Delta}\downarrow & & \downarrow{\scriptstyle\Delta} \\ (CH_G^* \otimes CH_C^*) \otimes (M \otimes CH_C^*) & \longrightarrow & M \otimes CH_C^*. \end{array}$$

Here the bottom map comes from the products $CH_C^* \otimes CH_C^* \to CH_C^*$ and $CH_G^* \otimes M \to M$. Combined with the fact that L is a sub-CH_C^*-comodule of CH_G^*, this diagram shows that $LM \subset M$ is a CH_C^*-comodule, as we want. So M_i is a sub-CH_C^*-comodule of M for every $i \geq 0$.

Therefore, for each $i \geq 0$, $\mathrm{gr}_i M = M_i/M_{i+1}$ is a CH_C^*-comodule. By definition of M_i, $\mathrm{gr}_i M$ is also a module over the ring $K := \mathrm{im}(CH_G^* \to CH_C^*)$. We have shown that K is a Hopf subalgebra of CH_C^*. Our assumption on M implies that the K-module and CH_C^*-comodule structures on $\mathrm{gr}_i M$ are compatible. By Lemma 12.9 (for $H = CH_C^*$ and $K = \mathrm{im}(CH_G^* \to CH_C^*)$), $\mathrm{gr}_i M$ is a free B-module and the composite

$$P_C(\mathrm{gr}_i M) \hookrightarrow \mathrm{gr}_i M \twoheadrightarrow Q_B(\mathrm{gr}_i M)$$

is injective, for each $i \geq 0$.

Since $L \subset CH_G^{>0}$, the submodule $M_i = L^i M \subset M$ is concentrated in degrees at least i, and so the intersection of all the submodules M_i is zero. Therefore, from $\mathrm{gr}_i M$ being a free B-module for each $i \geq 0$, it follows that M is a free B-module. Thus (a) is proved. Freeness of $\mathrm{gr}_i M$ as an B-module for each $i \geq 0$ also implies that the extensions of B-modules given by our filtration of M are split, and so $Q_B(\mathrm{gr}_i M) \to Q_B(M/M_{i+1})$ is injective. To prove (b), let u be a nonzero homogeneous element of $P_C M$. Then there is a unique $i \geq 0$ such that $u \in M_i$ and $u \notin M_{i+1}$. Then u gives a nonzero element of $P_C(\mathrm{gr}_i M) \subset \mathrm{gr}_i M$. By the previous paragraph, the image of u in $Q_B(\mathrm{gr}_i M)$ is not 0. By the injectivity just shown, the image of u in $Q_B(M/M_{i+1})$ is not zero. A fortiori, the image of u in $Q_B M$ is not zero. Thus (b) is proved. □

Apply the lemma to $M = \mathrm{CEss}(CH_G^*) \subset CH_G^*$. Then part (b) implies that $e_{\mathrm{prim}}^{CH}(G) \leq e_{\mathrm{indec}}^{CH}(G)$, as we want.

Next, we show that $e_{\mathrm{indec}}^{CH}(G) \leq e^{CH}(G)$. This is the most surprising step: we are bounding the central essential ideal in CH_G^*, which can be considered the most mysterious part of CH_G^*, in terms of $\mathrm{im}(CH_G^* \to CH_C^*)$, which is much easier to estimate. This step uses the regularity theorem on the Chow ring (Theorem 6.5).

The ideal $\sum_{\substack{C \subsetneq V \\ V \text{ elem ab}}} \mathrm{tr}_{C_G(V)}^G CH_{C_G(V)}^*$ kills $\mathrm{CEss}(CH_G^*)$, by the projection formula $x \,\mathrm{tr}_{C_G(V)}^G(y) = \mathrm{tr}_{C_G(V)}^G(x|_{C_G(V)}\, y) = 0$ for $x \in \mathrm{CEss}(CH_G^*)$, $y \in CH_{C_G(V)}^*$ (Lemma 2.15(i)). By Carlson's theorem for Chow rings (Corollary 8.14), it follows that for any $u \in \ker(CH_G^* \to CH_C^*)$, there is a natural number m such that $u^{p^m} \cdot \mathrm{CEss}(CH_G^*) = 0$.

We now apply that fact to the local cohomology (see Section 3.1 for definitions) of certain CH_G^*-modules with respect to the maximal ideal $\mathfrak{m} = CH_G^{>0}$.

Lemma 12.11

$$Q_B \mathrm{CEss}(CH_G^*) = H_{\mathfrak{m}}^0(Q_B \mathrm{CEss}(CH_G^*)) = H_{\mathfrak{m}}^0(Q_B CH_G^*).$$

Proof The 0th local cohomology of a CH_G^*-module is the $\mathfrak{m}$-torsion subspace. So the first statement means that every element of $Q_B \mathrm{CEss}(CH_G^*)$ is $\mathfrak{m}$-torsion. That is, by the definition in Section 3.1, we have to show that for any homogeneous elements $x \in Q_B \mathrm{CEss}(CH_G^*)$ and $f \in CH_G^{>0}$, there is an $r > 0$ such that $f^r x = 0$. Since B is a Duflot algebra, there is an element $g \in B^{>0}$ such that f and g have the same restriction to CH_C^*. That is, $f - g$ is in the ideal $L = \ker(CH_G^* \to CH_C^*)$. As shown before this lemma, it follows that there is a natural number m with $(f-g)^{p^m} x = 0$. (We showed this for x in $\mathrm{CEss}(CH_G^*)$, which implies the same statement for the quotient space $Q_B \mathrm{CEss}(CH_G^*)$.) So $f^{p^m} x = g^{p^m} x$, which is zero in $Q_B \mathrm{CEss}(CH_G^*)$ since g is in $B^{>0}$. Thus we have shown that $Q_B \mathrm{CEss}(CH_G^*) = H_{\mathfrak{m}}^0(Q_B \mathrm{CEss}(CH_G^*))$.

For the second equality we want, consider the exact sequence

$$0 \to \mathrm{CEss}(CH_G^*) \to CH_G^* \to \prod_{\substack{C \subsetneq V \\ V \text{ elem ab}}} CH_{C_G(V)}^*.$$

These are CH_G^*-modules and CH_C^*-comodules in a compatible way. Therefore, applying Lemma 12.10 to the images and cokernels of the maps shown, all those images and cokernels are free B-modules. So the maps split, and we have an exact sequence of B-indecomposables:

$$0 \to Q_B\mathrm{CEss}(CH_G^*) \to Q_B CH_G^* \to \prod_{\substack{C \subsetneq V \\ V \text{ elem ab}}} Q_B CH_{C_G(V)}^*.$$

This trivially gives an exact sequence of $\mathfrak{m}$-torsion submodules:

$$0 \to H_{\mathfrak{m}}^0 Q_B\mathrm{CEss}(CH_G^*) \to H_{\mathfrak{m}}^0 Q_B CH_G^* \to \prod_{\substack{C \subsetneq V \\ V \text{ elem ab}}} H_{\mathfrak{m}}^0 Q_B CH_{C_G(V)}^*.$$

So the lemma follows if we can show that for every elementary abelian subgroup V of G that strictly contains C, we have $H_{\mathfrak{m}}^0 Q_B CH_{C_G(V)}^* = 0$. Let $H = C_G(V)$. We can assume that V is the whole group $Z(H)[p]$; if not, just enlarge V. Since CH_V^* is a CH_G^*-module and a CH_C^*-comodule in a compatible way, it is free as a B-module, by Lemma 12.10. Here CH_V^* is a graded polynomial ring over $\mathbf{F}_p$ on $r > c$ generators, and B is a graded polynomial ring on c generators $f_1, \ldots, f_c$. So $Q_B CH_V^*$ is a complete intersection ring, and hence Cohen-Macaulay. It has dimension $r - c > 0$.

Let W be a faithful representation of G of dimension n over k. We know that CH_V^* is generated in bounded degrees as a module over the ring $R = \mathbf{F}_p[c_1 W, \ldots, c_n W]$. Since CH_V^* is of finite type over $\mathbf{F}_p$, it is in fact finite over R. So the quotient ring $Q_B CH_V^*$ is finite over R. By the Noether normalization lemma, $Q_B CH_V^*$ is finite over some graded polynomial ring $S \subset R$ of dimension $r - c$ $(= \dim(Q_B CH_V^*))$. Since $Q_B CH_V^*$ is Cohen-Macaulay, it is a finitely generated free module over S. Since S has positive dimension, there is a nonzero element $h \in S^{>0}$; then h is a non-zero-divisor on $Q_B CH_V^*$. So $f_1, \ldots, f_c, h|_{C_G(V)} \in CH_{C_G(V)}^*$ restrict to a regular sequence in CH_V^*. By the Duflot theorem for Chow rings (Theorem 3.20), which applies as stated since we made V equal to $Z(C_G(V))[p]$, $f_1, \ldots, f_c, h|_{C_G(V)}$ form a regular sequence in $CH_{C_G(V)}^*$. So the element h of $\mathfrak{m} = CH_G^{>0}$ restricts to a non-zero-divisor on $Q_B CH_{C_G(V)}^*$. So the $\mathfrak{m}$-torsion subspace $H_{\mathfrak{m}}^0 Q_B CH_{C_G(V)}^*$ is zero, as we want. □

We want to prove that $e_{\text{indec}}^{CH}(G) \le e^{CH}(G)$. That is, we want to show that $Q_B\mathrm{CEss}(CH_G^*)$ is concentrated in degrees at most $e^{CH}(G)$. By Lemma 12.11, it is equivalent to show that $H_{\mathfrak{m}}^0 Q_B CH_G^*$ is concentrated in degrees at most $e^{CH}(G)$. The Duflot algebra B is a polynomial ring $\mathbf{F}_p[u_1, \ldots, u_c]$ with $|u_i| = p^{a_i}$ for some natural numbers $a_1, \ldots, a_c$. We know that CH_G^* is free as a B-module by the Duflot theorem for Chow rings (Theorem 3.20). Therefore, $Q_B CH_G^*$ is the quotient of CH_G^* by a regular sequence.

If M is a graded CH_G^*-module and $f \in CH_G^d$ is a non-zero-divisor on M, then $\mathrm{reg}(M/fM) \le \mathrm{reg}(M) + d$, by the long exact sequence of local cohomology applied to the exact sequence

$$0 \to \Sigma^d M \to M \to M/fM \to 0.$$

Therefore,

$$\begin{aligned}
\operatorname{reg}(Q_B CH_G^*) &\leq \operatorname{reg}(CH_G^*) + \sum_{i=1}^{c}(p^{a_i} - 1) \\
&= \operatorname{reg}(CH_G^*) + e^{CH}(G) \\
&\leq e^{CH}(G),
\end{aligned}$$

by the regularity theorem for Chow rings (Theorem 6.5). By the definition of regularity in terms of local cohomology, this says in particular that $H^0_{\mathfrak{m}} CH_G^*$ is concentrated in degrees at most $e^{CH}(G)$. By the previous paragraph, we have shown that $e^{CH}_{\text{indec}}(G) \leq e^{CH}(G)$.

Thus we have shown that $e^{CH}_{\text{prim}}(G) \leq e^{CH}_{\text{indec}}(G) \leq e^{CH}(G) \leq n - c$. The interest of $e^{CH}_{\text{prim}}(G)$ is shown by the following lemma on Hopf algebras, which we will apply to $H = CH_C^*$ and $M = \text{CEss}(CH_G^*)$.

Lemma 12.12 *Let H be a graded connected Hopf algebra over a field F. Let M be a graded comodule over H that is bounded below. If the subspace of primitives $P_H M$ is concentrated in degrees at most e, then the composite*

$$M \to H \otimes_k M \to H \otimes_k M^{\leq e}$$

is injective.

Proof Let x be a homogeneous element of M that is not primitive, say of degree d. Then the image of x under the coproduct $M \to H \otimes M$ has nonzero component in $H^{d-j} \otimes M^j$ for some $j < d$. By induction on d, if we apply the coproduct map enough times, then x has nonzero image in $H \otimes \cdots \otimes H \otimes M^{\leq e}$. Since M is an H-comodule, we have the commutative diagram

$$\begin{array}{ccc}
M & \longrightarrow & H \otimes M \\
\downarrow & & \downarrow {\scriptstyle 1\otimes\Delta} \\
H \otimes M & \xrightarrow{\Delta\otimes 1} & H \otimes H \otimes M.
\end{array}$$

Applying this repeatedly, we conclude that x has nonzero image in $H \otimes M^{\leq e}$. ☐

Thus, our bound on $e^{CH}_{\text{prim}}(G)$ implies that, if we write $e = n - c$, then the restriction map

$$\text{CEss}(CH_G^*) \to CH_C^* \otimes \text{CEss}(CH_G^{\leq e})$$

is injective. Equivalently, the product map

$$CH_G^* \to CH_C^* \otimes CH_G^{\leq e} \times \prod_{\substack{C \subsetneq V \\ V \text{ elem ab}}} CH_{C_G(V)}^*$$

is injective.

For any subgroup H of the p-group G, it is easy to see that $n_H - c_H \leq n_G - c_G$. (A faithful representation V of G over k restricts to a faithful representation of H over k, and so $n_H \leq n_G$. The p-rank c_H of the center of H may be smaller than than of G, but if that happens then we can omit some irreducible summands from $V|_H$ and still get a faithful representation; so we always have $n_H - c_H \leq n_G - c_G$.) Therefore, applying the previous injectivity result to each of the subgroups $H = C_G(V)$, we find that

$$CH_G^* \to \prod_{\substack{V \subset G \\ V \text{ elem ab}}} CH_V^* \otimes CH_{C_G(V)}^{\leq e}$$

is injective. □

13

Calculations

An important test of our machinery is to compute the Chow ring for large classes of finite groups, not just for particular families of well-behaved groups. For example, the Chow groups of the symmetric group S_n are known for any n (Corollary 2.22 and [138, section 11]), but computing Chow rings for more general groups can be hard, even for relatively small groups. The main problem is to compute the Chow ring of a p-group, since the Chow ring (with $\mathbf{F}_p$ coefficients) of any finite group is a summand of the Chow ring of a Sylow p-subgroup. In that spirit, this chapter uses the tools we have developed to compute the Chow ring for all 14 groups of order 16 and all 5 groups of order p^3. Chapter 14 computes the Chow ring for all 15 groups of order 81, and for 13 of the 15 groups of order p^4 with $p \geq 5$. We also compute the Chow ring for some infinite classes of p-groups, including all p-groups with a cyclic subgroup of index p.

Section 13.3 relates the Chow ring of a p-group to the Chow ring of an associated 1-dimensional group. This method simplifies a surprising number of calculations. Leary used the same method to good effect in group cohomology [89, 90].

In Section 13.5, we compute the topological nilpotence degree $d_0(H_G^*)$ of Henn-Lannes-Schwartz and the analogous invariant $d_0(CH_G^*)$ for some small p-groups G. In a sense, computing the cohomology ring or the Chow ring reduces to calculations in degrees at most $d_0(H_G^*)$ or $d_0(CH_G^*)$. In our examples, these numbers turn out to be surprisingly small. Any improvement on the known upper bounds for $d_0(H_G^*)$ or $d_0(CH_G^*)$ would be useful for computations of group cohomology or the Chow ring.

Throughout this chapter, we consider each finite p-group G of exponent e as an algebraic group over any field k of characteristic not p that contains the eth roots of unity. We find that the groups in this chapter have the same mod p Chow ring over all such fields. We write ζ_m to mean a primitive mth root of unity in k.

13.1 The Chow rings of the groups of order 16

We compute the Chow ring for all the groups of order 16 in this section. (There are 14 of them.) We also compute the Chow rings of all 2-groups with a cyclic subgroup of index 2 (the quaternion, dihedral, semidihedral, and modular 2-groups). Previously, Yagita computed the Chow rings of the groups of order 8 [155], and Guillot computed the image of the Chow ring in the $\mathbf{F}_2$-cohomology ring for 10 of the 14 groups of order 16 [64, 65].

Our methods (the regularity Theorem 6.5 and the detection Theorem 12.7) reduce computing the Chow ring for a group of order 16 to computations of the Chow ring in degree at most 1. We write CH_G^* for the Chow ring of G with $\mathbf{F}_2$ coefficients. Since CH_G^1 is just $\mathrm{Hom}(G, k^*)/2$, computing the Chow ring of a group of order 16 is completely algorithmic, and in fact easy to do by hand.

We first compute the Chow ring for the 5 groups of order 8. These are the abelian groups $\mathbf{Z}/8$, $\mathbf{Z}/4 \times \mathbf{Z}/2$, and $(\mathbf{Z}/2)^3$, the quaternion group Q_8, and the dihedral group D_8. The Chow rings of abelian 2-groups are polynomial rings over $\mathbf{F}_2$ (Theorem 2.10). In fact, we compute the Chow ring for some infinite families of 2-groups, namely the quaternion group Q_{2^n} and the dihedral group D_{2^n} with $n \geq 3$. The quaternion groups are the only non-cyclic p-groups of rank 1 [1, proposition IV.6.6].

Lemma 13.1 *Let G be the quaternion group of order 2^n, $n \geq 3$. Then*

$$CH_G^* = \mathbf{F}_2[c_2 V, a, b]/(a^2, ab, b^2),$$

where $|a| = |b| = 1$ and V is the standard representation $G \hookrightarrow SL(2, k)$. The ring CH_G^* has dimension 1 and depth 1 (so it is Cohen-Macaulay), as follows from the Duflot lower bound for depth since G is p-central.

The Chow ring of the quaternion group is Cohen-Macaulay but not Gorenstein. This contrasts with the Benson-Carlson theorem: if the cohomology ring of a finite group is Cohen-Macaulay, then it is Gorenstein [12, vol. 2, theorem 5.18.1].

Proof By definition, G is the group

$$\langle x, y : x^{2^{n-1}} = 1, y^4 = 1, yxy^{-1} = x^{-1}, y^2 = x^{2^{n-2}} \rangle.$$

The group G has a faithful irreducible representation V of dimension 2 over k, $G \hookrightarrow SL(2, k)$, by $x \mapsto \begin{pmatrix} \zeta_{2^{n-1}} & 0 \\ 0 & \zeta_{2^{n-1}}^{-1} \end{pmatrix}$ and $y \mapsto \begin{pmatrix} 0 & 1 \\ -1 & 0 \end{pmatrix}$.

By Theorem 6.5, CH_G^* is generated as a module over $\mathbf{F}_2[c_1 V, c_2 V]$ by elements of degree at most 1, modulo relations in degree at most 2. Since the representation V has trivial determinant, we have $c_1 V = 0$. The abelianization of G is isomorphic to $(\mathbf{Z}/2)^2$. So $CH_G^1 = \mathrm{Hom}(G, k^*)/2$ is the $\mathbf{F}_2$-vector space

$\mathbf{F}_2\{a, b\}$, where we define the homomorphism $a\colon G \to k^*$ as $x \mapsto -1,\ y \mapsto 1$ and $b\colon G \to k^*$ as $x \mapsto 1,\ y \mapsto -1$. Therefore

$$CH^*_G = \mathbf{F}_2[c_2V]\{1, a, b\}/(\text{relations}),$$

where $|a| = |b| = 1$. For a ring R, we write $R\{e_1, \ldots, e_m\}$ to mean the free R-module with basis elements $e_1, \ldots, e_m$. By our computation of CH^1_G, any relations are in degree 2.

The center of G is the group $C = \langle x^{2^{n-2}} \rangle \cong \mathbf{Z}/2$. Since C has rank 1, CH^*_G has depth at least 1 by Corollary 3.21 (the analog of Duflot's theorem for Chow rings). Therefore CH^*_G is a free module over $\mathbf{F}_2[c_2V]$, and so we have computed it as a module:

$$CH^*_G = \mathbf{F}_2[c_2V]\{1, a, b\}.$$

As in the proof of Theorem 12.7, let $e^{CH}(G)$ denote the maximum degree of a generator of CH^*_C as a module over $\mathrm{im}(CH^*_G \to CH^*_C)$. We have $e^{CH}(G) \le n - c = 2 - 1 = 1$, where n is the dimension of a faithful complex representation of G and c is the p-rank of the center. By Theorem 12.7, since the quaternion group G is p-central, it follows that

$$CH^*_G \to CH^*_C \otimes CH^{\le 1}_G$$

is injective. This is another way in which the computation of the Chow ring of G reduces to computations in degrees at most 1. (In the simple example of the quaternion group, we could check this injectivity by hand, since we have already computed CH^*_G as a module. But our purpose is to show how to use our general results to compute Chow rings as easily as possible.)

That restriction map sends $a \mapsto 1 \otimes a$ and $b \mapsto 1 \otimes b$. Therefore a^2 maps to $1 \otimes a^2 = 0$ in $CH^*_C \otimes CH^{\le 1}_G$, and likewise ab and b^2 map to zero. Since the map is injective, it follows that $a^2 = ab = b^2 = 0$ in CH^*_G. Thus we have computed CH^*_G as a ring:

$$CH^*_G = \mathbf{F}_2[a, b, c_2V]/(a^2, ab, b^2). \qquad \square$$

Lemma 13.2 *Let G be the dihedral group of order 2^n, $n \ge 3$. Then*

$$CH^*_G = \mathbf{F}_2[c_1V, c_2V, a]/(a^2 = ac_1V),$$

where $|a| = 1$ and V is the standard representation $G \hookrightarrow GL(2, k)$. The ring CH^*_G has dimension 2 and depth 2 (so it is Cohen-Macaulay, although the Duflot bound gives only that the depth is at least 1).

Proof By definition, G is the group

$$\langle x, y : x^{2^{n-1}} = 1, y^2 = 1, yxy^{-1} = x^{-1} \rangle.$$

The group G has a faithful irreducible representation V of dimension 2 over k, $G \hookrightarrow GL(2,k)$, by $x \mapsto \begin{pmatrix} \zeta_{2^{n-1}} & 0 \\ 0 & \zeta_{2^{n-1}}^{-1} \end{pmatrix}$ and $y \mapsto \begin{pmatrix} 0 & 1 \\ 1 & 0 \end{pmatrix}$.

By Theorem 6.5, CH_G^* is generated as a module over $\mathbf{F}_2[c_1V, c_2V]$ by elements of degree at most 1, modulo relations in degree at most 2. The abelianization of G is isomorphic to $(\mathbf{Z}/2)^2$. So $CH_G^1 = \mathrm{Hom}(G, k^*)/2$ is $\mathbf{F}_2\{a, b\}$, where we define $a\colon G \to k^*$ as $x \mapsto -1$, $y \mapsto 1$ and $b\colon G \to k^*$ as $x \mapsto 1$, $y \mapsto -1$. By computing the determinant of V, we find that $c_1V = b$. Therefore

$$CH_G^* = \mathbf{F}_2[c_1V, c_2V]\{1, a\}/(\text{relations}),$$

where $|a| = 1$ and any module relations are in degree 2.

There are two conjugacy classes of maximal elementary abelian 2-subgroups in the dihedral group G, $A_1 = \langle x^{2^{n-2}}, y\rangle \cong (\mathbf{Z}/2)^2$ and $A_2 = \langle x^{2^{n-2}}, xy\rangle \cong (\mathbf{Z}/2)^2$. For A_1, define $t_1, u_1\colon G \to k^*$ by $t_1\colon x^{2^{n-2}} \mapsto -1$, $y \mapsto 1$ and $u_1\colon x^{2^{n-2}} \mapsto 1$, $y \mapsto -1$, so that $CH_{A_1}^* = \mathbf{F}_2[t_1, u_1]$. The restriction $CH_G^* \to CH_{A_1}^*$ sends $a \mapsto 0$ and $c_1V = b \mapsto u_1$. Also, the representation V restricted to A_1 is $T_1 \oplus (T_1 \otimes U_1)$, where we write T_1 and U_1 for the 1-dimensional representations of A_1 with Chern classes t_1 and u_1, respectively. So the total Chern class $c(V)|_{A_1}$ is equal to $(1+t_1)(1+t_1+u_1) = 1 + u_1 + t_1(t_1+u_1)$. In particular, $c_2V \mapsto t_1(t_1+u_1)$. From our description of CH_G^* as a module, we know that CH_G^2 is spanned as an $\mathbf{F}_2$-vector space by c_2V, c_1^2V, ac_1V. Since these restrict on A_1 as $c_2V \mapsto t_1(t_1+u_1)$, $c_1^2V \mapsto u_1^2$, $ac_1V \mapsto 0$ in $\mathbf{F}_2[c_1V, c_2V]$, the only possible module relation in CH_G^* is that ac_1V may be zero.

But restricting to the other elementary abelian 2-subgroup A_2 shows that $ac_1V \neq 0$ in CH_G^*. (If we define $t_2\colon x^{2^{n-2}} \mapsto -1$, $xy \mapsto 1$ and $u_2\colon x^{2^{n-2}} \mapsto 1$, $xy \mapsto -1$, then we compute that $a \mapsto u_2$, $c_1V = b \mapsto u_2$, and $c_2V \mapsto t_2(t_2+u_2)$, and so $ac_1V \mapsto u_2^2 \neq 0$.) Thus, for the dihedral group $G = D_{2^n}$, we have computed CH_G^* as a module:

$$CH_G^* = \mathbf{F}_2[c_1V, c_2V]\{1, a\},$$

where $|a| = 1$. In particular, the ring CH_G^* has depth 2, although the Duflot bound gives only that the depth is at least 1.

At the same time, we showed that CH_G^2 is detected on A_1 and A_2, and so we can compute the ring structure. We find that $a^2 - ac_1V$ restricts to zero on A_1 and A_2, and so it is zero in CH_G^*. Thus

$$CH_G^* = \mathbf{F}_2[c_1V, c_2V, a]/(a^2 = ac_1V),$$

where $|a| = 1$. □

Now we compute the Chow rings of the 14 groups of order 16. There are five abelian groups, $\mathbf{Z}/16$, $\mathbf{Z}/8 \times \mathbf{Z}/2$, $(\mathbf{Z}/4)^2$, $\mathbf{Z}/4 \times (\mathbf{Z}/2)^2$, and $(\mathbf{Z}/2)^4$, for which the Chow ring is a polynomial ring. There are two other product groups, $D_8 \times \mathbf{Z}/2$ and $Q_8 \times \mathbf{Z}/2$, for which the Chow ring is determined by the Chow Künneth formula, Lemma 2.12 (which applies to any product with an abelian group, among others). We have already computed the Chow ring for the dihedral group D_{16} and the quaternion group Q_{16}. The remaining groups are the modular group of order 16, the semidihedral group of order 16, the central product $D_8 * C_4$, the split metacyclic group $\mathbf{Z}/4 \ltimes \mathbf{Z}/4$, and one more.

Some of these are important examples of p-groups. The only nonabelian p-groups with a cyclic subgroup of index p are the modular p-group Mod_{p^n}, the dihedral group D_{2^n}, the semidihedral group SD_{2^n}, and the quaternion group Q_{2^n} [4, section 23.4]. Here the *modular p-group* Mod_{p^n} is the split extension

$$\mathrm{Mod}_{p^n} = \langle x, y : x^{p^{n-1}} = 1, y^p = 1, yxy^{-1} = x^{p^{n-2}+1} \rangle,$$

where we assume that $n \geq 3$ for p odd and $n \geq 4$ for $p = 2$ (since $\mathrm{Mod}_8 \cong D_8$). The *semidihedral group* SD_{2^n} is the split extension

$$SD_{2^n} = \langle x, y : x^{2^{n-1}} = 1, y^2 = 1, yxy^{-1} = x^{2^{n-2}-1} \rangle$$

for $n \geq 4$. We have computed the Chow rings of the quaternion and dihedral groups, and we now compute the Chow rings for the remaining 2-groups with a cyclic subgroup of index 2. The modular p-group for p odd is handled in Lemma 13.8.

We use the numbering of p-groups from the Small Groups library in GAP, a free group theory program [46]. This is also the numbering used in Green and King's calculations of the cohomology of p-groups [52].

Lemma 13.3 *Let G be the modular 2-group of order 2^n, $n \geq 4$. For $n = 4$, this is #6 of the groups of order 16 in the Small Groups library [52]. Then*

$$CH_G^* = \mathbf{F}_2[c_1V, c_2V, a]/(a^2, ac_1V),$$

where $|a| = 1$ and V is the standard representation $G \hookrightarrow GL(2, k)$. The ring CH_G^* has dimension 2 and depth 1, which agrees with the Duflot lower bound.

Proof The group G has a faithful irreducible representation V of dimension 2 over k, $G \hookrightarrow GL(2, k)$, by $x \mapsto \begin{pmatrix} \zeta_{2^{n-1}} & 0 \\ 0 & \zeta_{2^{n-1}}^{2^{n-2}+1} \end{pmatrix}$ and $y \mapsto \begin{pmatrix} 0 & 1 \\ 1 & 0 \end{pmatrix}$.

By Theorem 6.5, CH_G^* is generated as a module over $\mathbf{F}_2[c_1V, c_2V]$ by elements of degree at most 1, modulo relations in degree at most 2. The abelianization of G is isomorphic to $\mathbf{Z}/2^{n-2} \times \mathbf{Z}/2$. So $CH_G^1 = \mathrm{Hom}(G, k^*)/2$ is $\mathbf{F}_2\{a, b\}$, where we define $a\colon G \to k^*$ as $x \mapsto \zeta_{2^{n-2}}$, $y \mapsto 1$ and $b\colon G \to k^*$ as $x \mapsto 1$, $y \mapsto -1$. By computing the determinant of V, we find that $c_1V = a + b$. Therefore

$$CH_G^* = \mathbf{F}_2[c_1V, c_2V]\{1, a\}/(\text{relations}),$$

where $|a| = 1$ and any module relations are in degree 2.

The center of G is $\langle x^2\rangle \cong \mathbf{Z}/2^{n-2}$, and so $C := Z(G)[p]$ is $\langle x^{2^{n-2}}\rangle \cong \mathbf{Z}/2$. We have $e^{CH}(G) \le n - c = 2 - 1 = 1$, where n is the dimension of a faithful complex representation of G and c is the p-rank of the center. By Theorem 12.7,

$$CH_G^* \to \prod_{\substack{C\subset V\\ V\text{elem ab}}} CH_V^* \otimes CH_{C_G(V)}^{\le 1}$$

is injective. There are two conjugacy classes of elementary abelian subgroups V that contain C, namely C and $A := \langle x^{2^{n-2}}, y\rangle \cong (\mathbf{Z}/2)^2$. The centralizer of A is $C_G(A) = \langle x^2, y\rangle \cong \mathbf{Z}^{2^{n-2}} \times \mathbf{Z}/2$. Since we are considering Chow groups with $\mathbf{F}_2$ coefficients, we compute that $a \in CH_G^1$ restricts to zero in $CH^1_{C_G(A)}$ and hence in $CH_A^* \otimes CH^{\le 1}_{C_G(A)}$. Also, a restricts to $1 \otimes a$ and c_1V to $1 \otimes c_1V$ in $CH^*V \otimes CH_G^{\le 1}$. It follows that a^2 and ac_1V restrict to zero in both $CH_C^* \otimes CH_G^{\le 1}$ and in $CH_A^* \otimes CH^{\le 1}_{C_G(A)}$. By the injectivity statement above, it follows that $a^2 = ac_1V = 0$ in CH_G^2.

By our description of CH_G^* as a module, we know that CH_G^2 is spanned by a^2, ac_1V, c_2V as an $\mathbf{F}_2$-vector space, hence just by c_2V. Here c_2V restricts to t^2 in $CH_C^* = \mathbf{F}_2[t]$, and so c_2V is not zero. This completes the determination of CH_G^* as a module over $\mathbf{F}_2[c_1V, c_2V]$, since we knew that any module relations were in degree 2:

$$CH_G^* = \mathbf{F}_2[c_1V, c_2V]\{1\} \oplus \mathbf{F}_2[c_1V, c_2V]/(c_1V)\{a\}.$$

Since $a^2 = 0$, the ring structure is determined:

$$CH_G^* \cong \mathbf{F}_2[c_1V, c_2V, a]/(a^2, ac_1V),$$

where $|a| = 1$. □

It happens that the semidihedral and modular 2-groups have isomorphic mod 2 Chow rings. With our approach, the calculations in the two cases are almost identical.

Lemma 13.4 *Let G be the semidihedral group of order 2^n,*

$$SD_{2^n} = \langle x, y : x^{2^{n-1}} = 1, y^2 = 1, yxy^{-1} = x^{2^{n-2}-1}\rangle,$$

with $n \ge 4$. For $n = 4$, this is #8 of the groups of order 16 in the Small Groups library [52]. Then

$$CH_G^* = \mathbf{F}_2[c_1V, c_2V, a]/(a^2, ac_1V),$$

where $|a| = 1$ and V is the standard representation $G \hookrightarrow GL(2,k)$. The ring CH_G^* has dimension 2 and depth 1, which agrees with the Duflot lower bound.

Proof The group G has a faithful irreducible representation V of dimension 2 over k, $G \hookrightarrow GL(2,k)$, by $x \mapsto \begin{pmatrix} \zeta_{2^{n-1}} & 0 \\ 0 & \zeta_{2^{n-1}}^{2^{n-2}-1}\end{pmatrix}$ and $y \mapsto \begin{pmatrix} 0 & 1 \\ 1 & 0\end{pmatrix}$.

By Theorem 6.5, CH_G^* is generated as a module over $\mathbf{F}_2[c_1V, c_2V]$ by elements of degree at most 1, modulo relations in degree at most 2. The abelianization of G is isomorphic to $(\mathbf{Z}/2)^2$. So $CH_G^1 = \mathrm{Hom}(G, k^*)/2$ is $\mathbf{F}_2\{a, b\}$, where we define $a\colon G \to k^*$ as $x \mapsto -1,\ y \mapsto 1$ and $b\colon G \to k^*$ as $x \mapsto 1,\ y \mapsto -1$. By computing the determinant of V, we find that $c_1V = a + b$. Therefore

$$CH_G^* = \mathbf{F}_2[c_1V, c_2V]\{1, a\}/(\text{relations}),$$

where $|a| = 1$ and any module relations are in degree 2.

The center of G is $\langle x^{2^{n-2}} \rangle \cong \mathbf{Z}/2$, and so $C := Z(G)[p]$ is $\langle x^{2^{n-2}} \rangle \cong \mathbf{Z}/2$. We have $e^{CH}(G) \leq n - c = 2 - 1 = 1$, where n is the dimension of a faithful complex representation of G and c is the p-rank of the center. By Theorem 12.7,

$$CH_G^* \to \prod_{\substack{C \subset V \\ V \text{elem ab}}} CH_V^* \otimes CH_{C_G(V)}^{\leq 1}$$

is injective. There are two conjugacy classes of elementary abelian subgroups V that contain C, namely C and $A := \langle x^{2^{n-2}}, y \rangle \cong (\mathbf{Z}/2)^2$. The centralizer $C_G(A)$ is equal to A. We compute that $a \in CH_G^1$ restricts to zero in CH_A^1 and hence in $CH_A^* \otimes CH_{C_G(A)}^{\leq 1}$. Also, a restricts to $1 \otimes a$ and c_1V to $1 \otimes c_1V$ in $CH^*V \otimes CH_G^{\leq 1}$. It follows that a^2 and ac_1V restrict to zero in both $CH_C^* \otimes CH_G^{\leq 1}$ and in $CH_A^* \otimes CH_{C_G(A)}^{\leq 1}$. By the injectivity statement above, it follows that $a^2 = ac_1V = 0$ in CH_G^2.

By our description of CH_G^* as a module, we know that CH_G^2 is spanned by a^2, ac_1V, c_2V as an $\mathbf{F}_2$-vector space, hence just by c_2V. Here c_2V restricts to t^2 in $CH_C^* = \mathbf{F}_2[t]$, and so c_2V is not zero. This completes the determination of CH_G^* as a module over $\mathbf{F}_2[c_1V, c_2V]$, since we knew that any module relations were in degree 2:

$$CH_G^* = \mathbf{F}_2[c_1V, c_2V]\{1\} \oplus \mathbf{F}_2[c_1V, c_2V]/(c_1V)\{a\}.$$

Since $a^2 = 0$, the ring structure is determined:

$$CH_G^* \cong \mathbf{F}_2[c_1V, c_2V, a]/(a^2, ac_1V),$$

where $|a| = 1$. □

We now compute the Chow ring for the three remaining groups of order 16.

Lemma 13.5 *For any $n \geq 4$, let G be the central product group $D_8 * C_{2^{n-2}} = (D_8 \times \mathbf{Z}/2^{n-2})/(\mathbf{Z}/2)$. For $n = 4$, this is #13 of the groups of order 16 in the Small Groups library [52]. Then*

$$CH_G^* =$$

$$\mathbf{F}_2[c_1V, c_2V, a, b]/(a^2 = ac_1V, ab = c_1^2V + ac_1V + bc_1V, b^2 = bc_1V),$$

where $|a| = |b| = 1$ and V is the standard representation $G \hookrightarrow GL(2, k)$. The ring CH_G^* has dimension 2 and depth 2, whereas the Duflot bound gives only that the depth is ≥ 1.

Proof We can write

$$G = \langle x, y, z : x^4 = 1, y^2 = 1, yxy^{-1} = x^{-1}, z^{2^{n-3}} = x^2, zx = xz, zy = yz \rangle.$$

The abelianization of G is isomorphic to $(\mathbf{Z}/2)^2 \times \mathbf{Z}/2^{n-3}$. So $CH_G^1 = \mathrm{Hom}(G, k^*)/2$ is isomorphic to $\mathbf{F}_2\{a, b, c\}$, where

$$a\colon x \mapsto -1, y \mapsto 1, z \mapsto 1$$

$$b\colon x \mapsto 1, y \mapsto -1, z \mapsto 1$$

$$c\colon x \mapsto 1, y \mapsto 1, z \mapsto \zeta_{2^{n-3}}.$$

Let $H = \langle y, z \rangle \cong \mathbf{Z}/2 \times \mathbf{Z}/2^{n-2} \subset G$, and let $\beta\colon H \to k^*$ be the representation $y \mapsto 1, z \mapsto \zeta_2^{n-2}$. Then G has a faithful irreducible representation of dimension 2 over k given by $V = \mathrm{Ind}_H^G \beta$. In a suitable basis for V, G acts by $x \mapsto \begin{pmatrix} 0 & -1 \\ 1 & 0 \end{pmatrix}, y \mapsto \begin{pmatrix} 1 & 0 \\ 0 & -1 \end{pmatrix}, z \mapsto \begin{pmatrix} \zeta_{2^{n-2}} & 0 \\ 0 & \zeta_{2^{n-2}} \end{pmatrix}$.

By theorem 6.5, CH_G^* is generated as a module over $\mathbf{F}_2[c_1V, c_2V]$ by elements of degree at most 1, modulo relations in degree at most 2. By computing the determinant of V, we find that $c_1V = b + c$. Using our computation of CH_G^1, we have

$$CH_G^* = \mathbf{F}_2[c_1V, c_2V]\{1, a, b\}/(\text{relations}),$$

where $|a| = |b| = 1$ and any module relations are in degree 2.

The center of G is $\langle z \rangle \cong \mathbf{Z}/2^{n-2}$, and so $C := Z(G)[2]$ is $\langle z^{2^{n-3}} \rangle \cong \mathbf{Z}/2$. We have $e^{CH}(G) \leq n - c = 2 - 1 = 1$, where n is the dimension of a faithful complex representation of G and c is the p-rank of the center. By Theorem 12.7,

$$CH_G^* \to \prod_{\substack{C \subset V \\ V \text{elem ab}}} CH_V^* \otimes CH_{C_G(V)}^{\leq 1}$$

is injective. There are four conjugacy classes of elementary abelian subgroups V that contain C, namely C and $A_1 = \langle y, z^{2^{n-3}} \rangle$, $A_2 = \langle xz^{2^{n-4}}, z^{2^{n-3}} \rangle$, and $A_3 = \langle xy, z^{2^{n-3}} \rangle$, where $A_i \cong (\mathbf{Z}/2)^2$ for $i = 1, 2, 3$.

We know that CH_G^2 is spanned by $c_2V, c_1^2V, ac_1V, bc_1V$ as an $\mathbf{F}_2$-vector space. We compute that these four elements have linearly independent images under the restriction map

$$CH_G^2 \to CH_{A_1}^2 \oplus CH_{A_2}^2 \oplus CH_{A_3}^2.$$

So we have computed CH^2_G. In fact, we have computed CH^*_G as a module, since we showed that any module relations were in degree 2:

$$CH^*_G = \mathbf{F}_2[c_1V, c_2V]\{1, a, b\}.$$

Thus CH^*_G is Cohen-Macaulay.

Since we showed that CH^2_G injects into $CH^2_{A_1} \oplus CH^2_{A_2} \oplus CH^2_{A_3}$, we can compute a^2, ab, b^2 in terms of our basis for CH^2_G by restricting to these three elementary abelian subgroups. We find that $a^2 = ac_1V$, $ab = c_1^2V + ac_1V + bc_1V$, and $b^2 = bc_1V$. Thus the central product $G = D_8 * C_{2^{n-2}}$ has Chow ring

$$CH^*_G = \mathbf{F}_2[c_1V, c_2V, a, b]/(a^2 = ac_1V, ab = c_1^2V + ac_1V + bc_1V, b^2 = bc_1V),$$

where $|a| = |b| = 1$. □

The next group has Chow ring a polynomial ring, surprisingly. By contrast, its $\mathbf{F}_2$-cohomology ring has dimension 3 and depth only 2 [52]. This is also the first p-group we have seen for which the inequality $\operatorname{reg}(CH^*_G) \leq 0$ of Theorem 6.5 is strict; namely, CH^*_G has regularity -1.

Our notation for commutators is that $[x, y] = xyx^{-1}y^{-1}$.

Lemma 13.6 *Let*

$$G = \langle x, y, z : x^4 = 1, y^2 = 1, z^2 = 1, [x, y] = z, xz = zx, yz = zy\rangle,$$

#3 of the groups of order 16 in the Small Groups library [52]. Then

$$CH^*_G = \mathbf{F}_2[c_1W, c_2W, c_1\beta].$$

Here G has a faithful k-representation $W \oplus \beta$ of dimension 3, with W irreducible of dimension 2 and β of dimension 1. The ring CH^*_G has dimension 3 and depth 3, whereas the Duflot bound gives only that the depth is ≥ 2.

Proof Let $H = \langle x, z\rangle \cong \mathbf{Z}/4 \times \mathbf{Z}/2 \subset G$. Let $\alpha\colon H \to k^*$ be the representation $x \mapsto 1$, $z \mapsto -1$. Let $W = \operatorname{Ind}_H^G \alpha$. The kernel of the 2-dimensional representation W of G is $\langle x^2\rangle \cong \mathbf{Z}/2$. Therefore, the representation $W \oplus \beta$ of G is faithful if we define $\beta\colon G \to k^*$ by $x \mapsto \zeta_4$, $y \mapsto 1$, $z \mapsto 1$.

Even though G does not have a faithful representation of dimension 2, Theorem 6.5 works just as well as for the other groups of order 16, because $\sigma(\mathbf{F}_2[c_1W, c_2W, c_1\beta]) = 1$. We deduce that CH^*_G is generated as a module over $\mathbf{F}_2[c_1W, c_2W, c_1\beta]$ by elements of degree at most 1, modulo relations in degree at most 2.

The abelianization of G is isomorphic to $\mathbf{Z}/4 \times \mathbf{Z}/2$. So $CH^1_G = \operatorname{Hom}(G, k^*)/2$ is isomorphic to $\mathbf{F}_2\{a, b\}$, where $a\colon x \mapsto \zeta_4$, $y \mapsto 1$, $z \mapsto 1$ and $b\colon x \mapsto 1$, $y \mapsto -1$, $z \mapsto 1$. By computing determinants, we find that $c_1\beta = a$

and $c_1 W = 2a + b = b$. So we do not need any module generators in degree 1: we have

$$CH_G^* = \mathbf{F}_2[c_1 W, c_2 W, c_1 \beta]/(\text{relations}),$$

where any module relations are in degree 2.

So CH_G^2 is spanned by $c_1^2 W, c_2 W, c_1^2 \beta, c_1 W c_1 \beta$ as an $\mathbf{F}_2$-vector space. We compute that their restrictions to the elementary abelian subgroup $K = \langle x^2, y, z\rangle \cong (\mathbf{Z}/2)^3$ are linearly independent. So there are no module relations in dimension 2, and we have shown that

$$CH_G^* = \mathbf{F}_2[c_1 W, c_2 W, c_1 \beta].$$

(Alternatively, one can show that there are no relations among these generators without computing any restrictions. Indeed, G has p-rank 3, and so the ring CH_G^* has dimension 3 by Yagita's theorem (Theorem 8.10). But the quotient ring of $\mathbf{F}_2[c_1 W, c_2 W, c_1 \beta]$ by any nonzero ideal would have dimension at most 2.) □

We now find the Chow ring for the last of the 14 groups of order 16.

Lemma 13.7 *Let G be the split metacyclic group $\mathbf{Z}/4 \ltimes \mathbf{Z}/4$,*

$$G = \langle x, y : x^4 = 1, y^4 = 1, yxy^{-1} = x^{-1}\rangle,$$

#4 of the groups of order 16 in the Small Groups library [52]. Then

$$CH_G^* = \mathbf{F}_2[c_2 W, c_1 \beta, a]/(a^2),$$

where $|a| = 1$. Here G has a faithful k-representation $W \oplus \beta$ of dimension 3, with W irreducible of dimension 2 and β of dimension 1. The ring CH_G^* has dimension 2 and depth 2, which follows from the Duflot bound because G is p-central.

Proof Let $H = \langle x^2, y\rangle \cong \mathbf{Z}/2 \times \mathbf{Z}/4 \subset G$. Let $\alpha : H \to k^*$ be the representation $x^2 \mapsto -1, y \mapsto 1$. Let $W = \operatorname{Ind}_H^G \alpha$. The kernel of the 2-dimensional representation W of G is $\langle y^2\rangle \cong \mathbf{Z}/2$. Therefore, the representation $W \oplus \beta$ of G is faithful if we define $\beta : G \to k^*$ by $x \mapsto 1, y \mapsto \zeta_4$.

By Theorem 6.5, CH_G^* is generated as a module over $\mathbf{F}_2[c_1 W, c_2 W, c_1 \beta]$ by elements of degree at most 1, modulo relations in degree at most 2. The abelianization of G is isomorphic to $\mathbf{Z}/2 \times \mathbf{Z}/4$. So $CH_G^1 = \operatorname{Hom}(G, k^*)/2$ is isomorphic to $\mathbf{F}_2\{a, b\}$, where $a : x \mapsto -1, y \mapsto 1$ and $b : x \mapsto 1, y \mapsto \zeta_4$. By computing determinants, $c_1 \beta = b$ and $c_1 W = 2b = 0$. Thus

$$CH_G^* = \mathbf{F}_2[c_2 W, c_1 \beta]\{1, a\}/(\text{relations}),$$

where any module relations are in degree 2.

The center of G is $C = \langle x^2, y^2 \rangle \cong (\mathbf{Z}/2)^2$, and G is p-central. By the Duflot bound 3.21, it follows that CH_G^* is Cohen-Macaulay. That is:

$$CH_G^* = \mathbf{F}_2[c_2 W, c_1 \beta]\{1, a\}.$$

By Theorem 12.7,

$$CH_G^* \to CH_C^* \otimes CH_G^{\leq 1}$$

is injective. Under this map, $a \mapsto 1 \otimes a$ and so $a^2 \mapsto 1 \otimes a^2 = 0 \in CH_C^* \otimes CH_G^{\leq 1}$. Therefore $a^2 = 0$ in CH_G^*. This determines the ring structure:

$$CH_G^* = \mathbf{F}_2[c_2 W, c_1 \beta, a]/(a^2),$$

where $|a| = 1$. □

13.2 The modular p-group

In this section and Section 13.4, we compute the Chow rings for the groups of order p^3, for an odd prime number p. These computations were first made by Yagita. (Yagita's original proof had a gap: [155, lemma 2.8] assumes that the geometric and topological filtrations are the same. That has now been fixed [156, lemma 3.7].) Besides the abelian groups $\mathbf{Z}/p^3$, $\mathbf{Z}/p^2 \times \mathbf{Z}/p$, and $(\mathbf{Z}/p)^3$, there are two nonabelian groups of order p^3, the modular p-group Mod_{p^3} of exponent p^2 and the extraspecial group p_+^{1+2} of order p^3 and exponent p (the group of strictly upper triangular 3×3 matrices over $\mathbf{F}_p$).

More generally, we compute the Chow ring for the modular group of order p^n. Combined with the results of Section 13.1, this completes the calculation of the Chow ring for all p-groups with a cyclic subgroup of index p.

Lemma 13.8 *Let $G = \mathrm{Mod}_{p^n}$ be the modular p-group*

$$\langle x, y : x^{p^{n-1}} = 1, y^p = 1, yxy^{-1} = x^{p^{n-2}+1} \rangle,$$

for p an odd prime number and $n \geq 3$. Then

$$CH_G^* = \mathbf{F}_p[b, c_p V, x_1, \ldots, x_{p-1}]/(bx_i = 0, x_i x_j = 0)$$

where $|b| = 1$, $|x_i| = i$, and the relations involve all $i, j \in \{1, \ldots, p-1\}$. Here V is a faithful k-representation of G with dimension p. The ring CH_G^* has dimension 2 and depth 1, which agrees with the Duflot bound. Finally, $d_0(CH_G^*) = 1$.

Proof The abelianization of G is isomorphic to $\mathbf{Z}/p^{n-2} \times \mathbf{Z}/p$. So $CH_G^1 = \mathrm{Hom}(G, k^*)/p$ is equal to $\mathbf{F}_p\{a, b\}$, where we define $a \colon G \to k^*$ by $x \mapsto \zeta_{p^{n-2}}$, $y \mapsto 1$ and $b \colon x \mapsto 1$, $y \mapsto \zeta_p$.

The center of G is $\langle x^p \rangle \cong \mathbf{Z}/p^{n-2}$. So the group $C = Z(G)[p]$ is $\langle x^{p^{n-2}} \rangle \cong \mathbf{Z}/p$. There is only one conjugacy class of elementary abelian subgroups strictly containing C, namely $A = \langle x^{p^{n-2}}, y \rangle \cong (\mathbf{Z}/p)^2$.

Let H be the subgroup $\langle x \rangle \cong \mathbf{Z}/p^{n-1}$ of G, and let $\alpha \colon H \to k^*$ be the representation $x \mapsto \zeta_{p^{n-1}}$. For any integer i, the induced representation $V_i := \mathrm{Ind}_H^G \alpha^{\otimes i}$ of G has restriction to H given by

$$\begin{aligned}(\mathrm{Ind}_H^G \alpha^{\otimes i})|_H &= \alpha^{\otimes i} \oplus y\alpha^{\otimes i} \cdots \oplus y^{p-1}\alpha^{\otimes i} \\ &= \alpha^{\otimes i} \oplus \alpha^{\otimes (p^{n-2}+1)i} \oplus \cdots \alpha^{\otimes (p-1)(p^{n-2}+1)i}\end{aligned}$$

For $i \not\equiv 0 \pmod{p}$, these p summands are all non-isomorphic as representations of H, and so V_i is an irreducible representation of G of dimension p. Moreover, the formula for the restriction to H shows that the irreducible representations V_i are non-isomorphic for $1 \leq i \leq p^{n-2}$ with $i \not\equiv 0 \pmod{p}$. Using that the sum of the squares of the dimensions of the irreducible representations is equal to the order of G, we find that these $p^{n-2}(p-1)$ representations of dimension p together with the p^{n-1} representations of dimension 1 are all the irreducible representations of G over k.

For an integer j prime to the order n of a group G, the Adams operation $\psi^j V$ of a representation V of G over the field $\mathbf{Q}(\mu_n)$ is the representation obtained by applying the automorphism $\zeta_n \mapsto \zeta_n^j$ of $\mathbf{Q}(\mu_n)$ [77, section 4.6]. Moreover, every complex representation of G can be defined over $\mathbf{Q}(\mu_n)$. Therefore, induction of representations commutes with the Adams operation ψ^j when j is prime to the order of the groups involved. Since $\psi^j(\alpha) = \alpha^{\otimes j}$, it follows that $V_j \cong \psi^j(V_1)$ for $j \not\equiv 0 \pmod{p}$. Our assumption on the field k implies that the classification of representations over k is the same as over $\mathbf{Q}(\mu_n)$, and so we have the same isomorphism of representations over k. Therefore, the Chern classes of V_i in CH_G^* are polynomials in those of V_1. By Theorem 2.25, the Chow ring CH_G^* in degrees at most p is generated by the Chern classes of representations. Therefore, CH_G^* in degrees at most p is generated by $a, b, c_1 V_1, \ldots, c_p V_1$. We compute that $c_1 V_1 = a$, and so CH_G^* in degrees at most p is generated by $a, b, c_2 V_1, \ldots, c_p V_1$.

Lemma 13.9 *Let p be a prime number, and let k be a field of characteristic not p that contains the pth roots of unity. Let H be a normal subgroup of index p in an affine algebraic group G over k. Let $b \in CH_G^1$ be the pullback of a generator of $CH_{G/H}^1 \cong \mathbf{F}_p$. Let α be a 1-dimensional representation of H, and let V be the induced representation of G. Then, in the Chow ring of G with $\mathbf{F}_p$ coefficients,*

$$c_j V = \mathrm{tr}_H^G y_j$$

for $1 \leq j \leq p-2$, while

$$c_{p-1} V = \mathrm{tr}_H^G (t_1 \cdots t_{p-1}) - b^{p-1}.$$

Here $t_1 = c_1(\alpha)$ in CH^1_H, $t_1, \ldots, t_p$ are the conjugates of t_1 by the elements of $G/H \cong \mathbf{Z}/p$, and y_j for $1 \le j \le p-1$ is the sum of a set of representatives for the free action of $\mathbf{Z}/p$ on the set of monomials $t_{i_1} \ldots t_{i_j}$ with $1 \le i_1 < \cdots < i_j \le p$.

Proof It suffices to prove these formulas in the universal case where G is the wreath product $\mathbf{Z}/p \wr G_m$ and H is the subgroup $(G_m)^p$. In that case, they follow from the calculation of CH^*_G in Lemma 2.21. Namely, that calculation shows that CH^*_G is detected on the subgroups H and $\mathbf{Z}/p \times G_m \subset G$, and the equalities are easily checked on restriction to those two subgroups. Alternatively, Lemma 2.21 shows that the mod p Chow ring of $G = \mathbf{Z}/p \wr G_m$ injects into the $\mathbf{F}_p$-cohomology of G, and then these formulas in the Chow ring follow from Evens's analogous formulas in cohomology [41]. □

For the modular group G, with normal subgroup $H \cong \mathbf{Z}/p^{n-1}$, the quotient group $\mathbf{Z}/p$ acts trivially on $CH^1_H = \mathbf{F}_p\{t\}$, where $t\colon H \to k^*$ sends $x \mapsto \zeta_{p^{n-1}}$. By Lemma 13.9, we have

$$c_j(V_1) = \frac{1}{p}\binom{p}{j}\operatorname{tr}^G_H t^j$$

for $1 \le j \le p-2$ and

$$c_{p-1}(V_1) = \operatorname{tr}^G_H t^{p-1} - b^{p-1}.$$

Let $x_i = \operatorname{tr}^G_H t^i$ for $i = 1, \ldots, p-1$. By these formulas, the ring CH^*_G in degrees $\le p$ is generated by b, $x_1 = a, x_2, \ldots, x_{p-1}$, and $c_p(V_1)$.

Since the transfer $\operatorname{tr}^G_H\colon CH^*_H \to CH^*_G$ is CH^*_G-linear, we have $x_i w = 0$ for any $1 \le i \le p-1$ and any $w \in CH^*_G$ that restricts to zero on H. For example, $b|_H = 0$, and so $bx_i = 0$ for all i. Also, $x_j|_H = \sum_{g \in \mathbf{Z}/p} g(t^i) = pt^i = 0$, using that $G/H = \mathbf{Z}/p$ acts trivially on $CH^*_H = \mathbf{F}_p[t]$. So $x_i x_j = 0$ for all $i, j \in \{1, \ldots, p-1\}$. Thus CH^*_G is spanned in degrees $\le p$ as an $\mathbf{F}_p$-vector space by:

$$1 \quad \begin{matrix} b & b^2 & \cdots & b^{p-1} & b^p \\ x_1 & x_2 & \cdots & x_{p-1} & c_p V_1 \end{matrix}.$$

Let $A = \langle x^{p^{n-2}}, y\rangle \cong (\mathbf{Z}/p)^2 \subset G$. We have $CH^1_A = \mathbf{F}_p\{t, u\}$, where $t\colon A \to k^*$ takes $x^{p^{n-2}} \mapsto \zeta_p$, $y \mapsto 1$ and $u\colon A \to k^*$ takes $x^{p^{n-2}} \to 1$, $y \mapsto \zeta_p$. We compute that $x_1, \ldots, x_{p-1}$ restrict to zero in $CH^*_A = \mathbf{F}_p[t, u]$, while $b \mapsto u$ and $c_p(V_1) \mapsto t^p - tu^{p-1}$. So the only possible relation in degrees $\le p$ beyond those found so far is that some of $x_1, \ldots, x_{p-1}$ might be zero. In fact, $x_1, \ldots, x_{p-1}$ are all nonzero, which implies that a basis for CH^*_G in degrees $\le p$ is given by the elements listed in the previous paragraph.

One way to show that the classes $x_i = \operatorname{tr}^G_H t^i$ for $i = 1, \ldots, p-1$ are nonzero in CH^*_G is to observe that they have nonzero image in $H^*(G, \mathbf{Z})/p \subset H^*(G, \mathbf{F}_p)$, by Thomas's calculation of $H^*(G, \mathbf{Z})$ [135, p. 74].

(In the special case $G = \mathrm{Mod}_{p^3}$, this follows from the earlier calculation of $H^*(G, \mathbf{Z})$ by Lewis [97, theorem 5.2].) But it seems preferable to give a direct proof using Chow rings. We will prove that $x_1, \ldots, x_{p-1}$ are nonzero in CH_G^* using the idea of the topological nilpotence degree from Section 12.2. Also, we know from Theorem 12.7 that $d_0(CH_G^*)$ is at most $p-1$. Given that, our calculation will simultaneously show that $d_0(CH_G^*) = 1$.

To begin, the element x_1 is nonzero in CH_G^1, because $CH_G^1 = \mathrm{Hom}(G, k^*)/p$ is isomorphic to $(\mathbf{F}_p)^2$ and CH_G^1 is generated by b and x_1. Since $C \subset H \subset G$, we have a pullback diagram

$$\begin{array}{ccc} C \times H & \longrightarrow & H \\ \downarrow & & \downarrow \\ C \times G & \longrightarrow & G. \end{array}$$

Because pushforward commutes with pullback (Lemma 2.16), we have $(\mathrm{tr}_H^G t^i)|_{C\times G} = \mathrm{tr}_{C\times H}^{C\times G}(t^i|_{C\times H})$ for all $i \geq 0$. The restriction map from $CH_H^* = \mathbf{F}_p[t]$ to CH_C^* is an isomorphism. The pullback $CH_H^* \to CH_C^* \otimes_{\mathbf{F}_p} CH_H^*$ sends t to $t \otimes 1 + 1 \otimes t$. So the image of t^i in $CH_C^* \otimes_{\mathbf{F}_p} CH^{\leq 1}H$ is

$$t^i \otimes 1 + it^{i-1} \otimes t.$$

Here $\mathrm{tr}_H^G 1 = 0$, but $\mathrm{tr}_H^G t = x_1$, which is not zero. So the image of $x_i = \mathrm{tr}_H^G t^i$ in $CH_C^* \otimes_{\mathbf{F}_p} CH_G^{\leq 1}$ is $it^{i-1} \otimes x_1$. This is nonzero for $i = 1, \ldots, p-1$. So $x_1, \ldots, x_{p-1}$ are nonzero in CH_G^*, as we wanted.

We know that CH_G^* is generated by elements of bounded degree as a module over $\mathbf{F}_p[c_1 V_1, \ldots, c_p V_1]$, since V_1 is a faithful representation. By the formulas above, it follows that CH_G^* is generated by elements of bounded degree as a module over $\mathbf{F}_p[b, x_1, \ldots, x_{p-1}, c_p V_1]$. We showed that $x_i^2 = 0$, and so CH_G^* is generated by elements of bounded degree as a module over $\mathbf{F}_p[b, c_p V_1]$. Since CH_G^* has regularity ≤ 0 (Theorem 6.5), it follows that CH_G^* is generated as a module over $\mathbf{F}_p[b, c_p V_1]$ by elements of degree at most $\sigma(\mathbf{F}_p[b, c_p V_1]) = p-1$, modulo relations in degree at most p. Since we have computed CH_G^* in degrees at most p, we know CH_G^* in all degrees as a module:

$$CH_G^* = \mathbf{F}_p[b, c_p V_1]\{1, x_1, \ldots, x_{p-1}\}/(bx_1 = 0, \ldots, bx_{p-1} = 0).$$

We also showed that $x_i x_j = 0$ for all $i, j \in \{1, \ldots, p-1\}$, which determines the ring structure on CH_G^*:

$$CH_G^* =$$
$$\mathbf{F}_p[b, c_p V_1, x_1, \ldots, x_{p-1}]/(bx_i = 0, x_i x_j = 0 \text{ for all } i, j \in \{1, \ldots, p-1\}),$$

where $|b| = 1$ and $|x_i| = i$. □

13.3 Central extensions by G_m

In computing the cohomology or Chow ring of a p-group, a natural inductive approach is to consider a central extension by $\mathbf{Z}/p$,

$$1 \to \mathbf{Z}/p \to E \to Q \to 1.$$

In this section, we study the Chow ring of such an extension (for Q any affine k-group scheme) by considering the associated extension by the multiplicative group,

$$1 \to G_m \to K \to Q \to 1.$$

Explicitly, for z a generator of the subgroup $\mathbf{Z}/p$ in E, let $K = (E \times G_m)/\mathbf{Z}/p$, where the subgroup $\mathbf{Z}/p$ is generated by (z, ζ_p^{-1}). This maneuver can wonderfully simplify the problem. Throughout, we work over a field k of characteristic not p that contains the pth roots of unity.

There is a simple relation between the cohomology of the $\mathbf{Z}/p$-extension E and the G_m-extension K, which becomes even simpler for Chow rings. Since $K = (E \times G_m)/(\mathbf{Z}/p)$, E is a normal subgroup of K:

$$1 \to E \to K \to G_m \to 1,$$

where we have identified $G_m/(\mathbf{Z}/p)$ with G_m. Therefore, we have a principal G_m-bundle

$$G_m \to BE \to BK,$$

(This can be formulated concretely in terms of the finite-dimensional approximations to classifying spaces.) Let $u \in CH^1_K$ be the first Chern class of the homomorphism $K \to G_m$ above. The Leray-Serre spectral sequence of the G_m-bundle above reduces to a short exact sequence

$$0 \to H_K^*/(u) \to H_E^* \to \Sigma \ker(u \colon H_K^* \to H_K^*) \to 0.$$

For Chow rings, Lemma 2.4 gives the even simpler statement:

Theorem 13.10

$$CH_E^* \cong CH_K^*/(u).$$

Thus computing Chow rings of central extensions by $\mathbf{Z}/p$ reduces to the same problem for central extensions of G_m. The problem is still nontrivial; for example, we do not know whether finiteness of the groups CH_Q^* implies finiteness of the groups CH_K^*. Nonetheless, we have gained something. One point is that several different $\mathbf{Z}/p$-extensions of a group Q can determine G_m-extensions that are isomorphic as groups. For example, the two nonabelian groups of order p^3 for p odd are central extensions of $(\mathbf{Z}/p)^2$ by $\mathbf{Z}/p$, and they both induce the same central extension W of $(\mathbf{Z}/p)^2$ by G_m, which we consider in Section 13.4.

Moreover, there is the pleasant special case that a nontrivial $\mathbf{Z}/p$-extension may give a trivial G_m-extension $K = Q \times G_m$, in which case we know the Chow ring CH^*_K completely in terms of CH^*_Q. This gives the following result.

Theorem 13.11 *Let Q be an affine group scheme over k with a homomorphism $\alpha\colon Q \to G_m$. Let E be the subgroup of $Q \times G_m$ of elements (x,t) with $t^p\alpha(x) = 1$. Then E is a extension of Q by $\mathbf{Z}/p$, and*

$$CH^*_E \cong (CH^*_Q/(c_1\alpha))[u],$$

where $|u| = 1$.

Proof To see what is going on, consider the case $k = \mathbf{C}$. Then central extensions of Q by $\mathbf{Z}/p$ are classified by $H^2(BQ,\mathbf{Z})$, and central extensions of Q by G_m are classified by $H^3(BQ,\mathbf{Z})$. The short exact sequence $0 \to \mathbf{Z} \to \mathbf{Z} \to \mathbf{Z}/p \to 0$ of coefficient groups gives an exact sequence

$$H^2(BQ,\mathbf{Z}) \to H^2(BQ,\mathbf{F}_p) \to H^3(BQ,\mathbf{Z}).$$

Therefore, the central extensions by $\mathbf{Z}/p$ whose associated G_m-extension is trivial are exactly those coming from $H^2(BQ,\mathbf{Z}) = \mathrm{Hom}(Q, G_m)$. We compute that the central extension associated to a homomorphism $\alpha\colon Q \to G_m$ is the one defined in this theorem.

Over any field k satisfying our assumptions, we can check directly that the G_m-extension K associated to the given $\mathbf{Z}/p$-extension is trivial. That is, K is isomorphic to $Q \times G_m$, and so CH^*BK is a polynomial ring $(CH^*BQ)[u]$ with $|u| = 1$. As explained earlier in this section, BE is a principal G_m-bundle over BQ. By Lemma 2.4, CH^*BE is the quotient of CH^*BK by the ideal generated by one element of degree 1, which we compute to be $c_1\alpha - pu$. Mod p, this gives that $CH^*_E \cong (CH^*_Q/(c_1\alpha))[u]$. □

For example, let p be an odd prime number and let Q be the modular p-group $\mathbf{Z}/p \ltimes \mathbf{Z}/p^2$, whose Chow ring is computed in Lemma 13.8. Then the split metacyclic group $\mathbf{Z}/p^2 \ltimes \mathbf{Z}/p^2$ is an "integral $\mathbf{Z}/p$-extension" of $\mathbf{Z}/p \ltimes \mathbf{Z}/p^2$, meaning an extension associated to a 1-dimensional representation as in Theorem 13.11. That immediately computes the Chow ring of $\mathbf{Z}/p^2 \ltimes \mathbf{Z}/p^2$, as follows. Other approaches are possible, but this method is the quickest.

Lemma 13.12 *Let p be an odd prime number, and let G be a split metacyclic group $\mathbf{Z}/p^n \ltimes \mathbf{Z}/p^m$. Suppose that the image of the conjugation action $\mathbf{Z}/p^n \to (\mathbf{Z}/p^m)^*$ has order p. Let V be an irreducible p-dimensional representation of G over k. If $n = 1$, then*

$$CH^*_G \cong \mathbf{F}_p[b, c_pV, x_1, \ldots, x_{p-1}]/(bx_i = 0, x_ix_j = 0),$$

where $|b| = 1$, $|x_i| = i$, *and the relations involve all* $i, j \in \{1, \ldots, p-1\}$. *The ring* CH^*_G *has dimension 2 and depth 1, and* $d_0(CH^*_G) = 1$.

If $n \geq 2$, *then*

$$CH^*_G \cong \mathbf{F}_p[u, c_pV, x_1, \ldots, x_{p-1}]/(x_i x_j = 0),$$

where $|u| = 1$ *and* $|x_i| = i$. *In this case, the ring* CH^*_G *is Cohen-Macaulay of dimension 2. This follows from the Duflot bound, since G is p-central. Moreover,* $d_0(CH^*_G) = 1$. *In particular, this applies to the group* $\mathbf{Z}/p^2 \ltimes \mathbf{Z}/p^2$, *#4 of the groups of order* p^4 *in the Small Groups library [52].*

Another integral $\mathbf{Z}/p$-extension of the modular group of order p^3 is the group

$$E = \langle a, b, e : a^{p^2} = 1, b^p = 1, e^p = 1, e \text{ central}, aba^{-1} = b^{1+p}e\rangle$$

$$\cong \langle a, b, c : a^{p^2} = 1, b^p = 1, c^p = 1, c \text{ central}, [a, b] = c\rangle.$$

This is #3 of the groups of order p^4 in the Small Groups library. It has rank 3, and the Chow ring does not map onto $H^{\text{ev}}(BE, \mathbf{Z})$, which complicates some approaches to computing the Chow ring. But Theorem 13.11 makes it easy. We state the result as follows.

Lemma 13.13 *Let p be an odd prime number. Let G be #3 of the groups of order* p^4 *in the Small Groups library [52]:*

$$G = \langle a, b, c : a^{p^2} = 1, b^p = 1, c^p = 1, c \text{ central}, [a, b] = c\rangle.$$

Then

$$CH^*_G \cong \mathbf{F}_p[b, u, c_pV, x_2, \ldots, x_{p-1}]/(bx_i = 0, x_i x_j = 0),$$

where $|b| = |u| = 1$, $|x_i| = i$, *and the relations involve all* $i, j \in \{2, \ldots, p-1\}$. *The ring* CH^*_G *has dimension 3 and depth 2, which agrees with the Duflot bound. The group G has a unique conjugacy class of maximal elementary abelian subgroups, which is of rank 3. Finally,* $d_0(CH^*_G) = 2$.

13.4 The extraspecial group E_{p^3}

The nonabelian group $E_{p^3} = p^{1+2}_+$ of order p^3 and exponent p is more complicated than the other nonabelian group of order p^3, the modular p-group. We first compute the Chow ring of the associated central extension W of $(\mathbf{Z}/p)^2$ by G_m. (One sign of the importance of W is that any nontrivial central extension of $(\mathbf{Z}/p)^2$ by G_m over $\mathbf{C}$ is isomorphic to W as an algebraic group.) Leary computed the integral cohomology of BW and used it to compute the cohomology of E_{p^3} and related groups [90].

Given the Chow ring of W, it is easy to read off the Chow rings of both E_{p^3} and the modular p-group M_{p^3} by Theorem 13.10, although we have preferred to compute the Chow ring of M_{p^3} separately in Lemma 13.8 because that is a simpler calculation. (Both M_{p^3} and E_{p^3} are kernels of homomorphisms from W to G_m.) One reason to approach the Chow ring of E_{p^3} via the 1-dimensional group W is that we can compute the Chow ring of E_{p^3} without any dependence on Lewis's calculation of the integral cohomology of E_{p^3} [97]. This happens because the Chow ring of W turns out to be simpler than than of E_{p^3}. In the notation of Section 12.2, $d_0(CH^*_W)$ is 1 while $d_0(CH^*_{E_{p^3}})$ is 2, for $p \geq 5$ (Theorem 13.23). This makes it easier to compute CH^*_W first, and then read off the Chow ring of E_{p^3}. Finally, computing the Chow ring of W determines the Chow ring for an infinite family of p-subgroups of W, the central products $E_{p^3} * C_{p^{n-2}}$ (Lemma 13.16).

Lemma 13.14 *Let p be an odd prime number, and let k be a field of characteristic not p that contains the pth roots of unity. Let W be the central product $p_+^{1+2} * G_m = (p_+^{1+2} \times G_m)/(\mathbf{Z}/p)$ over k. Here $G = p_+^{1+2}$ is the group*

$$G = \langle x, y, z : x^p = y^p = z^p = [x, z] = [y, z] = 1, [x, y] = z \rangle$$

and the subgroup $\mathbf{Z}/p$ is generated by (z, ζ_p^{-1}). Then

$$CH^*_W \cong \mathbf{F}_p[a, b, e_1, \ldots, e_{p-2}, e_{p-1}, c_pV]/(ae_i = 0 \text{ for } 1 \leq i \leq p-2,$$
$$be_i = 0 \text{ for } 1 \leq i \leq p-1, ae_{p-1} = ab^{p-1} - a^p,$$
$$e_ie_j = 0 \text{ for } 1 \leq i \leq p-2 \text{ and } 1 \leq j \leq p-1, e_{p-1}^2 = -a^{p-1}b^{p-1} + a^{2p-2}).$$

The ring CH^*_G has dimension 2 and depth 1, which agrees with the Duflot bound. The topological nilpotence degree $d_0(CH^*_W)$ is equal to 1. Finally, for k a subfield of $\mathbf{C}$, CH^*_W maps isomorphically to $H^*(BW, \mathbf{Z})/p$, which is concentrated in even degrees.

Proof The abelianization of W is isomorphic to $(\mathbf{Z}/p)^2 \times G_m$. So $CH^1_W = \mathbf{F}_p\{a, b, e_1\}$, where we define $a, b, e_1 \colon G \to G_m$ by $a \colon x \mapsto \zeta_p, y \mapsto 1, \lambda \mapsto 1$ (for λ in G_m), $b \colon x \mapsto 1, y \mapsto \zeta_p, \lambda \mapsto 1$, and $e_1 \colon x \mapsto 1, y \mapsto 1, \lambda \mapsto \lambda^p$.

The center of W is the isomorphic image of $1 \times G_m \subset G \times G_m$. So every irreducible representation of W restricts to G_m as a $\lambda \mapsto \lambda^n$ for some integer j, which we call the weight. Representations of W of weight j can be identified with representations of G on which z acts by the scalar ζ_p^j. By the representation theory of G, it follows that W has a unique irreducible representation of each weight $j \not\equiv 0 \pmod p$, and it has dimension p. Also, W has p^2 irreducible representations of each weight $j \equiv 0 \pmod p$, all of dimension 1.

Let H be the subgroup $(\langle x, z\rangle \times G_m)/\langle(z, \zeta_p^{-1})\rangle$. This is a normal subgroup of index p in W. Let $\beta\colon H \to G_m$ be a representation. Then

$$(\mathrm{Ind}_H^W \beta)|_H \cong \beta \oplus y(\beta) \oplus \cdots \oplus y^{p-1}(\beta),$$

where a generator y of W/H acts by conjugation on H. If β has weight $j \not\equiv 0 \pmod p$, then these 1-dimensional representations are all distinct, and so $\mathrm{Ind}_H^W \beta$ is irreducible. It must be the unique irreducible representation V_j of W of weight j.

For any integer i, the Adams operation ψ^i on the representation ring $R(W)$ takes a representation of W of dimension n and weight j to a virtual representation of dimension n and weight ij [80]. Therefore, for integers $i \not\equiv 0 \pmod p$ and $j \not\equiv 0 \pmod p$, ψ^i of the unique irreducible representation V_j of W of weight j must be the irreducible representation V_{ij} of weight ij. Therefore, the Chern classes of V_i in CH_W^* are polynomials in those of $V = V_1$. It follows that the ring of Chern classes of all representations in CH_W^* is generated by $a, b, e_1, c_1V, \ldots, c_pV$, where we define $V = \mathrm{Ind}_H^W \beta$ for a weight-1 homomorphism $\beta\colon H \to G_m$. By Theorem 2.25, the ring CH_W^* in degrees $\le p$ is generated by these classes. Also, the representation V of W has $c_1V = e_1$, and so CH_W^* is generated in degrees $\le p$ by $a, b, e_1, c_2V, \ldots, c_pV$.

We have $CH_H^1 = \mathbf{F}_p\{t, u\}$, where we define $t, u\colon H \to k^*$ by $t\colon x \mapsto \zeta_p, \lambda \mapsto 1$ (for λ in G_m) and $u\colon x \mapsto 1, \lambda \mapsto \lambda$. The restriction map $CH_W^* \to CH_H^*$ sends $a \mapsto t$, $b \mapsto 0$, and $e_1 \mapsto pu = 0$. Also,

$$V|_H \cong U \oplus (T \otimes U) \oplus \cdots \oplus (T^{\otimes p-1} \otimes U),$$

where T and U are 1-dimensional representations of H with Chern classes t and u. So the total Chern class of V restricts to

$$\begin{aligned} c(V)|_H &= (1+u)(1+t+u)\cdots(1+(p-1)t+u) \\ &= 1 - t^{p-1} + (u^p - t^{p-1}u). \end{aligned}$$

By Lemma 13.9, for $1 \le j \le p-2$ we have $c_jV = \mathrm{tr}_H^W y_j$ where y_j is the sum of a set of orbit representatives for $\mathbf{Z}/p$ acting on the set of products $w_{i_1} \cdots w_{i_j}$, where $0 \le i_1 < \cdots < i_j \le p-1$ and $w_0 = u$, $w_1 = t+u, \ldots, w_{p-1} = (p-1)t+u$. Since $b \in CH_W^1$ restricts to 0 on H, we have $b\,\mathrm{tr}_H^W(w) = 0$ for any $w \in CH_H^*$, and so $bc_jV = 0$ for $1 \le j \le p-2$. But V is also induced from a 1-dimensional representation of the subgroup $A = (\langle y, z\rangle \times G_m)/\langle(z, \zeta_p^{-1})\rangle$, and a restricts to 0 on A; so we also have $ac_j(V) = 0$ for $1 \le j \le p-2$. Since a restricts to t on H, we have $\mathrm{tr}_H^W(t^iu^j) = a^i\,\mathrm{tr}_H^W(u^j)$. Let $e_i = \mathrm{tr}_H^W(u^i)$ for $1 \le i \le p-1$. It follows by induction on j, from the more complicated formula for c_jV above, that $c_j(V) = (1/p)\binom{p}{j}e_j$ for $1 \le j \le p-2$ and that $ae_j = 0$ for $1 \le j \le p-2$. (In particular, $e_1 = c_1V$, as shown earlier.)

Lemma 13.9 also relates $c_{p-1}V$ to transfers:

$$\begin{aligned} c_{p-1}V &= \mathrm{tr}_H^W((t+u)\cdots((p-1)t+u)) - b^{p-1} \\ &= \mathrm{tr}_H^W(u^{p-1} - t^{p-1}) - b^{p-1} \\ &= e_{p-1} - b^{p-1}. \end{aligned}$$

The last step uses that $\mathrm{tr}_H^W(t^{p-1}) = a^{p-1}\mathrm{tr}_H^W(1) = 0$ in CH_W^*, since $[W: H] = p$ and we are using $\mathbf{F}_p$ coefficients. From the relations between Chern classes and the classes e_i, the ring CH_W^* is generated in degrees $\leq p$ by $a, b, e_1, e_2, \ldots, e_{p-1}, c_pV$.

Since $b \in CH_W^1$ restricts to zero on H, the transfers e_i satisfy $be_i = 0$ for $1 \leq i \leq p-1$. For $i = p-1$, this says that $bc_{p-1}V + b^p = 0$. Since the representation V is also induced from $A = (\langle y, z\rangle \times G_m)/\langle(z, \zeta_p^{-1})\rangle$ (the kernel of the homomorphism $a\colon W \to G_m$), the same argument shows that $ac_{p-1}V + a^p = 0$ in CH_W^*. It follows that $ae_{p-1} = ab^{p-1} - a^p$.

Also,

$$\begin{aligned} e_i|_H &= \mathrm{tr}_H^W(u^i)|_H \\ &= u^i + (t+u)^i + \cdots + ((p-1)t+u)^i, \end{aligned}$$

which is zero for $1 \leq i \leq p-2$. Therefore $e_ie_j = 0$ for $1 \leq i \leq p-2$ and $1 \leq j \leq p-1$. On the other hand, the formula for $e_i|_H$ gives that $e_{p-1}|_H = -t^{p-1}$. Therefore

$$\begin{aligned} e_{p-1}^2 &= (\mathrm{tr}_H^W u^{p-1})^2 \\ &= \mathrm{tr}_H^W(u^{p-1}e_{p-1}|_H) \\ &= \mathrm{tr}_H^W(u^{p-1}(-t^{p-1})) \\ &= -a^{p-1}\mathrm{tr}_H^W u^{p-1} \\ &= -a^{p-1}e_{p-1} \\ &= -a^{p-1}b^{p-1} + a^{2p-2}. \end{aligned}$$

To summarize, we have a ring homomorphism from

$$\begin{aligned} &\mathbf{F}_p[a, b, e_1, \ldots, e_{p-2}, e_{p-1}, c_pV]/(ae_i = 0 \text{ for } 1 \leq i \leq p-2, \\ &\qquad be_i = 0 \text{ for } 1 \leq i \leq p-1,\ ae_{p-1} = ab^{p-1} - a^p, \\ &e_ie_j = 0 \text{ for } 1 \leq i \leq p-2 \text{ and } 1 \leq j \leq p-1,\ e_{p-1}^2 = -a^{p-1}b^{p-1} + a^{2p-2}) \end{aligned}$$

to CH_W^*. We know that CH_W^* is generated by elements of bounded degree as a module over the Chern classes of the faithful representation V, hence over the ring above. The relations show that the elements $e_1, \ldots, e_{p-1}$ are integral over the ring $\mathbf{F}_p[a, b, c_pV]$, and so CH_W^* is generated by elements of bounded degree over $\mathbf{F}_p[a, b, c_pV]$. Since CH_W^* has regularity ≤ 0 (Theorem 6.5), it follows that CH_W^* is generated by elements of degree at most $\sigma(\mathbf{F}_p[a, b, c_pV]) = p - 1$ as a module over $\mathbf{F}_p[a, b, c_pV]$, modulo relations in degree p.

The relations above imply that we have a map of $\mathbf{F}_p[a, b, c_pV]$-modules:

$$\mathbf{F}_p[a, b, c_pV]\{1, e_1, \ldots, e_{p-1}\}/(ae_i = 0 \text{ for } 1 \le i \le p-2,$$
$$be_i = 0 \text{ for } 1 \le i \le p-1, ae_{p-1} = ab^{p-1} - a^p) \to CH_W^*.$$

This is surjective in degrees at most p (since CH_W^* is generated by Chern classes in that range), hence in all degrees (since all module generators for CH_W^* are in degree at most $p - 1$).

It remains to show that this map is injective in degrees at most p. A basis for the domain as an $\mathbf{F}_p$-vector space in degrees at most p is given by monomials in a and b in degrees at most p, e_{p-1}, c_pV, and the elements e_i for $1 \le i \le p-2$. The elements e_i for $1 \le i \le p-2$ restrict to zero on the $p+1$ elementary abelian p-subgroups V of rank 2 in W, for example because $e_i^2 = 0$ for $1 \le i \le p-2$. We will show that the other elements listed have linearly independent images in $\prod_V CH_V^*$.

Let $C = Z(W)[p] = \langle z \rangle \cong \mathbf{Z}/p$. Then c_pV has nonzero restriction to $CH_C^* = \mathbf{F}_p[u]$, whereas $a, b, e_1, \ldots, e_{p-1}$ all restrict to zero on C. So c_pV is linearly independent of the other elements listed above in $\prod_V CH_V^*$. Next, the kernels of restriction from $\mathbf{F}_p[a, b]$ to the $p+1$ elementary abelian p-subgroups V of rank 2 in W are the ideals $(a), (a+b), \ldots, (a+(p-1)b)$, and (b). So any element of $\mathbf{F}_p[a, b]$ that restricts to zero on all these subgroups must be a multiple of $b\prod_{i=0}^{p-1}(a+ib) = a^pb - ab^p$. In particular, the monomials of degree at most p in a, b are linearly independent in $\prod_V CH_V^*$. Next, in degree $p-1$, suppose that $f_{p-1}(a, b) + e_{p-1}$ restricts to zero in $\prod_V CH_V^*$. Multiplying by b, we deduce that $bf_{p-1}(a, b) \mapsto 0$, since $be_{p-1} = 0$. By the injectivity we proved in degree p, it follows that $f_{p-1} = 0$, and so $e_{p-1} \mapsto 0$ in $\prod_V CH_V^*$. But then multiplying by a gives that $ab^{p-1} - a^p \mapsto 0$ in $\prod_V CH_V^*$, contradicting the injectivity in degree p.

To finish the computation of CH_W^*, it remains to show that $e_i \ne 0$ in CH_W^* for $1 \le i \le p-2$. As in the proof of Lemma 13.8, we could prove this by observing that e_i has nonzero image in $H^*(BG, \mathbf{Z})/p$, by Lewis's calculation of $H^*(BG, \mathbf{Z})$ [97], but we prefer to give a direct proof using Chow rings. Again, we use the idea of the topological nilpotence degree from Section 12.2. We know that $e_1 \ne 0$ in $CH_W^1 = \text{Hom}(W, G_m)/p$.

Since the central subgroup $C \cong \mathbf{Z}/p$ is contained in the subgroup H of W, we have a pullback diagram

$$\begin{array}{ccc} C \times H & \longrightarrow & H \\ \downarrow & & \downarrow \\ C \times W & \longrightarrow & W. \end{array}$$

Because pushforward commutes with pullback (Lemma 2.16), we have $(\mathrm{tr}_H^W u^i)|_{C\times W} = \mathrm{tr}_{C\times H}^{C\times W}(u^i|_{C\times H})$ for all $i \geq 0$. The restriction map from CH_H^* to $CH_C^* = \mathbf{F}_p[u]$ sends u to u. The pullback $CH_H^* \to CH_C^* \otimes_{\mathbf{F}_p} CH_H^*$ sends u to $u \otimes 1 + 1 \otimes u$. So the image of u^i in $CH_C^* \otimes_{\mathbf{F}_p} CH^{\leq 1}H$ is

$$u^i \otimes 1 + iu^{i-1} \otimes u.$$

Here $\mathrm{tr}_H^W 1 = 0$, but $\mathrm{tr}_H^W u = e_1$, which is not zero. So the image of $x_i = \mathrm{tr}_H^W u^i$ in $CH_C^* \otimes_{\mathbf{F}_p} CH_W^{\leq 1}$ is $iu^{i-1} \otimes x_1$. This is nonzero for $i = 1, \ldots, p-2$. So $e_1, \ldots, e_{p-2}$ are nonzero in CH_W^*, as we wanted.

Since all module relations for CH_W^* over $\mathbf{F}_p[a, b, c_p V]$ are in degrees $\leq p$, there are no more relations. Thus we have computed CH_W^*. We have made this calculation without using Leary's calculation of $H^*(BW, \mathbf{Z})$ [90, theorem 2]. By inspecting that calculation, we see that $CH_W^* \to H^*(BW, \mathbf{Z})/p$ is an isomorphism. □

Corollary 13.15 *Let p be an odd prime number. Let G be the nonabelian group of order p^3 and exponent p, also called the extraspecial group p_+^{1+2} of order p^3 or the group of strictly upper-triangular 3×3 matrices over $\mathbf{F}_p$. Then*

$$CH_G^* \cong \mathbf{F}_p[a, b, e_2, \ldots, e_{p-2}, e_{p-1}, c_p V]/(ae_i = 0 \text{ for } 2 \leq i \leq p-2,$$
$$be_i = 0 \text{ for } 2 \leq i \leq p-1,\ ae_{p-1} = ab^{p-1} - a^p,$$
$$e_i e_j = 0 \text{ for } 2 \leq i \leq p-2 \text{ and } 2 \leq j \leq p-1,\ e_{p-1}^2 = -a^{p-1}b^{p-1} + a^{2p-2}).$$

Here $|a| = |b| = 1$, $|e_i| = i$ for $2 \leq i \leq p-1$, and $|c_p V| = p$. The ring CH_G^* has dimension 2 and depth 1 for $p \geq 5$, which agrees with the Duflot bound, whereas CH_G^* is Cohen-Macaulay of dimension 2 for $p = 3$. Finally, for $k \subset \mathbf{C}$, CH_G^* maps isomorphically to $H^{\mathrm{ev}}(BG, \mathbf{Z})/p$.

For comparison with Lewis's computation of the integral cohomology ring [97, theorem 6.26], which we state as Theorem 13.22, note that the relations above imply that $ab^p - a^p b = abe_{p-1} = 0$.

Proof The group $G = p_+^{1+2}$ is a normal subgroup of the group W of Lemma 13.14, with $1 \to G \to W \to G_m \to 1$. It follows that CH_G^* is the quotient ring of CH_W^* by the first Chern class of the homomorphism $W \to G_m$, which is e_1. We can also compute $H^*(BG, \mathbf{Z})$ in terms of $H^*(BW, \mathbf{Z})$, as mentioned before Theorem 13.10. Then the isomorphism from CH_G^* to $H^{\mathrm{ev}}(BG, \mathbf{Z})/p$ follows from the isomorphism from CH_W^* to $H^*(BW, \mathbf{Z})/p$ (Theorem 13.14). □

Here is another family of p-groups whose Chow ring reduces to the Chow ring of the 1-dimensional group W.

Lemma 13.16 *Let p be an odd prime number. Let G be the central product $p_+^{1+2} * C_{p^{n-2}} = (p_+^{1+2} \times \mathbf{Z}/p^{n-2})/(\mathbf{Z}/p)$ for $n \geq 4$:*

$$G = \langle x, y, z : x^p = y^p = z^{p^{n-2}} = [x,z] = [y,z] = 1, [x,y] = z^{p^{n-3}} \rangle.$$

For $= 4$, this is group #14 of order p^4 in the Small Groups library [52]. Then

$$CH_G^* \cong \mathbf{F}_p[a, b, e_1, e_2, \ldots, e_{p-2}, e_{p-1}, c_p V]/(ae_i = 0 \text{ for } 1 \leq i \leq p-2,$$
$$be_i = 0 \text{ for } 1 \leq i \leq p-1, ae_{p-1} = ab^{p-1} - a^p,$$
$$e_i e_j = 0 \text{ for } 1 \leq i \leq p-2 \text{ and } 1 \leq j \leq p-1, e_{p-1}^2 = -a^{p-1}b^{p-1} + a^{2p-2}).$$

The ring CH_G^* has dimension 2 and depth 1, which agrees with the Duflot bound. The topological nilpotence degree $d_0(CH_G^*)$ is equal to 1. Finally, for $k \subset \mathbf{C}$, CH_G^* maps isomorphically to $H^{\text{ev}}(BG, \mathbf{Z})/p$.

For $n = 3$, the presentation above defines the extraspecial group p_+^{1+2}, whose Chow ring looks the same but without the generator e_1 (Corollary 13.15). For any $n \geq 3$, Leary computed the integral cohomology of the group G in the lemma [90]. He observed that it is the unique isomorphism class of groups that is neither abelian nor metacyclic but can be written as a central extension of $(\mathbf{Z}/p)^2$ by $\mathbf{Z}/p^n$. For $n = 4$, this group is sometimes called the almost extraspecial group of order p^4.

Proof Write E_{p^3} for the extraspecial p-group of order p^3 and exponent p. Let W be the central product $(E_{p^3} \times G_m)/(\mathbf{Z}/p)$, and let $e_1 : W \to G_m$ be the homomorphism that is trivial on E_{p^3} and sends $\lambda \in G_m$ to λ^p. The group G is the kernel of the homomorphism $e_1^p : W \to G_m$. By Theorem 13.10, the restriction $CH_W^* \to CH_G^*$ is an isomorphism. Since CH_W^* is computed in Lemma 13.14, we know CH_G^*. The fact that CH_G^* maps isomorphically to $H^{\text{ev}}(BG, \mathbf{Z})/p$ follows from the statement that CH_W^* maps isomorphically to $H^*(BW, \mathbf{Z})/p$ (Lemma 13.14). □

13.5 Calculations of the topological nilpotence degree

Kuhn gave good upper bounds for the topological nilpotence degree of any finite group (Theorem 12.6, above). But so far, most exact calculations of the topological nilpotence degree have focused on mod 2 cohomology [69, sections II.4 and II.5], [85, appendix A]. In this section, we compute the topological nilpotence degree $d_0(H_G^*)$ of Henn-Lannes-Schwartz for some simple examples of p-groups with p odd: split metacyclic groups (including the nonabelian group of order p^3 and exponent p^2), the extraspecial group of order p^3 and exponent p, and some p-central groups of order p^6. In some cases, we also compute the topological nilpotence degree of the Chow ring.

Theorems 12.4 and 12.7 bound these invariants in terms of the smallest faithful complex representation of G. The experimental evidence in this section suggests the possibility of major improvements to the known upper bounds for $d_0(CH^*_G)$ and $d_0(H^*_G)$, at least for p odd. Proving such bounds would bring the cohomology ring and Chow ring of a finite group under much better control.

Another observation is that the Chow ring of a finite group G tends to be simpler than the cohomology ring. For example, a cyclic group $G = \mathbf{Z}/p^r$ with $r \geq 2$ has $d_0(CH^*_G) = 0$; equivalent statements are that the mod p Chow ring of G is reduced (that is, has no nilpotent elements except zero), or that it is detected on elementary abelian subgroups. The $\mathbf{F}_p$-cohomology of G is a little more complicated, in the sense that $d_0(H^*_G) = 1$. More generally, Guillot observed that the mod l Chow ring of a finite group G of Lie type is often reduced, so that $d_0(CH^*_G) = 0$ [61]. By the examples in this section, we cannot expect to have the Chow ring to have $d_0 = 0$ in much generality, but $d_0(CH^*_G)$ is often smaller than current bounds would predict.

As in Section 13.1, when we discuss the Chow ring of a p-group G of exponent e in this section, we consider G as an algebraic group over any field k of characteristic not p that contains the eth roots of unity.

Kuhn was able to improve the bound in Theorem 12.6 by 1 in some cases, as follows. Let G be a finite group, and let C be the maximal central elementary abelian subgroup of G. (In this book, we normally consider this subgroup C only when G is a p-group.) Let $\mathrm{CEss}(H^*_G)$ be the *central essential ideal* in the cohomology of G, the ideal in H^*_G of elements that restrict to zero on $C_G(V)$ for all elementary abelian subgroups V that strictly contain C. Kuhn showed that the central essential ideal is nonzero if and only if the depth of H^*_G is equal to the rank of C (that is, the Duflot lower bound is an equality) [86, theorem 2.30]. This was proved earlier by Green when G is a p-group [49]. Let $e_{\mathrm{prim}}(G)$ be the supremum of the degrees in which the H^*_C-primitive subspace $P_C\mathrm{CEss}(H^*_G)$ is not zero. We write $e_{\mathrm{prim}}(G) = -\infty$ if the central essential ideal is zero.

Theorem 13.17 *[86, proposition 2.8, corollary 2.14, corollary 2.20, theorem 2.22] Let G be a finite group. Then*

$$d_0(H^*_G) = \max\{e_{\mathrm{prim}}(C_G(V)) : V \subset G \text{ elementary abelian } p\text{-subgroup}\}.$$

Here $e_{\mathrm{prim}}(H) \leq e(H)$ for all finite groups H, with $e_{prim}(H) < e(H)$ when H is not p-central. Finally, $e(H) \leq e(G)$ for every subgroup H of a p-group G.

For any p-central finite group G, $d_0(H^*_G) = e_{\mathrm{prim}}(G) = e(G)$.

It follows that $d_0(H^*_G) \leq e(G)$ for any p-group G, and sometimes Theorem 13.17 gives a strict inequality. For $p = 2$, the theorem is optimal in some cases.

For example, let G be group #8 of order 32 in the Small Groups library,

$$G = \langle x, y | x^{-1}yx = x^2y^{-1}, x^{-1}y^2x = y^{-2}, x^4 = y^4 \rangle.$$

Then G has rank 2, the center has rank 1, and G has type [8] in the sense of Definition 12.5, for example by the computations of Green and King [52]. The group G has a unique elementary abelian subgroup V of rank 2, which is normal, with centralizer isomorphic to $Q_8 \times \mathbf{Z}/2$. Since that centralizer is 2-central, $e_{\text{prim}}(C_G(V)) = e(C_G(V)) = 3$. Since G is not 2-central, Theorem 13.17 gives that $e_{\text{prim}}(G) < e(G) = 7$. From [85, table 3] (where G is called group number 48), we see that $d_0(H_G^*) = \max(e_{\text{prim}}(G), e_{\text{prim}}(C_G(V))) = \max(6, 3) = 6$, showing the optimality of Theorem 13.17. The following calculations suggest that things may be better at odd primes p.

Let p be an odd prime number, and let G be a split metacyclic p-group. That is, G is a semidirect product $\mathbf{Z}/p^n \ltimes \mathbf{Z}/p^m$ for some positive integers m and n. We compute $d_0(H_G^*)$ and give a partial result on $d_0(CH_G^*)$.

We assume that G is not abelian. Then G has a presentation

$$G = \langle s, t | s^{p^n} = t^{p^m} = 1, sts^{-1} = t^{1+p^l} \rangle$$

where $m > l \geq \max(m - n, 1)$. Diethelm computed the $\mathbf{F}_p$-cohomology ring of G, as follows [33, theorem 1].

Theorem 13.18 *Let G be a nonabelian split metacyclic p-group with p odd. If $l > m - n$ in the notation above, then*

$$H_G^* \cong \mathbf{F}_p\langle a, b, x, y \rangle,$$

where $|a| = |b| = 1$ and $|x| = |y| = 2$. Here b and y are pulled back from nonzero classes on the cyclic group $G/\langle t \rangle$, while a and x have nonzero restrictions to the cyclic group $\langle t \rangle$.

If $l = m - n$, then

$$H_G^* \cong \mathbf{F}_p\langle a_1, \ldots, a_{p-1}, b, y, v, w \rangle / (a_i a_j = a_i y = a_i v = 0),$$

where $|a_i| = 2i - 1$, $|b| = 1$, $|y| = 2$, $|v| = 2p - 1$, and $|w| = 2p$. Here b and y are pulled back from nonzero classes on the cyclic group $G/\langle t \rangle$, while the elements a_i, v, w all have nonzero restriction to the cyclic group $\langle t \rangle$.

We first observe that the topological nilpotence degree $d_0(CH_G^*)$ is positive for every nonabelian split metacyclic p-group G. If $d_0(CH_G^*)$ were zero, then the ring CH_G^* would be reduced. Consider the 1-dimensional representation L of G over k given by $s \mapsto 1$ and $t \mapsto \zeta_{p^l}$. Then $z := c_1 L$ is nonzero in $CH_G^1 = \text{Hom}(G, \mathbf{C}^*)/p$ (Lemma 2.26). But z is nilpotent by Yagita's theorem (Theorem 8.10), because it restricts to zero on the unique maximal elementary abelian p-subgroup of G, $\langle s^{p^{n-1}}, t^{p^{m-1}} \rangle \cong (\mathbf{Z}/p)^2$. Since $CH_G^1 = H^2(BG, \mathbf{Z})/p$ always injects into H_G^2, the image of z in H_G^* is a nonzero element that restricts to zero on all elementary abelian subgroups V, and so we also have $d_0(H_G^*) > 0$. More precisely, since z comes from CH_G^1, its image in the group $H_V^1 \otimes H_{C_G(V)}^1$ is also zero, and so $d_0(H_G^*) \geq 2$.

Theorem 13.19 *Let p be an odd prime number. Let G be a nonabelian split metacyclic p-group. Then $d_0(H_G^*) = 2$.*

Proof First consider G of the first type in Theorem 13.18, where $l > m - n$. In this case, G is p-central, with $C := Z(G)[p]$ equal to $\langle s^{p^{n-1}}, t^{p^{m-1}} \rangle \cong (\mathbf{Z}/p)^2$. Since G is a p-central p-group, Kuhn showed that $d_0(H_G^*)$ is equal to $e(G)$, the largest degree of a generator for H_C^* as a module over H_G^* (Theorem 12.6). By Quillen's theorem (Theorem 8.4) together with the computation of H_G^* in Theorem 13.18, we know that $\text{im}(H_G^* \to H_C^*/\text{rad}(H_C^*)) = H_G^*/\text{rad}(H_G^*) = \mathbf{F}_p[x, y]$, where $|x| = |y| = 2$. By inspection, the generators a, b of $H_G^1 = \text{Hom}(G, \mathbf{Z}/p)$ restrict to zero on C. So the image of $H_G^* \to H_C^*$ is exactly the polynomial ring $\mathbf{F}_p[x, y]$. It follows that $d_0(H_G^*) = e(G) = 2$.

Next, again for an odd prime p, let G be a split metacyclic p-group that is not p-central. This means that $l = m - n$, in the notation above. It follows that the cyclic normal subgroup $\mathbf{Z}/p^m \subset G$ has $m \geq 2$. An example is the modular p-group $\mathbf{Z}/p \ltimes \mathbf{Z}/p^m$; for $m = 2$, this is the nonabelian group of order p^3 and exponent p^2.

For $l = m - n$, the group $C = Z(G)[p]$ is $\langle t^{p^{m-1}} \rangle \cong \mathbf{Z}/p$. Here $e(G) = 2p - 1$, since the computation of H_G^* in Theorem 13.18 implies that $\text{im}(H_G^* \to H_C^*) = \mathbf{F}_p[w]$ where $|w| = 2p$. Kuhn's upper bound (Theorem 13.17) implies that $d_0(H_G^*) \leq 2p - 2$, as we check below. In fact, the cohomology of G is much simpler than that; we will show that $d_0(H_G^*) = 2$. Perhaps Kuhn's upper bound can be improved for non-p-central p-groups in general.

The group G has a unique maximal elementary abelian subgroup, $A = \langle s^{p^{n-1}}, t^{p^{m-1}} \rangle \cong (\mathbf{Z}/p)^2$. The centralizer $C_G(A)$ is the subgroup $\langle s, t^p \rangle$ of G, which is a p-central split metacyclic group (possibly abelian) of the form $\mathbf{Z}/p^n \ltimes \mathbf{Z}/p^{m-1}$. Our earlier calculation gives that $e_{\text{prim}}(C_G(A)) \leq d_0(C_G(A)) \leq 2$. Therefore, by Theorem 13.17, $d_0(H_G^*) \leq \max(2, e_{\text{prim}}(G)) \leq e(G) - 1 = 2p - 2$. That is,

$$H_G^* \to H_C^* \otimes_{\mathbf{F}_p} H_G^{\leq 2p-2} \times H_A^* \otimes_{\mathbf{F}_p} H_{C_G(A)}^{\leq 2p-2}$$

is injective.

It remains to analyze the cohomology of G in degrees at most $2p - 2$. Namely, we want to show that the homomorphism

$$H_G^* \to H_C^* \otimes_{\mathbf{F}_p} H_G^{\leq 2} \times H_A^* \otimes_{\mathbf{F}_p} H_{C_G(A)}^{\leq 2}$$

is injective in degrees at most $2p - 2$.

Although other approaches are possible, we do this by constructing generators of H_G^* explicitly. By Theorem 13.18, a basis for H_G^* as an $\mathbf{F}_p$-vector space in degrees at most $2p-2$ is given by:

0	1	2	3	4	5	$\cdots$	$2p-3$	$2p-2$
1	a_1, b	$a_1 b, y$	a_2, yb	$a_2 b, y^2$	$a_3, y^2 b$	$\cdots$	$a_{p-1}, y^{p-2}b$	$a_{p-1}b, y^{p-1}$

Theorem 13.18 also gives that

$$H^*_{C_G(A)} = \mathbf{F}_p\langle \alpha, b, \chi, y\rangle,$$

where $|\alpha| = |b| = 1$, $|\chi| = |y| = 2$, b and y are pulled back from $C_G(A)/\langle t^p\rangle \cong \langle s\rangle \cong \mathbf{Z}/p^n$, and α and χ have nonzero restriction to $\langle t^p\rangle \cong \mathbf{Z}/p^{m-1}$. We can assume that b and y are pulled back from the classes with the same names in H^*_G.

Define $f_i = \mathrm{tr}^G_{C_G(A)}(\alpha\chi^{i-1})$, for $1 \le i \le p$. I claim that f_i and $y^{i-1}b$ form a basis for H^{2i-1}_G, for $1 \le i \le p$. By Theorem 13.18, this holds if f_i has nonzero restriction to $\langle t\rangle \cong \mathbf{Z}/p^m$. By the double coset formula (Lemma 2.15),

$$f_i|_{\langle t\rangle} = \mathrm{tr}^{\langle t\rangle}_{\langle t^p\rangle}(\alpha\chi^{i-1}).$$

Here $\alpha\chi^{i-1}$ is nonzero in $H^*_{\langle t^p\rangle} \cong H^*_{\mathbf{Z}/p^{m-1}} = \mathbf{F}_p\langle \alpha, \chi\rangle$. The element χ is the restriction of a generator χ of $H^2_{\langle t\rangle} = H^2_{\mathbf{Z}/p^m}$, while the transfer of α from $H^1_{\mathbf{Z}/p^{m-1}}$ to $H^1_{\mathbf{Z}/p^m}$ is not zero. It follows that the transfer of $\alpha\chi^{i-1}$ from $\mathbf{Z}/p^{m-1}$ to $\mathbf{Z}/p^m$ is not zero. Thus f_i has nonzero restriction to $\langle t\rangle$, and hence is a generator of H^*_G as claimed.

Therefore, a basis for H^*_G in degrees at most $2p-1$ is given by f_i and $y^{i-1}b$ in degree $2i-1$, and $f_i b$ and y^i in degree $2i$. By the calculation of $H^*_{C_G(A)}$, y^i and $y^i b$ have nonzero restriction to $C_G(A)$. Since $d_0(H^*_{C_G(A)}) \le 2$, y^i and $y^i b$ have nonzero images in $H^*_A \otimes H^{\le 2}_{C_G(A)}$. On the other hand, y and b restrict to zero on C, and so the images of y and b in $H^*_C \otimes_{\mathbf{F}_p} H^*_G$ are of the form $1 \otimes y$ (using that y comes from CH^1_G) and $1 \otimes b$. So y^i for $i \ge 2$ and $y^i b$ for $i \ge 1$ map to zero in $H^*_C \otimes H^{\le 2}_G$.

To finish the proof that $d_0(H^*_G) \le 2$, it suffices to show that the images of f_i and $f_i b$ in $H^*_C \otimes H^{\le 2}_G$ are linearly independent of the images of $y^{i-1}b$ and y^i, for $1 \le i \le p$. This is clear for $i = 1$ (by the basis for H^*_G). It remains to show that f_i and $f_i b$ have nonzero image in $H^*_C \otimes H^{\le 2}_G$ for $2 \le i \le p$. This is easy from the definition of f_i as a transfer. Assume first that $m \ge 3$; then the element $\alpha \in H^1_{C_G(A)}$, which is nonzero on $\langle t^p\rangle$, restricts to zero on $C = \langle t^{p^{m-1}}\rangle$. So

$$\begin{aligned} f_i|_{C\times G} &= \mathrm{tr}^{C\times G}_{C\times C_G(A)}(\alpha\chi^{i-1}|_{C\times C_G(A)}) \\ &= \mathrm{tr}^{C\times G}_{C\times C_G(A)}((1\otimes\alpha)(\chi\otimes 1 + 1\otimes\chi)^{i-1}) \\ &= \mathrm{tr}^{C\times G}_{C\times C_G(A)}(\chi^{i-1}\otimes\alpha) \end{aligned}$$

in $H_C^* \otimes H_G^{\leq 2}$. So

$$f_i|_{C\times G} = \chi^{i-1} \otimes f_1.$$

This is nonzero in $H_C^* \otimes H_G^{\leq 2}$ (and even in $H_C^* \otimes H_G^{\leq 1}$). Likewise,

$$\begin{aligned} f_i b|_{C\times G} &= (\chi^{i-1} \otimes f_1)(1 \otimes b) \\ &= \chi^{i-1} \otimes f_1 b, \end{aligned}$$

which is again nonzero in $H_C^* \otimes H_G^{\leq 2}$. For $m = 2$ (the case of the modular p-group $G = \mathbf{Z}/p \ltimes \mathbf{Z}/p^2$), a slightly more complicated calculation leads to the same conclusion. This completes the proof that $d_0(H_G^*) \leq 2$. We showed earlier that $d_0(H_G^*) \geq 2$ for every nonabelian split metacyclic p-group, and so equality holds. □

In some cases, one can bound $d_0(CH_G^*)$ in terms of $d_0(H_G^*)$ using the following lemma. Note that, since k contains the pth roots of unity, the Chow ring CH_G^* maps to etale cohomology $H_{\text{et}}^*(BG_{k_s}, \mathbf{F}_p)$, which can be identified with the usual cohomology H_G^* of G.

Lemma 13.20 *Let p be a prime number. Let G be a finite group with $CH_G^* \to H_G^*$ injective. Then $d_0(CH_G^*) \leq d_0(H_G^*)/2$.*

Proof By Henn-Lannes-Schwartz's interpretation of $a := d_0(H_G^*)$ (Theorem 12.2), we know that

$$H_G^* \to \prod_{V\subset G} H_V^* \otimes_{\mathbf{F}_p} H_{C_G(V)}^{\leq a}$$

is injective, where the product runs over all elementary abelian subgroups V of G. Since $CH_G^* \to H_G^*$ is injective, the Chow ring also injects into the product ring above. This homomorphism factors through $\prod_{V\subset G} CH_V^* \otimes_{\mathbf{F}_p} CH_{C_G(V)}^{\leq a/2}$, and so

$$CH_G^* \to \prod_{V\subset G} CH_V^* \otimes_{\mathbf{F}_p} CH_{C_G(V)}^{\leq a/2}$$

is injective. That means that $d_0(CH_G^*) \leq a/2$. □

One can use Lemma 13.20 and Theorem 13.19 to show that $d_0(CH_G^*) = 1$ when p is an odd prime number and $G = \mathbf{Z}/p^n \ltimes \mathbf{Z}/p^m$ is a split metacyclic p-group with im $(\mathbf{Z}/p^n \to (\mathbf{Z}/p^m)^*)$ of order p. Indeed, one can show that CH_G^* injects into H_G^* in that case; we do not present the details, because we have already shown more directly that $d_0(CH_G^*) = 1$ (Lemma 13.12). Theorem 12.7 gives only that $d_0(CH_G^*) \leq p - 1$ in this case, which suggests that Theorem 12.7 may be improvable more generally.

It seems reasonable to conjecture that the cycle map $CH^*BG \to H^{\text{ev}}(BG, \mathbf{Z})$ is an isomorphism for all p-groups G of rank at most 2. Then

Theorem 13.19 would imply that $d_0(CH_G^*) = 1$ for every nonabelian split metacyclic p-group G. For now, that remains an open question.

We now compute $d_0(H_G^*)$ for the nonabelian group of order p^3 and exponent p, the extraspecial group p_+^{1+2}. For $p = 3$, the result is due to Milgram and Tezuka [103].

Theorem 13.21 *Let p be an odd prime number, and let G be the nonabelian group of order p^3 and exponent p,*

$$G = \langle A, B | A^p = B^p = [A, B]^p = [A, [A, B]] = [B, [A, B]] = 1\rangle.$$

If $p = 3$, then $d_0(H_G^*) = 0$. (Equivalently, H_G^* is detected on elementary abelian subgroups.) If $p \geq 5$, then $d_0(H_G^*) = 4$.

Proof The center of G is the subgroup $\mathbf{Z}/p$ generated by the element $[A, B]$, and so $C = C(G)$ is isomorphic to $\mathbf{Z}/p$. For V an elementary abelian subgroup of G that strictly contains C, V is isomorphic to $(\mathbf{Z}/p)^2$ and the centralizer of V in G is equal to V (otherwise, G would be abelian). In particular, $e(C_G(V)) = 0$ for such subgroups V. Since G is not p-central, Theorem 13.17 gives that $d_0(H_G^*) = e_{\text{prim}}(G) < e(G)$. Here $e(G) \leq 2p - 1$ because there is an element ζ of H_G^{2p}, the Euler class of a p-dimensional irreducible complex representation of G, whose restriction to C is nonzero. So $d_0(H_G^*) \leq 2p - 2$. That is, the homomorphism

$$H_G^* \to H_C^* \otimes_{\mathbf{F}_p} H_G^{\leq 2p-2} \times \prod_{V \subset G} H_V^*$$

is injective.

We use Lewis's calculation of $H^*(BG, \mathbf{Z})$, as follows [97, theorem 6.26]. In fact, we use Leary's choice of generators, which seems more natural [90, theorem 3]. Although this is complicated, the mod p cohomology ring H_G^* is even more complicated and took 25 years longer to compute [91]. It is a good feature of the proof that we can prove a strong property of the mod p cohomology ring by working with integral cohomology.

Theorem 13.22 *Let p be an odd prime number. Let G be the nonabelian group of order p^3 and exponent p, the extraspecial group p_+^{1+2}. Then*

$$\begin{aligned}
&H^*(BG, \mathbf{Z}) \\
&= \mathbf{Z}\langle \alpha, \beta, \mu, \nu, \chi_2, \ldots, \chi_{p-1}, \zeta\rangle/(p\alpha = p\beta = p\mu = p\nu = p\chi_i = p^2\zeta = 0, \\
&\alpha\mu = \beta\nu, \alpha^p\mu = \beta^p\nu, \alpha\chi_i = \beta\chi_i = \mu\chi_i = \nu\chi_i = 0 \, for \, 2 \leq i, j < p - 1,
\end{aligned}$$

$$\chi_i\chi_j = 0 \textit{ for } 2 \le i \le p-2 \textit{ and } 2 \le j \le p-1,\ \chi_{p-1}^2 = \alpha^{2p-2} + \beta^{2p-2} - \alpha^{p-1}\beta^{p-1},$$

$$\alpha\chi_{p-1} = -\alpha^p,\ \beta\chi_{p-1} = -\beta^{p-1},\ \mu\chi_{p-1} = -\mu\beta^{p-1},\ \nu\chi_{p-1} = -\nu\alpha^{p-1},$$

$$\alpha\beta^p = \beta\alpha^p,\ \mu\nu = \begin{cases} \lambda\chi_3 & \textit{for some } \lambda \in \mathbf{F}_p^* \textit{ if } p \ge 5 \\ 3\lambda\zeta & \textit{for some } \lambda \in \{\pm 1\} \textit{ if } p = 3 \end{cases}\Bigg)$$

Here $|\alpha| = |\beta| = 2$, $|\mu| = |\nu| = 3$, $\chi_i = 2i$, *and* $|\zeta| = 2p$.

Explicitly, α is the first Chern class of the complex representation $A \mapsto \zeta_p$, $B \mapsto 1$, and β is the first Chern class of $A \mapsto 1$, $B \mapsto \zeta_p$. Also, following Lewis's notation, let C be the central element $B^{-1}A^{-1}BA$ and define elementary abelian subgroups $H_t = \langle AB^t, C\rangle$ for $0 \le t \le p-1$ and $H = \langle B, C\rangle$. Then β and μ both restrict to zero on H_0 and have nonzero restriction to H_t for $t \ne 0$ and to H [97, lemma 6.16]. Likewise, α and ν both restrict to zero on H and have nonzero restriction to H_t for all t. Finally, let γ be the first Chern class of the 1-dimensional representation of H given by $B \mapsto 1$, $C \mapsto \zeta_p$. Then Leary defines $\chi_i = \mathrm{tr}_H^G \gamma^i$ for $2 \le i < p-1$ and $\chi_{p-1} = \mathrm{tr}_H^G \gamma^{p-1} - \alpha^{p-1}$.

We continue the proof of Theorem 13.21. Theorem 13.22 implies that $H^*(BG, \mathbf{Z})$ is killed by p in degrees from 1 to $2p-1$. We also read off that a basis for $H^i(BG, \mathbf{Z})/p$ over $\mathbf{F}_p$ in degrees $\le 2p-1$ is given by the monomials $\alpha^i\beta^j$ and χ_i in even degrees, and the monomials $\alpha^i\beta^j\nu$ and $\beta^i\mu$ in odd degrees. I claim that $H^*(BG, \mathbf{Z})$ injects into $\prod_V H_V^*$ in odd degrees from 1 to $2p-1$, where the product runs over the elementary abelian subgroups V of rank 2 (namely, the groups H_t and H). We first note that the kernel of restriction from $\mathbf{F}_p[\alpha, \beta]$ to the Chow ring of H_t is the ideal $(\beta - t\alpha)$, and the kernel of restriction to H is (α). So any element of $\mathbf{F}_p[\alpha, \beta]$ that restricts to zero on all these subgroups must be a multiple of $\alpha\prod_{t=0}^{p-1}(\beta - t\alpha) = \alpha\beta^p - \alpha^p\beta$. In particular, the monomials $\alpha^i\beta^j$ with $i + j \le p$ are linearly independent in $\prod_V CH_V^*$ (and hence in $\prod_V H_V^*$). Likewise, the monomials $\alpha^i\beta^j$ with $i + j \le p-1$ are linearly independent in $\prod_{t=0}^{p-1} CH_{H_t}^*$.

Next, it is convenient to observe that any nonzero element of CH_V^* is a non-zero-divisor in H_V^*, for an elementary abelian p-group V. Since μ has nonzero restriction to H, it follows that $\beta^i\mu$ is nonzero on H for all $i \ge 0$, whereas all the elements $\alpha^i\beta^j\nu$ restrict to zero on H. Likewise, since ν has nonzero restriction to H_t for each $0 \le t \le p-1$, the monomials $\alpha^i\beta^j\nu$ with $i + j \le p-1$ are linearly independent in $\prod_{t=0}^{p-1} H_{H_t}^*$. Therefore, the elements $\alpha^i\beta^j\nu$ and $\beta^i\mu$ of degrees at most $2p+1$ are linearly independent in $\prod_V H_V^*$. (We only needed this in degrees at most $2p-1$.)

Thus we have shown that $H^*(BG, \mathbf{Z})$ injects into $\prod_V H_V^*$ in odd degrees from 1 to $2p-1$. We have also seen that the elements $\alpha^i\beta^j$ in degrees at most $2p$ are also linearly independent in $\prod_V H_V^*$. In degree $2p-2$, we have the stronger statement that the elements $\alpha^j\beta^{p-1-j}$ and χ_{p-1} are linearly independent in $\prod_V H_V^*$. To show this, suppose that $f_{p-1}(\alpha,\beta)+\chi_{p-1}$ restricts to zero in $\prod_V H_V^*$. We use the relations that $\beta(\beta^{p-1}+\chi_{p-1})=0$ and $\alpha(\alpha^{p-1}+\chi_{p-1})=0$ in $H^*(BG,\mathbf{Z})$. Multiplying by β, we deduce that $\beta f_{p-1}(\alpha,\beta)-\beta^p$ maps to zero in $\prod_V H_V^*$. By the linear independence of $\alpha^i\beta^j$ for $i+j\le p$ in $\prod_V H_V^*$, it follows that $f_{p-1}=\beta^{p-1}$. But multiplying by α, we find that $\alpha\beta^{p-1}-\alpha^p$ maps to zero in $\prod_V H_V^*$, contradicting the linear independence of those monomials. This completes the proof that the elements $\alpha^j\beta^{p-1-j}$ and χ_{p-1} are linearly independent in $\prod_V H_V^*$. For $p=3$, that completes the proof: H_G^* is detected on elementary abelian p-subgroups.

By contrast, the elements χ_j for $2\le j\le p-2$ restrict to zero on all elementary abelian subgroups V of rank 2 in G (hence to all proper subgroups of G). To check this, note that for $2\le j\le p-2$ we have $\chi_j=\operatorname{tr}_H^G(\gamma^j)$. For $V\ne H$, the double coset formula (Lemma 2.15) gives that

$$\operatorname{res}_V^G\operatorname{tr}_H^G(\gamma^j))=\operatorname{tr}_{H\cap V}^V\operatorname{res}_{H\cap V}^H(\gamma^j),$$

which is zero because all transfers from proper subgroups are zero in the cohomology of V. For $V=H$, we have

$$\begin{aligned}\operatorname{res}_H^G\operatorname{tr}_H^G(\gamma^j)&=\sum_{g\in G/H} g(\gamma^j)\\&=\sum_{a\in\mathbf{F}_p}(\gamma+a\beta)^j\\&=\sum_{m=0}^{j}\binom{j}{m}\gamma^{j-m}\beta^m\sum_{a\in\mathbf{F}_p}a^m.\end{aligned}$$

Since $\sum_{a\in\mathbf{F}_p}a^m$ is zero for $0\le m\le p-2$ (Lemma 7.5), this restriction is zero.

In particular, it follows that $\operatorname{tr}_H^G 1=\operatorname{tr}_H^G\gamma=0$ in CH_G^* and hence in H_G^*, because every nonzero element of $CH_G^1=\operatorname{Hom}(G,G_m)/p\cong(\mathbf{Z}/p)^2$ has nonzero restriction to some elementary abelian subgroup of G.

Nonetheless, we will show that each element χ_i with $2\le i\le p-2$ has nonzero image in $H_C^*\otimes H_G^{\le 4}$. Since $C\subset H\subset G$, we have a pullback diagram

$$\begin{array}{ccc}C\times H&\longrightarrow&H\\ \downarrow&&\downarrow\\ C\times G&\longrightarrow&G.\end{array}$$

Because pushforward commutes with pullback (cf. Lemma 2.16), we have $(\operatorname{tr}_H^G y)|_{C\times G}=\operatorname{tr}_{C\times H}^{C\times G}(y|_{C\times H})$ for all $y\in H_H^*$. We will apply this to $y=\gamma^i$, noting that $\chi_i = \mathrm{tr}_H^G(\gamma^i)$. The pullback $H_H^* \to H_C^* \otimes_{\mathbf{F}_p} H_H^*$ sends γ to $\gamma \otimes 1 + 1 \otimes \gamma$. So γ^i pulls back to $\sum_j \binom{i}{j}\gamma^{i-j} \otimes \gamma^j$. For f in H_C^* and g in H_H^*, we have

$$\mathrm{tr}_{C\times H}^{C\times G}(f \otimes g) = f \otimes \mathrm{tr}_H^G(g),$$

by the projection formula (cf. Lemma 2.15(i)). So

$$\chi_i|_{C\times G} = \sum_j \binom{i}{j}\gamma^{i-j} \otimes \mathrm{tr}_H^G(\gamma^j).$$

Here $\mathrm{tr}_H^G 1 = 0$ and $\mathrm{tr}_H^G \gamma = 0$, but $\mathrm{tr}_H^G(\gamma^2) = \chi_2 \neq 0$ in H_G^4 by Theorem 13.22. Since $\binom{i}{2}$ is not zero in $\mathbf{F}_p$ for $2 \leq i \leq p-2$, we have shown that χ_i has nonzero image in $H_C^* \otimes H_G^{\leq 4}$ for $2 \leq i \leq p-2$.

Thus the homomorphism

$$H^*(BG, \mathbf{Z})/p \to H_C^* \otimes_{\mathbf{F}_p} H_G^{\leq 4} \times \prod_{V\subset G} H_V^*$$

is injective in degrees at most $2p-1$. Since $H^*(BG,\mathbf{Z})$ is killed by p in degrees at most $2p-1$, the Bockstein β has $\ker(\beta) = \mathrm{im}(\beta)$ on H_G^* in degrees between 1 and $2p-2$. Let x be any element of H_G^* of degree between 1 and $2p-2$ that maps to zero under the homomorphism above. Then the integral class βx must be zero by the injectivity above. So we can write $x = \beta y$ for some y in H_G^*. Thus x itself is integral, and so the injectivity above implies that $x = 0$. Thus we have shown that

$$H_G^* \to H_C^* \otimes_{\mathbf{F}_p} H_G^{\leq 4} \times \prod_{V\subset G} H_V^*$$

is injective in degrees at most $2p-2$. Combining this with the fact that $d_0(H_G^*) \leq 2p-2$, we conclude that $d_0(H_G^*) \leq 4$.

Finally, let us show that $d_0(H_G^*) \geq 4$ for $p \geq 5$. Since $2 \leq p-2$, the element χ_2 is nonzero in H_G^4 by Theorem 13.22, but it restricts to zero in H_V^* for all elementary abelian subgroups V. It remains to show that χ_2 pulls back to zero in $H_C^* \otimes H_G^{\leq 3}$ (that is, χ_2 is H_C^*-primitive). This follows from the formula above for the pullback of χ_2 to $H_C^* \otimes H_G^*$, using that $\mathrm{tr}_H^G 1 = \mathrm{tr}_H^G \gamma = 0$. □

We also record the topological nilpotence degree of the Chow ring in this case. The result is much better than the bound $d_0(CH_G^*) \leq p-1$ given by Theorem 12.7.

Theorem 13.23 *Let p be an odd prime number. Let G be the nonabelian group of order p^3 and exponent p, the extraspecial group p_+^{1+2}. Then $d_0(CH_G^*)$ is zero for $p = 3$, and is 2 for $p \geq 5$.*

Proof The calculation of CH_G^* in Corollary 13.15 shows that CH_G^* injects into $H^*(BG,\mathbf{Z})/p$ and hence into H_G^*. By Lemma 13.20, it follows that $d_0(CH_G^*) \leq$

$d_0(H_G^*)/2$. By Theorem 13.21, it follows that $d_0(CH_G^*)$ is zero for $p = 3$ and is at most 2 for $p \geq 5$. For $p \geq 5$, the proof of Theorem 13.21 shows that the element χ_2 comes from CH^2BG and is nonzero and H_C^*-primitive in H_G^*. So χ_2 is nonzero and CH_C^*-primitive in CH_G^*. Therefore, $d_0(CH_G^*)$ is equal to 2. □

So far, we have only seen p-groups with $d_0(H_G^*)$ comparable in size to the p-rank of G. In the case $p = 2$, Kuhn gave examples of 2-central 2-groups with $d_0(H_G^*)$ fairly large: using the numbering of the Small Groups library, group #32 of order 32 has p-rank 2 and type [4, 4], and hence $d_0 = 6$, and group #245 of order 64, the 2-Sylow subgroup of the simple group $U_3(4)$, has p-rank 2 and type [8, 8], and hence $d_0 = 14$ [85, theorem 2.9 and appendix A]. We now exhibit a p-central group for p odd with d_0 fairly large. The group we use belongs to a family of p-central groups considered by Browder-Pakianathan and Weigel [23, 151]. The group in Theorem 13.24 is not unusual; many similar groups also have d_0 large, as the proof will show.

Theorem 13.24 *Let p be an odd prime number. Let G be the group*

$$G = \langle e_1, e_2, e_3, f_1, f_2, f_3 : e_i^p = f_i,\ f_i^p = 1,\ f_i \text{ central},$$
$$[e_1, e_2] = f_1, [e_2, e_3] = f_2, [e_3, e_1] = f_3\rangle.$$

Then G is a p-central group of p-rank 3 and order p^6, and $d_0(H_G^*) = 2p + 1$.

Proof The group G is a central extension

$$1 \to (\mathbf{Z}/p)^3 \to G \to (\mathbf{Z}/p)^3 \to 1,$$

with the normal subgroup $C := (\mathbf{Z}/p)^3$ generated by the elements f_i. The fact that G has order equal to p^6 (not smaller) is a special case of [23, proposition 2.5]. From the relations $e_i^p = f_i$, we see that every element of order p in G is in C, and so G is p-central.

By inspection of the relations, the commutator subgroup $[G, G]$ is equal to C. So the restriction map $H_G^1 \to H_C^1$ is zero. It follows that the image of $H_G^* \to H_C^*$ is a polynomial ring of the form $\mathbf{F}_p[y_1^{p^{a_1}}, y_2^{p^{a_2}}, y_3^{p^{a_3}}]$ for some natural numbers a_1, a_2, a_3, as discussed before Definition 12.5. Here y_1, y_2, y_3 form a basis for $\beta(H_C^1) \subset H_C^2$. Since G is p-central, Theorem 12.6 says that $d_0(H_G^*) = \sum_{i=1}^{3}(2p^{a_i} - 1)$. So we have $d_0(H_G^*) \geq (2p - 1) + 1 + 1 = 2p + 1$ if we can show that H_G^2 does not map onto $\beta(H_C^1) \subset H_C^2$.

The fact that $H_G^2 \to H_C^2$ does not map onto $\beta(H_C^1)$ is a special case of Browder-Pakianathan [23, theorem 2.10], using that the alternating bilinear product $[e_1, e_2] = e_1$, $[e_2, e_3] = e_2$, $[e_3, e_1] = e_3$ on $(\mathbf{F}_p)^3$ is not a Lie algebra; that is, the alternating 3-form

$$J(x, y, z) = [[x, y], z] + [[y, z], x] + [[z, x], y]]$$

on $(\mathbf{F}_p)^3$ is not zero. (This would be true for most choices of relations, not just the ones we chose.)

To compute $d_0(H_G^*)$ exactly (and also prove by hand that $H_G^2 \to H_C^2$ does not map onto $\beta(H_C^1)$), we use the Hochschild-Serre spectral sequence for the group extension $1 \to C \to G \to Q \to 1$, where $C \cong Q \cong (\mathbf{Z}/p)^3$.

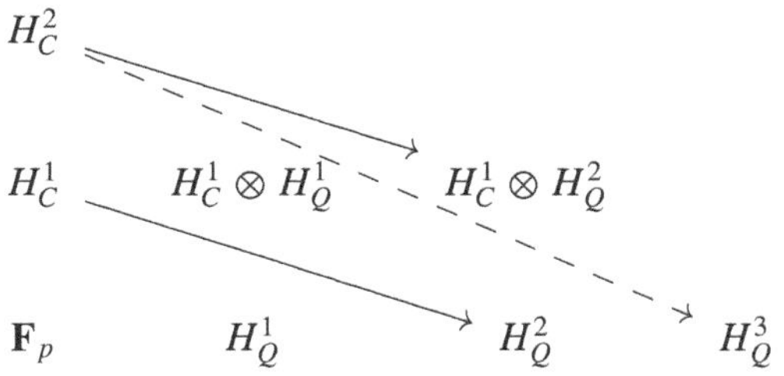

Write $H_C^* = \mathbf{F}_p\langle x_i, y_i\rangle$ and $H_Q^* = \mathbf{F}_p\langle u_i, v_i\rangle$, where $1 \leq i \leq 3$, $|x_i| = |u_i| = 1$, $y_i = \beta x_i$, and $v_i = \beta u_i$. The differential $d_2\colon H_C^1 \to H_Q^2$ has the form $d_2(x_1) = v_1 + u_1u_2$, $d_2(x_2) = v_2 + u_2u_3$, and $d_2(x_3) = v_3 + u_3u_1$, by the relations defining G. The differential d_2 is zero on $y_i = \beta x_i$, and $d_3(\beta x_i) = \beta d_2(x_i)$ [12, vol. 2, theorem 4.8.1]. So $d_3(y_1) = v_1u_2 - u_1v_2$, $d_3(y_2) = v_2u_3 - u_2v_3$, and $d_3(y_3) = v_3u_1 - u_3v_1$. These expressions are elements of $E_3^{*,0}$, the quotient of H_Q^* by the image of d_2. In this quotient ring, we have $v_1 = -u_1u_2$, $v_2 = -u_2u_3$, and $v_3 = -u_3u_1$. It follows that $d_3(y_1) = d_3(y_2) = d_3(y_3) = u_1u_2u_3$. (This gives another proof that $H_G^2 \to H_C^2$ is not surjective.)

Since d_3 vanishes on $y_2 - y_1$ and $y_3 - y_1$, those elements are permanent cycles in the spectral sequence. That is, those elements of $\beta(H_C^1)$ are in the image of H_G^2. We will show that y_1^p is in the image of H_G^*; that will imply that $d_0(H_G^*) = e(G) \leq (2p-1) + (2-1) + (2-1) = 2p+1$, as we want.

Let $H \subset G$ be the subgroup of index p such that $\ker(H_G^1 \to H_H^1)$ is spanned by u_3. A similar calculation to the one above shows that for the central extension $1 \to C \to H \to H/C \to 1$, $d_3(y_i)$ is zero for all $1 \leq i \leq 3$. That is, the restriction $H_H^2 \to H_C^2$ maps onto βH_C^1. In particular, there is an element z in H_H^2 that restricts to y_1 on C. Then the norm $N_H^G(z)$ is an element of H_G^{2p} that restricts to y_1^p on C by Lemma 8.1(v), using that C is central in G. The proof is complete. □

14

Groups of Order p^4

In this chapter, we determine the Chow ring for the 15 groups of order 81 and for 13 of the 15 groups of order p^4 with $p \geq 5$. Although we have tried to give arguments that apply directly to Chow rings, in the hardest cases we use calculations of group cohomology by Leary and Yagita. We prove some general results about when the Chow ring injects into cohomology.

As in Chapter 13, we consider each finite p-group G of exponent e as an algebraic group over any field k of characteristic not p that contains the eth roots of unity. We find that the groups in this chapter have the same mod p Chow ring over all such fields.

14.1 The wreath product $\mathbf{Z}/3 \wr \mathbf{Z}/3$

Lemma 14.1 *Let G be the wreath product $\mathbf{Z}/3 \wr \mathbf{Z}/3$, also known as the Sylow 3-subgroup of the symmetric group S_9 or as #7 of the groups of order 81 in the Small Groups library [52]. Then*

$$CH_G^* = \mathbf{F}_3[y_1, y_2, w, \delta, u]/(\delta^2 = y_1^2 y_2^2 + y_2^3 - y_1^3 w,\ y_1 u = y_2 u = \delta u = 0),$$

where $|y_i| = i$, $|w| = 3$, $|\delta| = 3$, and $|u| = 1$. The ring CH_G^* has dimension 3 and depth 2, whereas the Duflot bound gives only that the depth is at least 1. The topological nilpotence degree $d_0(CH_G^*)$ is zero; that is, CH_G^* is detected on elementary abelian subgroups., Finally, $CH_G^* \to H_G^*$ is injective.

Proof The paper [138, sections 8 and 9] gives an additive description of the mod p Chow ring of a wreath product $G = \mathbf{Z}/p \wr H$, assuming that k contains the pth roots of unity and that H is one of the groups in Lemma 2.21. Namely, let elements e_i for i in a set I be a basis for CH_H^* as an $\mathbf{F}_p$-vector space. Let $u \in CH_G^1$ be the first Chern class of the representation $\mathbf{Z}/p \to k^*$ sending the generator to a primitive pth root of unity. Then a basis for CH_G^* as an $\mathbf{F}_p$-vector space is given by the transfers $\mathrm{tr}^G_{H^p}(e_{i_1} \otimes \cdots \otimes e_{i_p})$ for $(i_1, \ldots, i_p)$ running over a set of representatives for the free action of $\mathbf{Z}/p$ on $I^p - \Delta_I$, together with the norms $u^j N^G_{H^p}(e_i \otimes 1 \otimes \cdots \otimes 1)$ for $i \in I$ and $j \geq 0$.

Let $p = 3$ and $H = \mathbf{Z}/3$. Write $CH^*_H = \mathbf{F}_3[v]$ with $|v| = 1$. We deduce that an $\mathbf{F}_3$-basis for CH^*_G is given by the elements $\mathrm{tr}^G_{H^3}(v^a \otimes v^b \otimes v^c)$ with $0 \leq a \leq b$ and $a < c$, together with the elements $u^b N^G_{H^3}(v^a \otimes 1 \otimes 1)$ with $a \geq 0$ and $b \geq 0$. Let $w = N^G_{H^3}(v \otimes 1 \otimes 1)$ in CH^3_G. By multiplicativity of the norm, we can write the second set of elements as $w^a u^b$ with $a, b \geq 0$.

By Lemma 2.21 and its proof,

$$CH^*_G \to CH^*_{H^3} \times CH^*_{\mathbf{Z}/3 \times H}$$

is injective, and CH^*_G maps onto the invariants $(CH^* H^3)^{\mathbf{Z}/3}$. This ring of invariants is $\mathbf{F}_3[x_1, x_2, x_3]^{\mathbf{Z}/3}$. By thinking of $\mathbf{Z}/3$ as the alternating group A_3, we see that this ring of invariants is Cohen-Macaulay, as discussed in Section 5.2:

$$\mathbf{F}_3[x_1, x_2, x_3]^{\mathbf{Z}/3} = \mathbf{F}_3[e_1, e_2, e_3]\{1, \Delta\},$$

where e_i is the ith elementary symmetric polynomial and $\Delta = \prod_{i<j}(x_i - x_j)$ is the square root of the discriminant. By the formula for the discriminant of a cubic polynomial [88, exercise VI.12], we have

$$\begin{aligned}
&\mathbf{F}_3[x_1, x_2, x_3]^{\mathbf{Z}/3} \\
&\quad = \mathbf{F}_3[e_1, e_2, e_3, \Delta]/(\Delta^2 = e_1^2 e_2^2 - 4e_2^3 - 4e_1^3 e_3 - 27e_3^2 + 18e_1e_2e_3) \\
&\quad = \mathbf{F}_3[e_1, e_2, e_3, \Delta]/(\Delta^2 = e_1^2 e_2^2 - e_2^3 - e_1^3 e_3)
\end{aligned}$$

The restriction of the norm w to $CH^*_{H^3}$ is $x_1x_2x_3 = e_3$. We can choose other elements of CH^*_G that restrict to e_1, e_2, and Δ. Namely, let $y_1 = \mathrm{tr}^G_{H^3}(1 \otimes 1 \otimes v)$ and $y_2 = \mathrm{tr}^G_{H^3}(1 \otimes v \otimes v)$; then $y_1|_{H^3} = x_1 + x_2 + x_3 = e_1$ and $y_2|_{H^3} = x_1x_2 + x_2x_3 + x_3x_1 = e_2$. Also, $\Delta = (x_1^2x_2 + x_2^2x_3 + x_3^2x_1) - (x_1x_2^2 + x_2x_3^2 + x_3x_1^2)$. So the element $\delta := \mathrm{tr}^G_{H^3}(1 \otimes v^2 \otimes v) - \mathrm{tr}^G_{H^3}(1 \otimes v \otimes v^2)$ of CH^3_G restricts to Δ.

Given any element of CH^*_G, we can subtract off a polynomial in y_1, y_2, e, δ of degree at most 1 in δ and get an element that restricts to zero on H^3. From the $\mathbf{F}_3$-basis for CH^*_G above, we read off that any element of CH^*_G that restricts to zero on the subgroup H^3 is an $\mathbf{F}_3$-linear combination of the elements $w^a u^b$ with $a \geq 0$ and $b > 0$. We therefore have a surjection

$$\mathbf{F}_3[y_2, y_2, w, \delta, u] \twoheadrightarrow CH^*_G.$$

Since u restricts to zero in the Chow ring of H^3, the projection formula gives that $u\, \mathrm{tr}^G_{H^3}(v^a \otimes v^b \otimes v^c) = 0$ for all a, b, c. So $y_1 u = y_2 u = \delta u = 0$ in CH^*_G.

Next, by our description of the invariant ring $(CH^*_{H^3})^{\mathbf{Z}/3}$, we know that

$$\delta^2 - y_1^2 y_2^2 + y_2^3 + y_1^3 w$$

in CH^*_G restricts to zero on the subgroup H^3. But the pullback diagram

$$\begin{array}{ccc} H & \longrightarrow & H^3 \\ \downarrow & & \downarrow \\ \mathbf{Z}/3 \times H & \longrightarrow & \mathbf{Z}/3 \ltimes H^3 \end{array}$$

implies that

$$\begin{aligned} (\mathrm{tr}^G_{H^3}(s))|_{\mathbf{Z}/3\times H} &= \mathrm{tr}_H^{\mathbf{Z}/3\times H}(s|_H) \\ &= 0 \end{aligned}$$

for every $s \in CH^*_{H^3}$. So the elements y_1, y_2, δ restrict to zero on the diagonal subgroup $\mathbf{Z}/3 \times H$. So the element $\delta^2 - y_1^2 y_2^2 + y_2^3 + y_1^3 w$ of CH^6_G restricts to zero on both $\mathbf{Z}/3 \times H$ and H^3, and hence is zero.

Thus we have a surjection

$$\mathbf{F}_3[y_1, y_2, w, \delta, u]/(\delta^2 - y_1^2 y_2^2 + y_2^3 + y_1^3 w = 0, y_1 u = y_2 u = \delta u = 0) \twoheadrightarrow CH^*_G.$$

By our $\mathbf{F}_3$-basis for CH^*_G, this map is an isomorphism.

We compute that this ring has depth 2, although the center $\mathbf{Z}/3$ of G only has rank 1. Finally, it is a general fact about wreath products that CH^*_G is detected on elementary abelian subgroups (Lemma 2.21). It follows that $CH^*_G \to H^*_G$ is injective. □

14.2 Geometric and topological filtrations

For several calculations of Chow rings of groups of order p^4, we will need the basic properties of the geometric and topological filtrations of the representation ring. Chapter 15 discusses these filtrations in more detail, and gives examples of p-groups for every prime number p such that the two filtrations differ.

For an affine group scheme G over a field k, a representation of G determines a vector bundle on the classifying space BG. The geometric filtration of the representation ring $R(G)$ comes from the filtration of the algebraic K-group $K_0 BG$ by codimension of support. When $k = \mathbf{C}$, the topological filtration comes from the filtration of the topological K-group $K^0 BG$ by codimension of support. Concretely, on a finite cell complex X approximating BG, the codimension of support of an element u in $K^0 X$ means the largest codimension of a closed subcomplex S such that u restricts to zero in $K^0(X - S)$.

The topological filtration of the representation ring is also the filtration given by the Atiyah-Hirzebruch spectral sequence [6, 7, 9]:

Theorem 14.2 *For any topological space X, there is a spectral sequence from ordinary cohomology to topological K-theory,*

$$E_2^{ij} = H^i(X, K^j(\mathit{point})) \Rightarrow K^{i+j}X.$$

The sequence is strongly convergent for X a finite CW complex or the classifying space of a compact Lie group. It is periodic of order 2 in the vertical direction and concentrated in even rows, since $K^j(\mathit{point})$ is isomorphic to $\mathbf{Z}$ for j even and to 0 for j odd. For $X = BG$ with G a compact Lie group, K^0BG is isomorphic to the completion $R(G)^\wedge$ of the complex representation ring with respect to powers of the augmentation ideal, while $K^1BG = 0$.

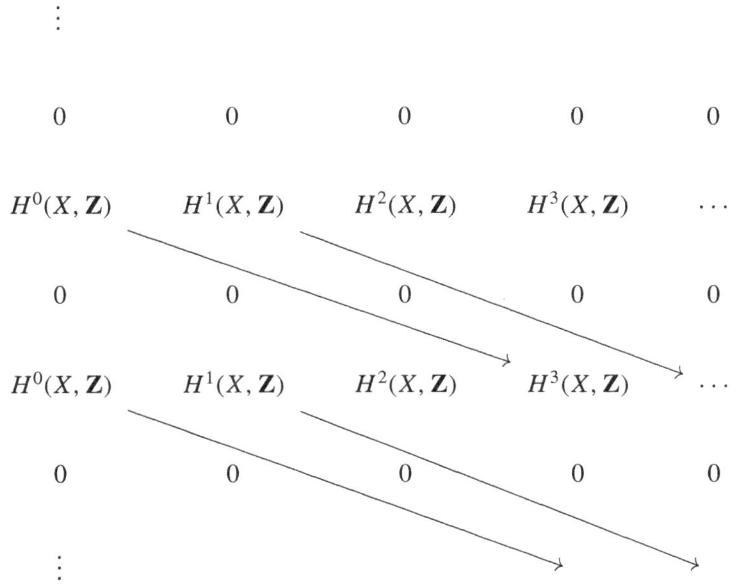

We have $F^i_{\mathrm{geom}}R(G) \subset F^{2i}_{\mathrm{top}}R(G)$, and hence there is a natural map $\mathrm{gr}^i_{\mathrm{geom}}R(G) \to \mathrm{gr}^{2i}_{\mathrm{top}}R(G)$. The Atiyah-Hirzebruch spectral sequence shows that $\mathrm{gr}^{2i}_{\mathrm{top}}R(G)$ is a subquotient of $H^{2i}(BG, \mathbf{Z})$ and that the topological filtration is concentrated in even degrees (that is, $F^{2i-1}_{\mathrm{top}}R(G) = F^{2i}_{\mathrm{top}}R(G)$). Similarly, we can describe the geometric filtration of $R(G)$ as the filtration of K_0BG associated to the spectral sequence from motivic cohomology to algebraic K-theory, $E_2^{pq} = H^p(BG, \mathbf{Z}(-q/2)) \Rightarrow K_{-p-q}BG$ [95, theorem 2.9], which we write out as Theorem 15.12. (The groups contributing to $K_0BG = R(G)^\wedge$ are the Chow groups $CH^iBG = H^{2i}(BG, \mathbf{Z}(i))$.) The relation between Chow groups and the geometric filtration is simpler than what happens in topology: the natural map $CH^i BG \to \mathrm{gr}^i_{\mathrm{geom}} R(G)$ is surjective for all i and is an isomorphism p-locally for $i \leq p$ (Theorem 2.25).

For completeness, we also define the γ-filtration of the representation ring $R(G)$ [44, chapter III]. For a representation E of G, write $\lambda^j(E)$ for the jth exterior power of E, and let $\lambda_t(E) = \sum_{j\geq 0} \lambda^j(E)t^j$ in the power series ring $R(G)[[t]]$. This definition extends to an operation $\lambda_t \colon R(G) \to R(G)[[t]]$ by $\lambda_t(E - F) = \lambda_t(E)\lambda_t(F)^{-1}$. Grothendieck defined operations γ^j on $R(G)$ by $\lambda_{t/(1-t)} E = \sum_{j\geq 0} \gamma^j(E)$. The γ-filtration is defined by: $F^0_\gamma R(G) = R(G)$, $F^1_\gamma R(G)$ is the kernel of the rank homomorphism $R(G) \to \mathbf{Z}$, and $F^m_\gamma R(G)$ for $m \geq 1$ is the subgroup generated by elements $\gamma^{j_1}(x_1) \cdots \gamma^{j_n}(x_n)$ with

$$x_1, \ldots, x_n \in F^1_\gamma R(G) \text{ and } \sum j_i \geq m.$$

One basic result on the γ-filtration is that $F^m_\gamma R(G)$ is contained in $F^m_{\mathrm{geom}} R(G)$.

14.3 Groups of order p^4 for $p \geq 5$

Theorem 14.3 *Let G be a finite group and p a prime number at least 5. Suppose that G has p-rank at most 2 and that G (or just its Sylow p-subgroup) has a faithful complex representation of the form $W \oplus X$ where W has dimension at most p and X is a sum of 1-dimensional representations. For e the exponent of G, consider G as an algebraic group over any subfield k of $\mathbf{C}$ that contains the eth roots of unity. Then the mod p Chow ring CH^*_G maps isomorphically to $H^{\mathrm{ev}}(BG, \mathbf{Z})/p$.*

Using Theorem 14.3, we can compute the Chow ring for 13 of the 15 groups of order p^4, for $p \geq 5$. Indeed, there are five abelian groups of order p^4 and two products of a nonabelian group of order p^3 with $\mathbf{Z}/p$. In those cases, we know the Chow rings by Section 13.2 and the Chow Künneth formula (Lemma 2.12).

Of the remaining eight groups, five have rank 2 and three have rank 3, for $p \geq 5$. The five groups of rank 2 (numbers 4, 6, 9, 10, 14 in the Small Groups library [52]) are handled by Theorem 14.3. In more detail, groups #4 and #6 are the split metacyclic groups $\mathbf{Z}/p^2 \ltimes \mathbf{Z}/p^2$ and $\mathbf{Z}/p \ltimes \mathbf{Z}/p^3$, whose Chow rings are given in Theorem 13.12. Groups #9 and #10 are discussed below. Group #14 is the central product $p^{1+2}_+ * C_{p^2}$, whose Chow ring is computed in Lemma 13.16. Of the three groups of rank 3 (numbers 3, 7, 8), we computed the Chow ring of group #3 in Lemma 13.13, while Section 14.5 suggests an approach to the open cases, groups #7 and #8.

The groups of order p^4 played a significant role in the history of group cohomology. C. B. Thomas conjectured that for a p-group G of rank at most 2,

$H^{\text{ev}}(BG, \mathbf{Z})$ should be generated by Chern classes of complex representations [135, chapter 8]. (The assumption on the rank is natural, since the statement fails for abelian groups of rank 3.) The conjecture was verified for various classes of p-groups of rank 2, such as metacyclic p-groups [71, 134]. But Leary-Yagita showed that $H^{\text{ev}}(BG, \mathbf{Z})$ is not generated by Chern classes for certain p-groups $G(n, \epsilon)$ of rank 2 [93]:

$$\begin{aligned} G(n, \epsilon) = \langle x, y, z : x^p &= y^p = z^{p^{n-2}} \\ &= [y, z] = 1, [x, z^{-1}] = y, [y, x] = z^{\epsilon p^{n-3}} \rangle, \end{aligned}$$

where $p \geq 5, n \geq 4$, the group has order p^n, and for fixed p and n there are two isomorphism classes of such groups, depending on whether ϵ is 1 or a quadratic non-residue modulo p. These include groups #9 and #10 of order p^4 for $p \geq 5$. Nonetheless, Yagita proved a strong substitute for Thomas's conjecture: for $p \geq 5$, every p-group of rank at most 2 has $H^{\text{ev}}(BG, \mathbf{Z})$ generated by transferred Euler classes [152]. In particular, Yagita computed the integral cohomology of groups #9 and #10, which gives the Chow rings by Theorem 14.3.

Proof of Theorem 14.3 It suffices to prove the theorem for G a p-group, by the properties of transfer for Chow rings and cohomology. By Yagita, since G has rank at most 2 and $p \geq 5$, $H^{\text{ev}}(BG, \mathbf{Z})$ is generated by transferred Euler classes [152]. The assumption on k implies that the representation theory of G over k is the same as over $\mathbf{C}$. It follows that the homomorphism $CH^*BG \to H^{\text{ev}}(BG, \mathbf{Z})$ is surjective.

We use the geometric and topological filtrations of the representation ring, as discussed in Section 14.2. (We consider the geometric filtration of the representation ring of G over k, and the topological filtration of the representation ring over $\mathbf{C}$; these are filtrations of the same ring $R(G)$, by our assumption on k.) For a k-representation V of G of dimension n and any $i > 0$, the element $\gamma^i(V - n)$ of the representation ring lies in $F^i_{\text{geom}}R(G)$, and its class in $\text{gr}^i_{\text{geom}}R(G)$ is the image of the Chern class c_iV in CH^iBG [44, chapter III]. So the image of $\gamma^i(V - n)$ in $\text{gr}^{2i}_{\text{top}}R(G)$ is represented by the Chern class c_iV in $H^{2i}(BG, \mathbf{Z})$, which is a permanent cycle in the Atiyah-Hirzebruch spectral sequence.

For the group G of rank 2, since $CH^iBG \to H^{2i}(BG, \mathbf{Z})$ is surjective for all i, the map $\text{gr}^i_{\text{geom}}R(G) \to \text{gr}^{2i}_{\text{top}}R(G)$ is surjective for all natural numbers i. Moreover, the two filtrations both define the same topology on $R(G)$, that associated to the powers of the augmentation ideal [9, corollary 2.3]. Therefore, this surjectivity implies that the two filtrations of $R(G)$ are actually the same. In particular, the map $\text{gr}^i_{\text{geom}}R(G) \to gr^{2i}_{\text{top}}R(G)$ is an isomorphism for all i. We have $CH^iBG \cong \text{gr}^i_{\text{geom}}R(G)$ for $i \leq p$ (Theorem 2.25), and so we have isomorphisms

$$CH^i BG \cong \mathrm{gr}^i_{\mathrm{geom}} R(G) \cong \mathrm{gr}^{2i}_{\mathrm{top}} R(G)$$

for $i \le p$. Since $\mathrm{gr}^{2i}_{\mathrm{top}} R(G)$ is a subquotient of $H^{2i}(BG, \mathbf{Z})$, it follows that the map $CH^i BG \to H^{2i}(BG, \mathbf{Z})$ is injective for $i \le p$. Since it is also surjective, it is an isomorphism for $i \le p$.

In particular, $CH^i_G = CH^i(BG)/p$ maps isomorphically to $H^{2i}(BG, \mathbf{Z})/p$ for $i \le p$, and so CH^i_G injects into $H^{2i}_G = H^{2i}(BG, \mathbf{F}_p)$ for $i \le p$. Now we use the assumption that G has a faithful complex representation of the form $W \oplus X$, where W has dimension at most p and X is a sum of 1-dimensional representations. By Theorem 12.7, the restriction map

$$CH^*_G \to \prod_{\substack{V \subset G \\ V \text{ elem ab}}} CH^*_V \otimes CH^{\le p-1}_{C_G(V)}$$

is injective. Our assumptions on G (rank at most 2 and faithful representation $W \oplus X$) pass to the subgroups $C_G(V)$, and so we know that $CH^{\le p-1}_{C_G(V)} \to H^{\le 2p-2}_{C_G(V)}$ is injective, for every elementary abelian subgroup V of G. Likewise, CH^*_V injects into H^*_V. So CH^*_G injects into H^*_G. We also know that CH^*_G maps onto $H^{\mathrm{ev}}(BG, \mathbf{Z})/p \subset H^*_G$, and so CH^*_G maps isomorphically to $H^{\mathrm{ev}}(BG, \mathbf{Z})/p$. □

14.4 Groups of order 81

Let G be #7, #8, #9, or #10 of the groups of order 81 in the Small Groups library [52]. Leary computed the integral cohomology of these four groups in a unified way: they are all normal subgroups of the same 1-dimensional group $\widetilde{G}$,

$$1 \to G \to \widetilde{G} \to S^1 \to 1$$

[92, corollary 9 and theorem 10]. (In more detail, $\widetilde{G}$ is a central extension of the extraspecial 3-group 3^{1+2}_+ by S^1.) For groups #8, #9, and #10, Leary showed that $H^{\mathrm{ev}}(BG, \mathbf{Z})$ is generated by Chern classes of complex representations. (That fails for #7, the wreath product $\mathbf{Z}/3 \wr \mathbf{Z}/3$, which has rank 3; but we have already computed the Chow ring in that case as Lemma 14.1.) In all three cases, G has a faithful irreducible complex representation of dimension 3. Given these facts, the proof of Theorem 14.3 shows that the cycle map

$$CH^* BG \to H^{\mathrm{ev}}(BG, \mathbf{Z})$$

is an isomorphism for groups #8, #9, and #10. We record the mod 3 Chow rings of these groups, using Leary's calculation of their integral cohomology.

This completes the calculation of the Chow ring for the 15 groups of order 81. First, there are five abelian groups of order 81 and two products of a nonabelian group of order 27 with $\mathbf{Z}/3$. In those cases, we know the Chow rings by Section 13.2 and the Chow Künneth formula (Lemma 2.12). For $p = 3$, two of the remaining eight groups have rank 3 and six have rank 2. The groups of rank 3 are the wreath product $\mathbf{Z}/3 \wr \mathbf{Z}/3$ (group #7 in the Small Groups library) and group #3, whose Chow rings are computed in Lemmas 14.1 and 13.13. The groups of rank 2 are the split metacyclic group $\mathbf{Z}/9 \ltimes \mathbf{Z}/9$ (group #4), the modular group $\mathbf{Z}/3 \ltimes \mathbf{Z}/27$ (group #6), the central product $3^{1+2}_{+} * C_9$ (group #14), and groups #8, #9, and #10. The Chow rings are computed in Lemmas 13.12, 13.16, and 14.4.

To simplify the statement, we first give the ring $R = H^{\text{ev}}(\widetilde{G}, \mathbf{Z})/3$:

$$\begin{aligned} R = \mathbf{F}_3[\alpha, \beta, \delta_1, \delta_2, \delta_3, \zeta]/(&\alpha\delta_1 + \alpha\beta = 0, \delta_1^2 + \delta_1\beta = 0, \alpha\delta_2 = 0, \\ &\delta_1\delta_2 = 0, \alpha^2\beta + \delta_1\beta^2 = 0, \alpha\zeta = 0, \delta_1\zeta = 0, \\ &\delta_2^3 + \zeta^2 + \delta_3\delta_1\beta^2 + \delta_3\beta^3 - \delta_2^2\beta^2 - \delta_2\beta^4 + \delta_1\beta^5 + \beta^6 = 0), \end{aligned}$$

where $|\alpha| = |\beta| = 2$, $|\delta_i| = 2i$, and $|\zeta| = 6$. To correct two typos in the published paper [92]: the relation for δ_1^2 in Theorem 10 should be $\delta_1^2 = 3\delta_2 - \delta_1\beta$ (as the proof shows), and the character table for $G(n, e)$ (before Proposition 2) should contain the entry $\eta(2 + \eta^{e3^{n-3}})$ for $e = \pm 1$, where η is any primitive 3^{n-2}th root of unity. For what follows, we repeat our convention that the commutator $[A, B]$ means $ABA^{-1}B^{-1}$.

Lemma 14.4 *Let G be #8 of the groups of order 81 in the Small Groups library [52], which Leary calls $G(4, 1)$,*

$$G = \langle A, B, C : A^3 = B^9 = C^3 = [B, C] = 1, [B, A] = C, [C, A] = B^3 \rangle.$$

Then CH^*_G is the quotient of the ring R above (with grading $|\alpha| = |\beta| = 1$, $|\delta_i| = i$, and $|\zeta| = 3$) by the ideal generated by $\delta_1 - \beta$. The ring CH^*_G has dimension 2 and depth 1, which agrees with the Duflot bound. The topological nilpotence degree $d_0(CH^*_G)$ is equal to 1. The group G has 2 conjugacy classes of maximal elementary abelian subgroups, both of rank 2.

Let G be #9 of the groups of order 81 in the Small Groups library [52], which Leary calls $G(4, -1)$, the Sylow 3-subgroup of the simple group $U_3(8) = PSU(3, \mathbf{F}_8)$:

$$G = \langle A, B, C : A^3 = B^9 = C^3 = [B, C] = 1, [B, A] = C, [C, A] = B^{-3} \rangle.$$

Then CH^*_G is the quotient of the ring R above by the ideal generated by $\delta_1 + \beta$. The ring CH^*_G is Cohen-Macaulay of dimension 2 (so it has depth 2), whereas the Duflot bound gives only that the depth is at least 1. The ring CH^*_G is reduced, and is detected on elementary abelian subgroups; that is, $d_0(CH^*_G) = 0$. The *group G has 4 conjugacy classes of maximal elementary abelian subgroups, all of rank 2.*

Let G be #10 of the groups of order 81 in the Small Groups library [52], which Leary calls $G'(4)$:

$$G = \langle A, B, C : B^9 = C^3 = [B, C] = 1, [B, A] = C, [C, A] = B^{-3} = A^3 \rangle.$$

Then CH_G^* is the quotient of the ring R above by the ideal generated by $\delta_1 + \beta + \alpha$. The ring CH_G^* has dimension 2 and depth 1, which agrees with the Duflot bound. We have $d_0(CH_G^*) = 2$. The group G has a unique conjugacy class of maximal elementary abelian subgroups, which is of rank 2.

14.5 A 1-dimensional group

Let p be a prime number at least 5. Let G be group #7 of order p^4 in the Small Groups library,

$$G = \langle h_1, h_2, h_3, h_4 : h_i^p = 1 \text{ for all } i,\ h_4 \text{ central},$$
$$[h_1, h_2] = h_3, [h_1, h_3] = h_4, [h_2, h_3] = 1 \rangle.$$

Then G has rank 3 and (using that $p \geq 5$) exponent p. The center of G is $\langle h_4 \rangle \cong \mathbf{Z}/p$.

Let $\widetilde{B} = (G \times G_m)/(\mathbf{Z}/p)$, where the subgroup $\mathbf{Z}/p$ is generated by (h_4, ζ_p^{-1}). Since $G/Z(G)$ is the extraspecial group E_{p^3}, $\widetilde{B}$ is a central extension

$$1 \to G_m \to \widetilde{B} \to E_{p^3} \to 1.$$

Any nontrivial central extension of E_{p^3} by G_m over $\mathbf{C}$ is isomorphic to $\widetilde{B}$ as an algebraic group, which explains why it is useful to consider the Chow ring of $\widetilde{B}$. In fact, groups #7, #8, #9, and #10 of order p^4 all occur as kernels of homomorphisms from $\widetilde{B}$ to G_m. As a result, the Chow rings of these four groups would follow immediately, using Theorem 13.10, if we knew the Chow ring of $\widetilde{B}$. That would complete the computation of the Chow ring for all groups of order p^4, by the discussion after Theorem 14.3. These groups are different in other ways; for example, groups #7 and #8 have rank 3 while groups #9 and #10 have rank 2.

Leary gave partial results on the integral cohomology $H^*(B\widetilde{B}, \mathbf{Z})$. In particular, he showed that $H^{\text{ev}}(B\widetilde{B}, \mathbf{Z})$ is generated by transferred Chern classes (using only transfers from finite-index subgroups) [89, corollary 3.10]. This is related to Yagita's construction of generators for the rank-2 subgroups #9 and #10 of $\widetilde{B}$ [152, theorem 5.29]. The general results of this book say a lot about the Chow ring of $B\widetilde{B}$, but we leave the full computation as an open problem.

We will just prove a relation between the Chow ring and cohomology for $\widetilde{B}$. The proof is similar to that of Theorem 14.3, but we have to argue separately because $\widetilde{B}$ is a 1-dimensional group of p-rank 3 rather than a finite group of p-rank 2.

Lemma 14.5 *Let p a prime number at least 5, k a subfield of $\mathbf{C}$ that contains the p^2 roots of unity, and let $\widetilde{B}$ be the above 1-dimensional group over k. Then the mod p Chow ring $CH^*_{\widetilde{B}}$ maps isomorphically to $H^{\text{ev}}(B\widetilde{B}, \mathbf{Z})/p$.*

Proof Because G has exponent p, every element of order p is $\widetilde{B}$ is contained in the subgroup $G \subset \widetilde{B}$. So every elementary abelian p-subgroup of $\widetilde{B}$ is contained in G. Some examples of elementary abelian subgroups of G are $A = \langle h_2, h_3, h_4\rangle \cong (\mathbf{Z}/p)^3$ and, for $i \in \mathbf{Z}/p$, $H_i = \langle h_1 h_2^i, h_4\rangle \cong (\mathbf{Z}/p)^2$. Here A is normal in G, while the subgroups H_i are not normal. Using that, it is straightforward to check that A and the p subgroups H_i are the only maximal elementary abelian subgroups of G up to conjugation. The centralizer $C_G(A)$ is equal to A, and likewise $C_G(H_i) = H_i$ for each i. It follows that A and the subgroups H_i are the only maximal elementary abelian p-subgroups of $\widetilde{B}$ up to conjugation, and we have $C_{\widetilde{B}}(A) = (A \times G_m)/(\mathbf{Z}/p) \cong (\mathbf{Z}/p)^2 \times G_m$ and $C_{\widetilde{B}}(H_i) = (H_i \times G_m)/(\mathbf{Z}/p) \cong \mathbf{Z}/p \times G_m$.

Since some results in this book are proved only for finite groups, it is convenient to observe that the mod p Chow ring of $\widetilde{B}$ is isomorphic to the mod p Chow ring of a certain finite subgroup. Let $\alpha\colon \widetilde{B} \to G_m$ be the homomorphism that pulls back on $G \times G_m$ to $(g, \lambda) \mapsto \lambda^p$; then $\widetilde{B}$ is the kernel of α. Let K be the kernel of α^p; then K is the central product $G * C_{p^2}$, of order p^5. Theorem 13.10 gives that the restriction map $CH^*_{\widetilde{B}} \to CH^*_K$ is an isomorphism. This allows us to apply some results on the Chow rings of finite groups to the Chow ring of $\widetilde{B}$.

By Leary, $H^{\text{ev}}(B\widetilde{B}, \mathbf{Z})$ is generated by transferred Euler classes [89, corollary 3.10]. The assumption on k implies that the representation theory of $\widetilde{B}$ over k is the same as over $\mathbf{C}$. It follows that the homomorphism $CH^*B\widetilde{B} \to H^{\text{ev}}(B\widetilde{B}, \mathbf{Z})$ is surjective.

We use the geometric and topological filtrations of the representation ring, as discussed in Section 14.2. The above surjectivity implies that $\text{gr}^i_{\text{geom}} R(\widetilde{B}) \to \text{gr}^{2i}_{\text{top}} R(\widetilde{B})$ is surjective for all natural numbers i. Moreover, the two filtrations both define the same topology on $R(\widetilde{B})$, that associated to the powers of the augmentation ideal [9, corollary 2.3]. Therefore, this surjectivity implies that the two filtrations of $R(\widetilde{B})$ are actually the same. In particular, the map $\text{gr}^i_{\text{geom}} R(\widetilde{B}) \to gr^{2i}_{\text{top}} R(\widetilde{B})$ is an isomorphism for all i. We have $CH^iB\widetilde{B} \cong \text{gr}^i_{\text{geom}} R(\widetilde{B})$ for $i \leq p$ (Theorem 2.25), and so we have isomorphisms

$$CH^iB\widetilde{B} \cong \text{gr}^i_{\text{geom}} R(\widetilde{B}) \cong \text{gr}^{2i}_{\text{top}} R(\widetilde{B})$$

for $i \leq p$. Since $\mathrm{gr}^{2i}_{\mathrm{top}} R(\widetilde{B})$ is a subquotient of $H^{2i}(B\widetilde{B}, \mathbf{Z})$, it follows that the map $CH^i B\widetilde{B} \to H^{2i}(B\widetilde{B}, \mathbf{Z})$ is injective for $i \leq p$. Since it is also surjective, it is an isomorphism for $i \leq p$.

In particular, $CH^i_{\widetilde{B}} = CH^i(B\widetilde{B})/p$ maps isomorphically to $H^{2i}(B\widetilde{B}, \mathbf{Z})/p$ for $i \leq p$, and so $CH^i_{\widetilde{B}}$ injects into $H^{2i}_{\widetilde{B}} = H^{2i}(B\widetilde{B}, \mathbf{F}_p)$ for $i \leq p$. Now we observe that $\widetilde{B}$ has a faithful irreducible representation of dimension p over k, which can be defined by inducing from the abelian subgroup of index p in $\widetilde{B}$. By Theorem 12.7, applied to the finite subgroup K of $\widetilde{B}$ with the same Chow ring, the restriction map

$$CH^*_{\widetilde{B}} \to \prod_{\substack{V \subset \widetilde{B} \\ V \text{ elem ab}}} CH^*_V \otimes CH^{\leq p-1}_{C_{\widetilde{B}}(V)}$$

is injective. Using the description of the maximal elementary abelian subgroups of $\widetilde{B}$, we check that the centralizer of each elementary abelian subgroup is either abelian or the whole group $\widetilde{B}$. Therefore, $CH^{\leq p-1}_{C_{\widetilde{B}}(V)} \to H^{\leq 2p-2}_{C_{\widetilde{B}}(V)}$ is injective, for every elementary abelian subgroup V of $\widetilde{B}$. Likewise, CH^*_V injects into H^*_V. So $CH^*_{\widetilde{B}}$ injects into $H^*_{\widetilde{B}}$. We also know that $CH^*_{\widetilde{B}}$ maps onto $H^{\mathrm{ev}}(B\widetilde{B}, \mathbf{Z})/p \subset H^*_{\widetilde{B}}$, and so $CH^*_{\widetilde{B}}$ maps isomorphically to $H^{\mathrm{ev}}(B\widetilde{B}, \mathbf{Z})/p$. □

15

The Geometric and Topological Filtrations of the Representation Ring

Let G be a complex algebraic group, for example a finite group. The complex representation ring $R(G)$ has two natural filtrations, the topological and geometric filtrations, defined using the topological or algebro-geometric codimension of support of a virtual bundle on BG. One might expect any naturally occurring filtration of the representation ring to have a purely algebraic interpretation, but no one knows how to do that for the geometric and topological filtrations; see the comments below on the counterexamples to Atiyah's conjecture.

In this chapter, we give examples of p-groups for any prime number p such that the geometric and topological filtrations differ. For $p = 2$, such examples were recently given by Yagita [156, corollary 5.7]. Related to that, we give examples of finite groups for which the cycle map from the mod p Chow ring to mod p cohomology is not injective. The examples build upon Vistoli's description of the Chow ring of the classifying space of $PGL(p)$ for a prime number p [143].

15.1 Summary

See Section 14.2 for the definitions of the geometric and topological filtrations of the representation ring $R(G)$.

For any prime number p, we give p-local examples of groups for which the geometric and topological filtrations differ. Our simplest example is the group $(SL(p)^2/(\mathbf{Z}/p)) \times \mathbf{Z}/p$ for any prime p. Seeking examples among finite groups, we find that the two filtrations differ for the 2-group $2^{1+4}_{+} \times \mathbf{Z}/2$. For p odd, we find a more complicated p-group for which the two filtrations differ: $G = (H_1 \times H_2)/(\mathbf{Z}/p) \times \mathbf{Z}/p$, where $H_1 = \mathbf{Z}/p \ltimes (\mathbf{Z}/p^2)^{p-1} \subset SL(p)$ and $H_2 = \mathbf{Z}/p \ltimes (\mathbf{Z}/p)^2 \subset SL(p)$.

Atiyah conjectured that the topological filtration of the representation ring was equal to the γ-filtration, [6]. See Section 14.2 for the definition of the

γ-filtration, which is purely algebraic. Atiyah's conjecture was disproved by Weiss, Thomas, and (for p-groups) Leary and Yagita [93]. Since

$$F^i_\gamma R(G) \subset F^i_{\text{geom}} R(G) \subset F^{2i}_{\text{top}} R(G),$$

the statement here (that the geometric and topological filtrations can differ) strengthens Leary-Yagita's result. Both the geometric and topological filtrations are preserved by transfers, in contrast to the γ-filtration. That makes it hard to distinguish the geometric and topological filtrations.

15.2 Positive results

We show that the geometric and topological filtrations of the representation ring agree in low degrees. The result (Lemma 15.2) is optimal, as we see in Theorems 15.7 and 15.13.

Write $\mathbf{Z}_p$ for the ring of p-adic integers.

Lemma 15.1 *Let G be a finite group scheme over a field k, and let p be a prime number invertible in k. Then the cycle map $(CH^2 BG)_{(p)} \to H^4_{\text{et}}(BG, \mathbf{Z}_p(2))$ is injective, with image the $\mathbf{Z}_p$-submodule generated by Chern classes of representations. For $k = \mathbf{C}$, $CH^2 BG \to H^4(BG, \mathbf{Z})$ is injective, with image the subgroup generated by Chern classes of complex representations.*

By contrast, for every prime number p there are p-groups G such that $CH^3 BG_{\mathbf{C}} \to H^6(BG, \mathbf{Z})$ is not injective. The example $G = 2^{1+4}_+ \times \mathbf{Z}/2$ was given in [137, section 5], and we give examples for any p in Lemma 15.3 and Theorem 15.7.

Proof This was proved for $k = \mathbf{C}$ in [138, corollary 3.5], by arguments of Bloch and Colliot-Thélène using the Merkurjev-Suslin theorem. For convenience, we give the argument in terms of the (more general, and now known) Beilinson-Lichtenbaum conjecture.

The etale cohomology $H^*_{\text{et}}(k, \mathbf{Z}_p(2))$ splits off as a summand from $H^*_{\text{et}}(BG, \mathbf{Z}_p(2))$, using the two morphisms $EG \to BG \to \operatorname{Spec} k$ whose composition induces an isomorphism on etale cohomology. Likewise, the motivic cohomology of k splits off as a summand from that of BG.

Let G have order p^r times a number prime to p. We have a transfer map from the etale cohomology of EG to that of BG, since $EG \to BG$ is approximated by finite morphisms of smooth varieties. The composition of pullback and pushforward is multiplication by $|G|$ on the etale cohomology of BG. Therefore the summand of $H^3_{\text{et}}(BG, \mathbf{Z}_p(2))$ that pulls back to zero in $H^3_{\text{et}}(k, \mathbf{Z}_p(2))$ is killed by $|G|$ and hence by p^r.

Consider the commutative diagram

$$\begin{array}{ccccc} H^2_M(BG, \mathbf{z}/p^2(2)) & \xrightarrow[\beta_r]{} & H^3_M(BG, \mathbf{Z}(2)) & \xrightarrow[p^r]{} & H^3_M(BG, \mathbf{Z}(2)) \\ \downarrow & & \downarrow & & \downarrow \\ H^2_{\text{et}}(BG, \mathbf{Z}/p^2(2)) & \xrightarrow[\beta_r]{} & H^3_{\text{et}}(BG, \mathbf{Z}_p(2)) & \xrightarrow[p^r]{} & H^3_{\text{et}}(BG, \mathbf{Z}_p(2)). \end{array}$$

By the previous paragraph, any element v of $H^3_{\text{et}}(BG, \mathbf{Z}_p(2))$ in the summand besides $H^3_{\text{et}}(k, \mathbf{Z}_p(2))$ is the Bockstein β_r of some element w of $H^2_{\text{et}}(BG, \mathbf{Z}/p^r(2))$. By the Beilinson-Lichtenbaum conjecture (Theorem 6.9), w is in the image of an element x of motivic cohomology, $H^2_M(BG, \mathbf{Z}/p^r(2))$. Then $\beta_r x$ is in $H^3_M(BG, \mathbf{Z}(2))$ and maps to the given element v in $H^3_{\text{et}}(BG, \mathbf{Z}_p(2))$.

Now let z be an element of $(CH^2BG)_{(p)} = H^4(BG, \mathbf{Z}_{(p)}(2))$ that maps to zero in $H^4_{\text{et}}(BG, \mathbf{Z}_p(2))$. Consider the commutative diagram

$$\begin{array}{ccccccc} H^3_M(BG, \mathbf{Z}_{(p)}(2)) & \longrightarrow & H^3_M(BG, \mathbf{Z}/p^2(2)) & \xrightarrow[\beta_r]{} & H^4_M(BG, \mathbf{Z}_{(p)}(2)) & \xrightarrow[p^r]{} & H^4_M(BG, \mathbf{Z}_{(p)}(2)) \\ \downarrow & & \downarrow & & \downarrow & & \downarrow \\ H^3_{\text{et}}(BG, \mathbf{Z}_p(2)) & \longrightarrow & H^3_{\text{et}}(BG, \mathbf{Z}/p^2(2)) & \xrightarrow[\beta_r]{} & H^4_{\text{et}}(BG, \mathbf{Z}_p(2)) & \xrightarrow[p^r]{} & H^4_{\text{et}}(BG, \mathbf{Z}_p(2)). \end{array}$$

We know that z is killed by $|G|$, hence by p^r. So z is the Bockstein β_r of some element y of $H^3_M(BG, \mathbf{Z}/p^r(2))$. We can assume that y is in the summand other than $H^3_M(k, \mathbf{Z}/p^r(2))$. Let t be the image of y in $H^3_{\text{et}}(BG, \mathbf{Z}/p^r(2))$. The Bockstein of t in $H^4_{\text{et}}(BG, \mathbf{Z}_p(2))$ is zero, by our assumption on z. So t is the image of some element v of $H^3_{\text{et}}(BG, \mathbf{Z}_p(2))$. Again, we can assume that v is in the summand other than $H^3_{\text{et}}(k, \mathbf{Z}_p(2))$. By the previous paragraph, v is in the image of some element q of $H^3_M(BG, \mathbf{Z}_p(2))$. Map q into $H^3_M(BG, \mathbf{Z}/p^r(2))$; by injectivity of $H^3_M(BG, \mathbf{Z}/p^r(2)) \to H^3_{\text{et}}(BG, \mathbf{Z}/p^r(2))$ (another application of Beilinson-Lichtenbaum, Theorem 6.9), the image of q is equal to y in $H^3_M(BG, \mathbf{Z}/p^r(2))$. Therefore the original element z (the Bockstein of y) in $(CH^2BG)_{(p)}$ is zero, as we want.

We know that CH^2BG is generated by Chern classes (Theorem 2.25). That completes the proof for an arbitrary field k. The statements about $k = \mathbf{C}$ follow from the isomorphism between completed etale cohomology and ordinary cohomology with $\mathbf{Z}_p$ coefficients [104, theorem III.3.12]. □

Lemma 15.2 *Let G be a finite group, viewed as an algebraic group over* $\mathbf{C}$. *Then $F^i_{\text{geom}}R(G) = F^{2i}_{\text{top}}R(G)$ for $i \leq 3$.*

Proof We have $\text{gr}^0_{\text{geom}}R(G) = CH^0BG = \mathbf{Z}$ and $\text{gr}^0_{\text{top}}R(G) = H^0(BG, \mathbf{Z})$. Since CH^0BG maps isomorphically to $H^0(BG, \mathbf{Z})$, we have $F^1_{\text{geom}}R(G) =$

$F^2_{\text{top}}R(G)$. Next, $\text{gr}^1_{\text{geom}}R(G) = CH^1BG$ and $\text{gr}^2_{\text{top}}R(G)$ is a subgroup of $H^2(BG, \mathbf{Z})$, by Theorem 2.25 and inspection of the Atiyah-Hirzebruch spectral sequence (Theorem 14.2). But CH^1BG (= $\text{Hom}(G, \mathbf{C}^*)$, by Lemma 2.26) maps isomorphically to $H^2(BG, \mathbf{Z})$. It follows that $\text{gr}^1_{\text{geom}}R(G) \to \text{gr}^2_{\text{top}}R(G)$ is an isomorphism, so that $F^2_{\text{geom}}R(G) = F^4_{\text{top}}R(G)$. At the same time, we have proved the well-known fact that $\text{gr}^2_{\text{top}}R(G)$ is equal to $H^2(BG, \mathbf{Z})$.

Next, we have a commutative diagram

$$\begin{array}{ccc} CH^2BG & \longrightarrow & \text{gr}^2_{\text{geom}}R(G) \\ \downarrow & & \downarrow \\ H^4(BG, \mathbf{Z})_{\text{perm}} & \longrightarrow & \text{gr}^4_{\text{top}}R(G), \end{array}$$

where $H^4(BG, \mathbf{Z})_{\text{perm}}$ denotes the subgroup of permanent cycles in $H^4(BG, \mathbf{Z})$ for the Atiyah-Hirzebruch spectral sequence, that is, the kernel of all differentials. The top horizontal map is an isomorphism by Theorem 2.25, and the right vertical map is surjective since $F^2_{\text{geom}}R(G) = F^4_{\text{top}}R(G)$. On the other hand, the left vertical map is injective by Lemma 15.1, and the bottom map is an isomorphism by inspection of the Atiyah-Hirzebruch spectral sequence. (The only possible differential into $H^4(BG, \mathbf{Z})$ would come from $H^1(BG, \mathbf{Z})$, which is zero.) Therefore, the map from upper left to lower right group is injective and surjective, hence an isomorphism. It follows that all the maps shown are isomorphisms. In particular, $\text{gr}^2_{\text{geom}}R(G)$ maps isomorphically to $\text{gr}^4_{\text{top}}R(G)$, and so $F^3_{\text{geom}}R(G) = F^6_{\text{top}}R(G)$. □

15.3 Examples at odd primes

In this section, we show that the geometric and topological filtrations of the representation ring can differ p-locally for p odd. Our examples are products with $\mathbf{Z}/p$, as suggested by the following lemma.

Lemma 15.3 *Let p be a prime number. Let G be a complex algebraic group such that the image of $(CH^2BG)_{(p)} \to H^4(BG, \mathbf{Z}_{(p)})$ is not a summand. Then $CH^3B(G \times \mathbf{Z}/p) \to H^6(B(G \times \mathbf{Z}/p), \mathbf{Z})$ is not injective. Also, if p is odd, then*

$$F^4_{\text{geom}}R(G \times \mathbf{Z}/p) \subsetneqq F^8_{\text{top}}R(G \times \mathbf{Z}/p).$$

Note that the map $CH^2BG \to H^4(BG, \mathbf{Z})$ is injective, with image the subgroup generated by Chern classes c_2 of representations of G, by Lemma 15.1. So we can rephrase the assumption by saying that the Chern subgroup of $H^4(BG, \mathbf{Z})$ is not a summand, p-locally.

Proof The assumption implies that $CH^2(BG)/p \to H^4(BG, \mathbf{F}_p)$ is not injective. Let $H = G \times \mathbf{Z}/p$. By the Künneth formulas for Chow groups (Lemma 2.12) and for cohomology, it follows that $(CH^3BH)_{(p)} \to H^6(BH, \mathbf{Z}_{(p)})$ is not injective. Now suppose that the prime number p is odd. Then the natural surjection $CH^3BH \to \mathrm{gr}^3_{\mathrm{geom}} R(H)$ is an isomorphism p-locally, by Riemann-Roch [138, proof of corollary 3.2]; more generally, this holds in degrees at most p. The group $\mathrm{gr}^6_{\mathrm{top}} R(H)$ is a subquotient of $H^6(BH, \mathbf{Z})$, since the topological filtration of $R(H)$ is the filtration associated to the Atiyah-Hirzebruch spectral sequence (Theorem 14.2). Therefore, the map $\mathrm{gr}^3_{\mathrm{geom}} R(H) \to \mathrm{gr}^6_{\mathrm{top}} R(H)$ is not injective p-locally. So

$$F^4_{\mathrm{geom}} R(H) \subsetneq F^8_{\mathrm{top}} R(H). \qquad \square$$

The simplest complex algebraic group I know for which the Chern subgroup of $H^4(BG, \mathbf{Z})$ is not a summand is $G = SO(4)$. The group $SO(4)$ was used to give examples of torsion algebraic cycles with unexpected behavior in [137]. Since $SO(4)$ is isomorphic to $SL(2)^2/(\mathbf{Z}/2)$, it makes sense to try the group $G = SL(p)^2/(\mathbf{Z}/p)$ to produce similar examples at an odd prime p, and indeed we will see that this works. To be precise, we define G by dividing $SL(p)^2$ by the diagonal subgroup $\{(a, a) : a \in \mu_p = Z(SL(p))\}$.

Theorem 15.4 *Let p be a prime number, and let $G = SL(p)^2/(\mathbf{Z}/p)$. Then the Chern subgroup of $H^4(BG, \mathbf{Z})$ is not a summand. Moreover, for p odd, the geometric and topological filtrations of the representation ring of $G \times \mathbf{Z}/p$ are different. Explicitly,*

$$F^4_{\mathrm{geom}} R(G \times \mathbf{Z}/p) \subsetneq F^8_{\mathrm{top}} R(G \times \mathbf{Z}/p).$$

In fact, for $p = 2$, the geometric and topological filtrations of the representation ring of $(SL(2)^2/(\mathbf{Z}/2)) \times \mathbf{Z}/2 = SO(4) \times \mathbf{Z}/2$ are also different. The proof is slightly more elaborate for $p = 2$, and so we prove that later as Theorem 15.13.

Proof The integral cohomology of $BPGL(p)$ in degrees at most 5 is

$$\begin{array}{cccccc} \mathbf{Z} & 0 & 0 & \mathbf{Z}/p & \mathbf{Z} & 0 \\ 0 & 1 & 2 & 3 & 4 & 5, \end{array}$$

This follows from Vistoli [143, theorem 3.6], who computed the integral cohomology of $BPGL(p)$ in all degrees; since we only want the cohomology in degrees at most 5, a direct computation is not hard. Next, the exact sequence $1 \to SL(p) \to G \to PGL(p) \to 1$ given by projecting $G = SL(p)^2/(\mathbf{Z}/p)$ to the second factor gives a spectral sequence $E_2 =$

$H^*(BPGL(p), H^*(BSL(p), \mathbf{Z})) \Rightarrow H^*(BG, \mathbf{Z})$:

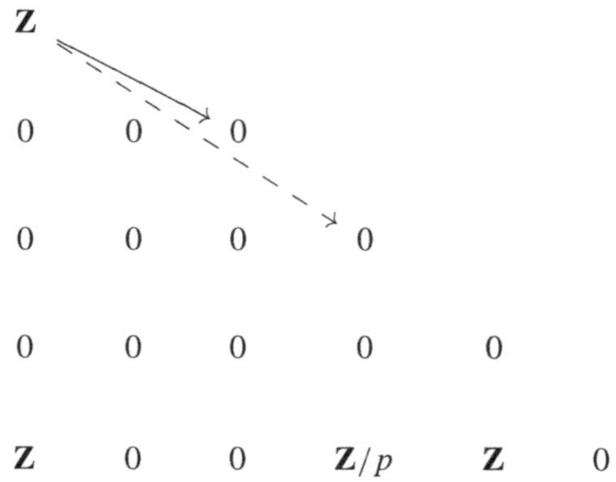

There are no possible differentials in this range. It follows that $H^4(BG, \mathbf{Z})$ is isomorphic to $\mathbf{Z}^2$ and the restriction map $H^4(BG, \mathbf{Z}) \to H^4(BSL(p), \mathbf{Z}) = \mathbf{Z}c_2V$ is surjective, where V denotes the standard representation of $SL(p)$. We know that the image of $CH^2BG \to H^4(BG, \mathbf{Z}) \cong \mathbf{Z}^2$ is a subgroup of finite index, and that this is the subgroup spanned by Chern classes (Theorem 2.25). So if we can show that the restriction map $CH^2BG \to CH^2BSL(p) = \mathbf{Z}c_2V$ lands in $\mathbf{Z}pc_2V$, then CH^2BG is not a summand of $H^4(BG, \mathbf{Z})$, as we want.

We need the following fact.

Lemma 15.5 *Let W be an irreducible representation of the complex group $SL(p)$ with nontrivial central character. Then* $\dim(W) \equiv 0 \pmod p$.

Proof By Schur's lemma, the center μ_p of $SL(p)$ acts on W by a homomorphism $\alpha \in \mathrm{Hom}(\mu_p, G_m) = \mathbf{Z}/p$, which we call the central character of W. We know that the determinant of W in $\mathrm{Hom}(SL(p), G_m)$ is trivial, since $\mathrm{Hom}(SL(p), G_m) = 0$. So the restriction $\det(W)|_{\mu_p}$ is trivial. But this restriction is equal to $\det(W|_{\mu_p}) = \alpha^{\dim(W)}$. So if α is nontrivial, then $\dim(W) \equiv 0 \pmod p$. □

Now let W be any irreducible representation of $G = SL(p)^2/(\mathbf{Z}/p)$, where we divide out by the diagonal central subgroup $\mathbf{Z}/p$. Then we can write $W = A \otimes B$, where A and B are irreducible representations of the two factors $SL(p)$ with inverse central characters. Restricting to the first factor of $SL(p)$, we have $W|_{SL(p)} \cong A^{\oplus \dim(B)}$. If A has nontrivial central character, then so does B, and so $\dim(B)$ is a multiple of p. We have $c_1A = 0$ in $CH^1BSL(p) = 0$, and so $c_2(W)|_{SL(p)} = \dim(B)c_2A \equiv 0 \pmod p$ in $CH^2BSL(p) = \mathbf{Z}c_2V$. On the other hand, if A has trivial central character, then we can view A as a representation of $PGL(p)$, and we can apply the following lemma.

Lemma 15.6 *The pullback map $H^4(BPGL(p), \mathbf{Z}) \to H^4(BSL(p), \mathbf{Z}) = \mathbf{Z}c_2V$ is zero modulo p.*

Proof This follows from Vistoli's description of $H^*(BPGL(p), \mathbf{Z})$ [143, section 14]. In more detail, an element of $H^*(BPGL(p), \mathbf{Z})$ maps by restriction to the maximal torus to an S_p-invariant element of the ring $\mathbf{Z}[y_1 - y_0, \ldots, \ldots, y_{p-1} - y_{p-2}] \subset \mathbf{Z}[y_0, \ldots, y_{p-1}]$, where the symmetric group S_p permutes the variables $y_0, \ldots, y_{p-1}$. The S_p-invariants in degree 2 are the subgroup generated by an element γ_2. Likewise, an element of $H^*(BSL(p), \mathbf{Z})$ restricts to an S_p-invariant element of the quotient ring $\mathbf{Z}[y_0, \ldots, y_{p-1}]/(y_0 + \cdots + y_{p-1} = 0)$. Vistoli shows (in the equation "$\gamma_k = p^{k-1}\gamma_k'$") that the image of γ_2 in the latter invariant ring is p times the class of $c_2V = \sum_{0 \le i < j \le p-1} y_i y_j$ for p odd, up to sign. For $p = 2$, a shorter calculation shows that the image of γ is 4 times the class of c_2V up to sign, which is even better. □

Therefore, if A is a representation of $SL(p)$ with trivial central character, the class $c_2(A) \in CH^2BSL(p) = \mathbf{Z}c_2V$ is zero modulo p. This completes the proof that the restriction to the first factor $SL(p)$ of the second Chern class of any representation $A \otimes B$ of $G = SL(p)^2/(\mathbf{Z}/p)$ is zero in $CH^2(BSL(p))/p$. By our earlier discussion, it follows that $CH^2BG \subset H^4(BG, \mathbf{Z})$ is not a summand. Theorem 15.4 is proved. (The fact that the geometric and topological filtrations are different follows from Lemma 15.3.) □

Thus, for every odd prime p, we have an example of a complex algebraic group for which the geometric and topological filtrations of the representation ring are different, namely $(SL(p)^2/(\mathbf{Z}/p)) \times \mathbf{Z}/p$. We now want to give examples among finite p-groups, for any odd prime number p. (We will give examples among 2-groups later.) The idea is to find a big enough finite p-subgroup S of $SL(p)^2/(\mathbf{Z}/p)$ to make $CH^2BS \subset H^4(BS, \mathbf{Z})$ not a summand. Then the geometric and topological filtrations for the p-group $S \times \mathbf{Z}/p$ are different, by Lemma 15.10.

Theorem 15.7 *Let G be an odd prime number, and let S be the p-group $(H_1 \times H_2)/(\mathbf{Z}/p)$, where $H_1 = \mathbf{Z}/p \ltimes (\mathbf{Z}/p^2)^{p-1} \subset SL(p)$ and H_2 is the extraspecial p-group of order p^3 and exponent p, $H_2 = \mathbf{Z}/p \ltimes (\mathbf{Z}/p)^2 \subset SL(p)$. Then the geometric and topological filtrations of the representation ring of $S \times \mathbf{Z}/p$ are different. Explicitly,*

$$F^4_{\text{geom}} R(S \times \mathbf{Z}/p) \subsetneqq F^8_{\text{top}} R(S \times \mathbf{Z}/p).$$

To clarify the definition of H_1, let T denote the subgroup of diagonal matrices in $SL(p)$, so that $T \cong (G_m)^{p-1}$. Then the Weyl group $N(T)/T$ is isomorphic to the symmetric group S_p. In particular, using the subgroup $\mathbf{Z}/p \subset S_p$, $SL(p)$ contains an extension of $\mathbf{Z}/p$ by T, which is a semidirect product $\mathbf{Z}/p \ltimes T$

for p odd. Writing $A[n]$ for the subgroup of an abelian group A killed by n, it follows that $SL(p)$ contains $\mathbf{Z}/p \ltimes T[p^2] \cong \mathbf{Z}/p \ltimes (\mathbf{Z}/p^2)^{p-1}$, which is the subgroup we call H_1.

Here both H_1 and H_2 contain the center $\mathbf{Z}/p$ of $SL(p)$, and we define S by dividing by the diagonal subgroup $\{(a, a) : a \in \mathbf{Z}/p = Z(SL(p))\}$. Then it is clear that S is a subgroup of $SL(p)^2/(\mathbf{Z}/p)$. The group S is the smallest p-group for odd p for which I was able to make the argument work, but there are probably smaller examples.

The following lemma reduces the problem to one about H_1.

Lemma 15.8 *Let p be an odd prime number. Let H_1 be a finite p-subgroup of $SL(p)$ that contains the center $\mathbf{Z}/p$, and let H_2 be the extraspecial p-group of order p^3 and exponent p, $H_2 = \mathbf{Z}/p \ltimes (\mathbf{Z}/p)^2 \subset SL(p)$. Let V be the standard representation of $SL(p)$. Suppose that $c_2 V \in CH^2 BH_1 \subset H^4(BH_1, \mathbf{Z})$ is not in the subgroup $p\,CH^2 BH_1 + CH^2 B(H_1/(\mathbf{Z}/p)) + H^4(BH_1, \mathbf{Z})[p]$, where $M[p]$ denotes the subgroup of an abelian group M killed by p. Then the p-group $S = (H_1 \times H_2)/(\mathbf{Z}/p)$ has the property that the subgroup $CH^2 BS$ of $H^4(BS, \mathbf{Z})$ is not a summand. Moreover, the geometric and topological filtrations of the representation ring are different for $S \times \mathbf{Z}/p$. Explicitly,*

$$F^4_{\text{geom}} R(S \times \mathbf{Z}/p) \subsetneqq F^8_{\text{top}} R(S \times \mathbf{Z}/p).$$

Proof If $CH^2 BS$ is a summand of $H^4(BS, \mathbf{Z})$, and if $x \in H^4(BS, \mathbf{Z})$ is an element such that px is in $CH^2 BS$, then x is the sum of an element of $CH^2 BS$ with an element of $H^4(BS, \mathbf{Z})[p]$, as one checks immediately.

We know that S is a subgroup of $SL(p)^2/(\mathbf{Z}/p)$. There is an element x of $H^4(B(SL(p)^2/(\mathbf{Z}/p)), \mathbf{Z})$ that restricts on the first factor $SL(p) \subset SL(p)^2/(\mathbf{Z}/p)$ to the generator $c_2 V$ of $H^4(BSL(p), \mathbf{Z})$, by the proof of Theorem 15.4. Apply the previous paragraph's observation to the restriction of x to $H^4(BS, \mathbf{Z})$. We know that $CH^2 B(SL(p)^2/(\mathbf{Z}/p))$ contains p times $H^4(B(SL(p)^2/(\mathbf{Z}/p)), \mathbf{Z})$, by pulling back from the two projections to $PGL(p)$, using that $CH^2(PGL(p)) = H^4(BPGL(p), \mathbf{Z})$ by Vistoli [143, corollary 3.5]. It follows that px belongs to $CH^2 B(SL(p)^2/(\mathbf{Z}/p))$, and so px restricted to S is in $CH^2 BS$. Thus, to show that $CH^2 BS$ is not a summand of $H^4(BS, \mathbf{Z})$, it suffices to show that x is not in the subgroup $CH^2 BS + H^4(BS, \mathbf{Z})[p]$ of $H^4(BS, \mathbf{Z})$.

We know that the restriction of x to H_1 is equal to $c_2 V$ in $CH^2 BH_1 \subset H^4(BH_1, \mathbf{Z})$. So it suffices to show that $c_2 V$ in $H^4(BH_1, \mathbf{Z})$ is not in

$$\operatorname{im}(CH^2 BS \to CH^2 BH_1) + H^4(BH_1, \mathbf{Z})[p].$$

We know that $CH^2 BS$ is generated by Chern classes (Theorem 2.25). We therefore consider the complex representations of S and their restrictions to H_1. Since H_2 is a p-group, every irreducible representation of H_2 has degree a power of p. Also, every irreducible representation R of $S = (H_1 \times H_2)/(\mathbf{Z}/p)$ is the tensor product of an irreducible representation of H_1 with an irreducible representation of H_2 having the inverse central character (because we divide out by the diagonal subgroup $\{(a, a) : a \in \mathbf{Z}/p\}$ of $H_1 \times H_2$). Finally, the extraspecial p-group H_2 has the property that the center of H_2 is contained in the commutator subgroup of H_2, and so every 1-dimensional representation of H_2 has trivial central character. So the restriction of R to H_1 is either a representation of $H_1/(\mathbf{Z}/p)$ or the direct sum of a multiple of p copies of a representation of H_1. In the second case, $c_2 R$ belongs to $pCH^2 BH_1$. (This uses that p is odd, so that $c_2(M^{\oplus p}) = pc_2 M + \binom{p}{2} c_1^2 M$ for a representation M, where $\binom{p}{2}$ is a multiple of p.) In the first case, $c_2 R$ belongs to the image of $CH^2 B(H_1/(\mathbf{Z}/p))$.

Thus, if $c_2 V$ in $H^4(BH_1, \mathbf{Z})$ is not in

$$pCH^2 BH_1 + CH^2 B(H_1/(\mathbf{Z}/p)) + H^4(BH_1, \mathbf{Z})[p],$$

then we have shown that $CH^2 BS$ is not a summand of $H^4(BS, \mathbf{Z})$. By Lemma 15.3, it follows that the geometric and topological filtrations of the representation ring are different for the p-group $S \times \mathbf{Z}/p$. □

We now prove the property of H_1 we want.

Lemma 15.9 *Let $H_1 = \mathbf{Z}/p \ltimes (\mathbf{Z}/p^2)^{p-1} \subset SL(p)$. Then the element $c_2 V \in CH^2 BH_1 \subset H^4(BH_1, \mathbf{Z})$ is not in the subgroup $pCH^2 BH_1 + CH^2 B(H_1/(\mathbf{Z}/p)) + H^4(BH_1, \mathbf{Z})[p]$.*

Proof The maximal torus in $GL(p)$ has Chow ring $\mathbf{Z}[y_0, \ldots, y_{p-1}]$. The maximal torus in $SL(p)$ has Chow ring a quotient of that ring, namely $\mathbf{Z}[y_0, \ldots, y_{p-1}]/(y_0 + \cdots + y_{p-1})$. In those terms, the subgroup $(\mathbf{Z}/p^2)^{p-1}$ of $H_1 \subset SL(p)$ has Chow ring

$$\mathbf{Z}[y_0, \ldots, y_{p-1}]/(p^2 y_i, y_0 + \ldots + y_{p-1}).$$

The total Chern class of V restricted to $(\mathbf{Z}/p^2)^{p-1}$ is $(1 + y_0) \cdots (1 + y_{p-1})$, and so

$$c_2 V|_{(\mathbf{Z}/p^2)^{p-1}} = \sum_{0 \le i < j \le p-1} y_i y_j.$$

Writing this in terms of $y_0, \ldots, y_{p-2}$ gives

$$\begin{aligned} c_2 V|_{(\mathbf{Z}/p^2)^{p-1}} &= \sum_{0 \le i < j \le p-2} y_i y_j - \left(\sum_{i=0}^{p-2} y_i \right)^2 \\ &= -\sum_{i=0}^{p-2} y_i^2 - \sum_{0 \le i < j \le p-2} y_i y_j. \end{aligned}$$

Write $R\{e_i : i \in I\}$ for the free module over a ring R with basis elements e_i. Then we can view c_2V as an element of $CH^2B(\mathbf{Z}/p^2)^{p-1} = \mathbf{Z}/p^2\{y_iy_j : 0 \leq i \leq j \leq p-2\}$.

We will show that c_2V in $H^4(BH_1, \mathbf{Z})$ is not in the subgroup $pCH^2BH_1 + CH^2B(H_1/(\mathbf{Z}/p)) + H^4(BH_1, \mathbf{Z})[p]$ by restricting to $H^4(BK, \mathbf{Z})$, where $K = (\mathbf{Z}/p^2)^{p-1}$. Since $H^4(BK, \mathbf{Z})$ contains CH^2BK as a summand (as is true for any abelian group), it suffices to show that the restriction of c_2V to CH^2BK is not in $pCH^2BK + CH^2B(H_1/(\mathbf{Z}/p)) + (CH^2BK)[p]$. Since $CH^2BK = \mathbf{Z}/p^2\{y_iy_j : 0 \leq i \leq j \leq p-2\}$ is a sum of copies of $\mathbf{Z}/p^2$, it suffices to show that the restriction of c_2V to $CH^2(BK)/p = \mathbf{F}_p\{y_iy_j : 0 \leq i \leq j \leq p-2\}$ is not in the image of $CH^2B(H_1/(\mathbf{Z}/p))$.

We know that CH^2 of $H_1/(\mathbf{Z}/p)$ is generated by Chern classes of complex representations, by Theorem 2.25. Since $H_1/(\mathbf{Z}/p) = \mathbf{Z}/p \ltimes (K/(\mathbf{Z}/p))$ has an abelian normal subgroup of index p, every complex irreducible representation of $H_1/(\mathbf{Z}/p)$ is either 1-dimensional or induced from a 1-dimensional representation of $K/(\mathbf{Z}/p)$ [124, proposition 24].

In terms of the inclusion of $K = (\mathbf{Z}/p^2)^{p-1}$ in the maximal torus of $SL(p)$, the subgroup $\mathbf{Z}/p \subset K$ is the center of $SL(p)$. The Chow ring of the quotient group $K/(\mathbf{Z}/p) \cong (\mathbf{Z}/p^2)^{p-2} \times \mathbf{Z}/p$ is $\mathbf{Z}[z_0, \ldots, z_{p-3}, w]/(p^2z_i = 0, pw = 0)$, where $z_i \mapsto y_{i+1} - y_i$ for $0 \leq i \leq p-3$ and $w \mapsto py_0$. The Chern class c_1 of any 1-dimensional representation of $H_1/(\mathbf{Z}/p) = \mathbf{Z}/p \ltimes (K/(\mathbf{Z}/p))$ must be an element of the subgroup of $\mathbf{Z}/p$-invariants in $CH^1B(K/(\mathbf{Z}/p))$, which we compute is generated by $py_0 + 2py_1 + \cdots + (p-1)py_{p-2}$. In particular, such classes pull back to zero in $(CH^*BK)/p$.

It remains to analyze the Chern class c_2 of a representation W of $H_1/(\mathbf{Z}/p) = \mathbf{Z}/p \ltimes (K/(\mathbf{Z}/p))$ induced from a 1-dimensional representation L of $K/(\mathbf{Z}/p)$. By Lemma 13.9, c_2W is the transfer of some element u in $CH^2B(K/(\mathbf{Z}/p))$ (plus a term that restricts to zero on $K/(\mathbf{Z}/p)$, in the case $p = 3$). Write $\text{tr} = 1 + \sigma + \cdots + \sigma^{p-1}$ acting on K and on $K/(\mathbf{Z}/p)$. We deduce that c_2W restricted to $K/(\mathbf{Z}/p)$ can be written as the trace $\text{tr}(u)$ for some $u \in CH^2BK/(\mathbf{Z}/p)$.

Thus it suffices to show that c_2V restricted to $(CH^2BK)/p$ is not in the image of the subgroup $\text{tr}(CH^2B(K/(\mathbf{Z}/p)))$. This is somewhat subtle, because c_2V is in the image of $CH^2B(K/(\mathbf{Z}/p))$. I don't have a general explanation for this, but we can check it by calculation. Namely, the image of the Chow ring of $K/(\mathbf{Z}/p)$ in the mod p Chow ring of K is the subring of $\mathbf{F}_p[y_0, \ldots, y_{p-1}]/(y_0 + \cdots + y_{p-1})$ generated by $y_1 - y_0, \ldots, y_{p-2} - y_{p-3}$. And we can rewrite c_2V, computed earlier as a polynomial in $y_0, \ldots, y_{p-2}$, as a polynomial in $y_1 - y_0, \ldots, y_{p-2} - y_{p-3}$. Explicitly:

$$c_2V|_K = -\sum_{i=1}^{p-2}(y_i - y_0)^2 - \sum_{1\leq i<j\leq p-2}(y_i - y_0)(y_j - y_0).$$

Let W_p denote the $\mathbf{F}_p$-vector space with basis $y_0, \ldots, y_{p-1}$, and let W_{p-1} be the subspace of W_p spanned by $y_1 - y_0, \ldots, y_{p-1} - y_{p-2}$. Let W_{p-2} denote the image of $CH^1B(K/(\mathbf{Z}/p)) \to (CH^1BK)/p$, which is the quotient of W_{p-1} by the line W_1 spanned by $y_0 + \cdots + y_{p-1} = 0$. These vector spaces are all representations of the symmetric group S_p, and hence of the subgroup $\mathbf{Z}/p$, by permuting $y_0, \ldots, y_{p-1}$.

Define a linear map $S^2W_p \to \mathbf{F}_p$ by mapping y_i^2 to 1 for all i and y_iy_j to 0 for all $i \neq j$, and write f for the restriction of this map to S^2W_{p-1}. Clearly the map f is $\mathbf{Z}/p$-invariant (and even S_p-invariant, although we do not use that). We compute that

$$f((y_0 + \cdots + y_{p-1})(y_{i+1} - y_i)) = 0$$

for all i, and so f factors through the surjection $S^2W_{p-1} \twoheadrightarrow S^2W_{p-2}$. Thus we have a $\mathbf{Z}/p$-invariant linear map $f\colon S^2W_{p-2} \to \mathbf{F}_p$.

Since f is $\mathbf{Z}/p$-invariant, we have $f(\mathrm{tr}(y)) = 0$ for all $y \in S^2W_{p-2} = \mathrm{im}(CH^2B(K/\mathbf{Z}/p) \to (CH^2BK)/p)$. Therefore, to show that c_2V is not in the image of the trace map on S^2W_{p-2}, it suffices to show that $f(c_2V) \in \mathbf{F}_p$ is not zero. We gave a formula above for c_2V as an element of S^2W_{p-2} (that is, as a polynomial in the elements $y_{i+1} - y_i$, modulo the relation $y_0 + \cdots + y_{p-1} = 0$):

$$c_2V = -\sum_{i=1}^{p-2}(y_i - y_0)^2 - \sum_{1 \leq i < j \leq p-2}(y_i - y_0)(y_j - y_0).$$

It follows that

$$\begin{aligned} f(c_2V) &= -2(p-2) - \binom{p-2}{2} \\ &= (p+1)(p-2)/2 \\ &\neq 0 \in \mathbf{F}_p. \end{aligned}$$

So c_2V in $(CH^2BK)/p$ is not in the image of $\mathrm{tr}(CH^2B(K/(\mathbf{Z}/p)))$.

This completes the proof that the subgroup CH^2BS of $H^4(BS, \mathbf{Z})$ is not a summand. By Lemma 15.3, the geometric and topological filtrations on the representation ring of $S \times \mathbf{Z}/p$ are different. □

15.4 Examples for $p = 2$

In order to show that the geometric and topological filtrations of the representation ring can differ 2-locally, it seems that we need a little more homotopy theory, notably Steenrod operations. We formulate the method p-locally for any prime p, although our applications only involve $p = 2$. Yagita has also shown that the geometric and topological filtrations can differ 2-locally, namely for the extraspecial group 2_+^{1+6} [156, corollary 5.7].

Lemma 15.10 *Let p be a prime number. Let G be a complex algebraic group such that $CH^p(BG)/p \to H^{2p}(BG, \mathbf{F}_p)$ is not injective. Then $CH^{p+1}B(G \times \mathbf{Z}/p) \to H^{2p+2}(B(G \times \mathbf{Z}/p), \mathbf{Z})$ is not injective. If in addition the abelianization of G is killed by p, then*

$$F^{p+2}_{\text{geom}} R(G \times \mathbf{Z}/p) \subsetneq F^{2p+4}_{\text{top}} R(G \times \mathbf{Z}/p).$$

Proof Let $u \in CH^pBG$ be an element that is nonzero in $CH^p(BG)/p$ but maps to zero in $H^{2p}(BG, \mathbf{F}_p)$. Let $K = G \times \mathbf{Z}/p$, and let $v \in CH^1BK$ be the pullback of a generator of $CH^1B\mathbf{Z}/p \cong \mathbf{Z}/p$. By the Künneth formulas for Chow groups (Lemma 2.12) and for cohomology, it follows that $(CH^{p+1}BK)_{(p)} \to H^{2p+2}(BK, \mathbf{Z}_{(p)})$ is not injective. Explicitly, uv is a nonzero element of the kernel.

The natural surjection $CH^iBK \to \text{gr}^i_{\text{geom}} R(K)$ is an isomorphism p-locally for $i \leq p$, by Riemann-Roch (Theorem 2.25). For $i = p + 1$, this map need not be an isomorphism, but we can compute the kernel, as follows.

Lemma 15.11 *For any smooth scheme X over the complex numbers and any prime number p, there is a natural operation*

$$\beta P^1 \colon H^3(X, \mathbf{Z})_{p\text{-power torsion}} \to CH^{p+1}X.$$

For any complex algebraic group G, we have

$$CH^{p+1}BG/\beta P^1 H^3(BG, \mathbf{Z}) \cong \text{gr}^{p+1}_{\text{geom}} R(G).$$

Proof The operation $\beta P^1 \colon H^3(X, \mathbf{Z})_{p\text{-power torsion}} \to CH^{p+1}X$ comes from Voevodsky's Steenrod operations (section 6.3) together with the Bloch-Kato conjecture proved by Voevodsky and Rost. This is a remarkable operation, since it produces algebraic cycles from purely topological input. As a first step, we write βP^1 for the composition

$$H^3(X, \mathbf{Z}(2)) \to H^3(X, \mathbf{Z}/p(2)) \xrightarrow{P^1} H^{2p+1}(X, \mathbf{Z}/p(p+1))$$

$$\xrightarrow{\beta} H^{2p+2}(X, \mathbf{Z}(p+1)) = CH^{p+1}X,$$

which clearly maps into the p-torsion subgroup $CH^{p+1}X[p]$.

Next, the Beilinson-Lichtenbaum conjecture (Theorem 6.9) gives an isomorphism from ordinary to motivic cohomology,

$$H^2(X, \mathbf{Z}/p^r) \cong H^2_M(Z, \mathbf{Z}/p^r(2)).$$

Thus, for a complex scheme X, βP^1 gives a map from the ordinary cohomology $H^2(X, \mathbf{Z}[1/p]/\mathbf{Z})$ to $CH^{p+1}[p]$ (identify the domain with $\varinjlim_r H^2(X, \mathbf{Z}/p^r) \cong \varinjlim_r H^2_M(X, \mathbf{Z}/p^r(2))$, apply the Bockstein to get to motivic cohomology $H^3_M(X, \mathbf{Z}(2))$, and apply βP^1 as before). Finally, this map vanishes on the image of $H^2(X, \mathbf{Z}[1/p])$ in $H^2(X, \mathbf{Z}[1/p]/\mathbf{Z})$, because it lands in a group killed by p. By the exact sequence

$$H^2(X, \mathbf{Z}[1/p]) \to H^2(X, \mathbf{Z}[1/p]/\mathbf{Z}) \to H^3(X, \mathbf{Z}) \to H^3(X, \mathbf{Z}[1/p]),$$

we have a natural map βP^1 from $H^3(X, \mathbf{Z})_{p\text{-power torsion}}$ to $(CH^{p+1}X)[p]$, as we want.

In view of the exact sequence of motivic cohomology groups

$$H^2(X, \mathbf{Z}/p^r(2)) \to H^3(X, \mathbf{Z}(2)) \underset{p^r}{\to} H^3(X, \mathbf{Z}(2)),$$

the image of βP^1 on the ordinary cohomology $H^3(X, \mathbf{Z})_{p\text{-power torsion}}$ is equal to the image of βP^1 on motivic cohomology $H^3(X, \mathbf{Z}(2))_{p\text{-power torsion}}$. We now specialize to the case $X = BG$, for a finite group G. (Any given motivic cohomology group of BG can be computed on a suitable finite-dimensional approximation U/G, where U is an open subset of a representation of G on which G acts freely.) Using transfers, we see that $H^3(BG, \mathbf{Z}(2)) = H^3(\mathbf{C}, \mathbf{Z}(2)) \oplus K$ where K is an abelian group killed by the order of G. Moreover, the operation βP^1 vanishes on $H^3(\mathbf{C}, \mathbf{Z}(2))$, because $CH^{p+1}(\text{Spec } \mathbf{C}) = 0$. So all that matters is the torsion part. We conclude that the image of βP^1 on the motivic cohomology $H^3(BG, \mathbf{Z}(2))$ is equal to the image of βP^1 on the ordinary cohomology $H^3(BG, \mathbf{Z})_{p\text{-power torsion}}$.

The motivic Atiyah-Hirzebruch spectral sequence from motivic cohomology to algebraic K-theory has the following form [95]:

Theorem 15.12 *Let X be a smooth scheme over a field k. Then there is a spectral sequence*

$$E_2^{ij} = H^i(X, \mathbf{Z}(-j/2)) \Rightarrow K_{-i-j}X.$$

The E_2 term looks like:

$H^0(X, \mathbf{Z}(0))$	0	0	0	0
0	0	0	0	0
$H^0(X, \mathbf{Z}(1))$	$H^1(X, \mathbf{Z}(1))$	$H^2(X, \mathbf{Z}(1))$	0	0
0	0	0	0	0
$H^0(X, \mathbf{Z}(2))$	$H^1(X, \mathbf{Z}(2))$	$H^2(X, \mathbf{Z}(2))$	$H^3(X, \mathbf{Z}(2))$	$H^4(X, \mathbf{Z}(2))$
$\vdots$				

The groups on the top diagonal are the Chow groups $CH^iX = H^{2i}(X, \mathbf{Z}(i))$. If we localize the spectral sequence at a prime number p, then all differentials are zero except $d_{a(2p-2)+1}$ for positive integers a, by the splitting of p-local algebraic K-theory into $p-1$ summands given by Adams operations [157, proposition 1.2].

So the first possible p-local differential is

$$d_{2p-1}\colon H^i(X, \mathbf{Z}_{(p)}(j)) \to H^{i+2p-1}(X, \mathbf{Z}_{(p)}(j+p-1)).$$

This is in fact equal to $-\beta P^1$; most of the argument was given by Yagunov [157]. Indeed, by naturality of the spectral sequence and its stability under suspensions, it suffices to compute this differential for the universal class on the motivic Eilenberg-MacLane spectrum $H\mathbf{Z}_{(p)}(j)$. Moreover, this differential is killed by p [95], and so the differential on the universal class is the Bockstein of some element of $H^{2p-2}(H\mathbf{Z}_{(p)}(j), \mathbf{Z}/p(j+p-1))$. By Voevodsky's calculation of the motivic Steenrod algebra over the complex numbers, such a class is a polynomial in the Bockstein and the operations P^i [146, theorem 3.49]. By comparison with topology, where the corresponding differential in the Atiyah-Hirzebruch spectral sequence is well known [8, proposition 7.2], this operation in degree $2p-2$ must be equal to $-P^1$. We conclude that the first p-local differential in the motivic Atiyah-Hirzebruch spectral sequence is equal to $-\beta P^1$.

The geometric filtration on K_0X, for a smooth scheme X over $\mathbf{C}$, is the filtration associated to this spectral sequence. (The groups contributing to K_0X in the spectral sequence are those on the top diagonal, the Chow groups.) For a finite group G, taking $X = BG$ over $\mathbf{C}$, the ring K_0BG is the completed representation ring of G, by Merkurjev [101, corollary 4.4], [138, theorem 3.1]. All differentials are zero on Chow groups (by the form of the spectral sequence), and there are no possible p-local differentials into CH^iBG for $i \leq p$; this proves again Theorem 2.25's isomorphism $(CH^iBG)_{(p)} \cong \mathrm{gr}^i_{\mathrm{geom}}R(G)_{(p)}$ for $i \leq p$. We can now add that the only possible p-local differential into $CH^{p+1}BG$ is

$$d_{2p-1}\colon H^3(BG, \mathbf{Z}(2)) \to H^{2p+2}(BG, \mathbf{Z}(p+1)) = CH^{p+1}BG.$$

This map is equal to $-\beta P^1$. As we have discussed, its image is equal to the image of the map $\beta P^1\colon H^3(BG, \mathbf{Z}) \to CH^{p+1}BG$. (For any complex algebraic group G, $H^*(BG, \mathbf{Z})$ is torsion in odd degrees. So we can think of βP^1 as being defined on all of $H^3(BG, \mathbf{Z})$, not just on the p-power torsion subgroup.) We conclude that

$$\mathrm{gr}^{p+1}_{\mathrm{geom}}R(G)_{(p)} \cong (CH^{p+1}BG)_{(p)}/\beta P^1 H^3(BG, \mathbf{Z}). \qquad \square$$

We return to the proof of Lemma 15.10. We know that the product group $K = G \times \mathbf{Z}/p$ has $(CH^{p+1}BK)_{(p)} \to H^{2p+2}(BK, \mathbf{Z}_{(p)})$ not injective. Explicitly, we defined a nonzero element uv of the kernel. Since $(gr^{2p+2}_{\mathrm{top}}R(K))_{(p)}$ is a subquotient of $H^{2p+2}(BK, \mathbf{Z}_{(p)})$, it follows that

$$(CH^{p+1}BK)_{(p)} \to (gr_{\text{top}}^{2p+2}R(K))_{(p)}$$

is not injective. Again, uv is a nonzero element of the kernel.

This map factors through $\text{gr}_{\text{geom}}^{p+1}R(K)_{(p)}$, and the subgroup of $(CH^{p+1}NK)_{(p)}$ that maps to zero in that group is $\beta P^1 H^3(BK, \mathbf{Z})$ by Lemma 15.11. So to show that $\text{gr}_{\text{geom}}^{p+1}R(K)_{(p)} \to \text{gr}_{\text{top}}^{2p+2}R(K)_{(p)}$ is not injective, it remains to show that $uv \in CH^{p+1}BK$ is not in the image of $H^3(BK, \mathbf{Z})$.

By the Künneth formula for integral cohomology, we have

$$H^3(BK, \mathbf{Z}) \cong H^3(BG, \mathbf{Z}) \oplus \text{Tor}_1^{\mathbf{Z}}(H^2(BG, \mathbf{Z}), H^2(B\mathbf{Z}/p, \mathbf{Z})).$$

We assumed that the abelianization of G is killed by p. It follows that

$$\beta\colon H^1(BG, \mathbf{F}_p) \to H^2(BG, \mathbf{Z})$$

is an isomorphism. Using that, we can describe the Tor term explicitly as follows. Let $x_1, \ldots, x_m$ be a basis for $H^1(BG, \mathbf{F}_p)$ and w a generator of $H^1(B\mathbf{Z}/p, \mathbf{F}_p) \cong \mathbf{F}_p$; we can assume that $\beta w = v$. Then

$$H^3(BK, \mathbf{Z}) = H^3(BG, \mathbf{Z}) \oplus \mathbf{F}_p\{\beta(x_1 w), \ldots, \beta(x_m w)\}.$$

The operation βP^1 sends $H^3(BG, \mathbf{Z}) \subset H^3(BK, \mathbf{Z})$ into $CH^{p+1}BG \subset CH^{p+1}BK$, by functoriality. We can also compute βP^1 on the elements $\beta(x_i w)$, using the formal properties of Voevodsky's Steenrod operations (Section 6.3):

$$\begin{aligned}\beta P^1(\beta(x_i w)) &= \beta P^1((\beta x_i)w - x_i \beta w)\\ &= \beta((\beta x_i)^p w - x_i(\beta w)^p)\\ &= (\beta x_i)^p \beta w - (\beta x_i)(\beta w)^p\\ &= (\beta x_i)^p v - (\beta x_i) v^p.\end{aligned}$$

Now we can see that uv is linearly independent of $\beta P^1 H^3(BK, \mathbf{Z})$, even in $CH^*(BK)/p$. We have $CH^*(BK)/p \cong CH^*(BG)/p[v]$ by the Chow Künneth formula, and we know that u is nonzero in $CH^p(BG)/p$. Also, the elements $\beta x_1, \ldots, \beta x_m$ are linearly independent in $CH^1(BG)/p = H^2(BG, \mathbf{Z})/p$. So if $uv = \beta P^1(s + \sum a_i \beta(x_i w))$ for some $s \in H^3(BG, \mathbf{Z})$ and $a_i \in \mathbf{F}_p$, then all a_i must be zero (otherwise the right side would have nonzero coefficient of v^p). But then the right side has zero coefficient of v, a contradiction.

Thus $uv \in (CH^{p+1}BK)[p]$ is not in the image of $H^3(BK, \mathbf{Z})$ and hence maps to a nonzero element of the kernel of $\text{gr}_{\text{geom}}^{p+1}R(K) \to \text{gr}_{\text{top}}^{2p+2}R(K)$. □

Theorem 15.13 *The Chern subgroup of $H^4(BSO(4), \mathbf{Z})$ is not a summand. Moreover, the geometric and topological filtrations of the representation ring* *of $SO(4) \times \mathbf{Z}/2$ are different. Explicitly,*

$$F^4_{\text{geom}} R(SO(4) \times \mathbf{Z}/2) \subsetneq F^8_{\text{top}} R(SO(4) \times \mathbf{Z}/2).$$

Likewise, the extraspecial 2-group

$$H = 2^{1+4}_{+} = \langle u_1, u_2, u_3, u_4 : u_i^2 = 1, [u_1, u_2] = [u_3, u_4] = v, v \text{ central},$$
$$[u_1, u_3] = [u_1, u_4] = [u_2, u_3] = [u_2, u_4] = 1 \rangle$$

has

$$F^4_{\text{geom}} R(H \times \mathbf{Z}/2) \subsetneq F^8_{\text{top}} R(H \times \mathbf{Z}/2).$$

Proof The abelianization of $SO(4)$ is trivial, hence killed by 2. By Lemma 15.10, the conclusion on $SO(4)$ follows if we can show that $CH^2(BSO(4))/2 \to H^4(BSO(4), \mathbf{F}_2)$ is not injective. That holds by a calculation with Steenrod operations: we have $H^4(BSO(4), \mathbf{Z}) \cong \mathbf{Z}^2$, but the Euler class in this group has nonzero image under the odd-degree mod 2 Steenrod operation Sq^3, and hence is not in the image of $CH^2BSO(4)$ [137, section 5]. On the other hand, $CH^2BSO(4)$ maps onto $H^4(BSO(4), \mathbf{Z})$ tensor $\mathbf{Q}$, because $H^*(BG, \mathbf{Q})$ is generated by Chern classes for every complex algebraic group G. It follows that $CH^2(BSO(4))/2 \to H^4(BSO(4), \mathbf{F}_2)$ is not injective.

The abelianization of the subgroup $H = 2^{1+4}_{+} \subset SO(4)$ is isomorphic to $(\mathbf{Z}/2)^4$, which is killed by 2. By Lemma 15.10, the conclusion on H follows if we can show that $CH^2(BH)/2 \to H^4(BH, \mathbf{F}_2)$ is not injective. Since c_2V in $CH^2BSO(4)$ maps to zero in $H^4(BSO(4), \mathbf{F}_2)$, its restriction to CH^2BH maps to zero in $H^4(BH, \mathbf{F}_2)$. Finally, c_2V is nonzero in $CH^2(BH)/2$, as shown in [137, section 5] by computing that c_2V has nonzero image in $MU^*BH \otimes_{MU^*} \mathbf{F}_2$. Thus $CH^2(BH)/2 \to H^4(BH, \mathbf{F}_2)$ is not injective. □

16

The Eilenberg-Moore Spectral Sequence in Motivic Cohomology

In this chapter, we construct an Eilenberg-Moore spectral sequence in motivic cohomology for schemes with group actions. In topology, for a space X with an action of a topological group G, there is a fibration

$$X \to X//G \to BG,$$

and the Eilenberg-Moore spectral sequence converges to the cohomology of the fiber X, given the cohomology of the base BG and total space $X//G = (X \times EG)/G$ [98, chapter 7].

We prove a spectral sequence of the same form for motivic cohomology, when X is a smooth scheme with an action of a split reductive group. The spectral sequence was defined by Krishna with rational coefficients [82, theorem 1.1]. Our method is essentially the same, but we give an integral statement as far as possible. Other related results include Merkurjev's construction of the Eilenberg-Moore spectral sequence of a group action for algebraic K-theory [101, theorem 4.3] and for K-cohomology [102, section 3a].

Our Eilenberg-Moore spectral sequence works only after inverting the torsion index, a positive integer associated to G. (If G is $GL(n)$ or a torus, then the torsion index is 1, and so the spectral sequence computes motivic cohomology integrally.) This is unavoidable: the spectral sequence does not hold in the same form as in topology without inverting the torsion index, by Remark 16.7. One could hope for some more general form of the Eilenberg-Moore spectral sequence in motivic homotopy theory.

The relevance of the Eilenberg-Moore spectral sequence for this book is that it clarifies the relation between the motivic cohomology of the classifying space BG and the finite-dimensional variety $GL(n)/G$, for any affine group scheme G with a faithful representation $G \to GL(n)$. A basic result in this book, Theorem 5.1, which relates the Chow groups of BG and $GL(n)/G$, is an easy special case of the spectral sequence (Theorem 16.6).

16.1 Motivic cohomology of flag bundles

In this section, we show that computing equivariant motivic cohomology with respect to a split reductive group G reduces to the case of a torus, after inverting the torsion index of G. This is close to several results in the literature: Edidin-Graham [38, theorem 6] is a similar statement for Chow groups, and Asok-Doran-Kirwan [5, proposition 3.9] gives the result here on motivic cohomology after tensoring with the rationals.

Theorem 16.1 *Let G be a split reductive group over a field k. Let B be a Borel subgroup of G, and let $t(G)$ be the torsion index of G. Then there are elements $e_1, \ldots, e_m$ of $CH^*(BB)[1/t(G)]$ that restrict to a basis for $CH^*(G/B)[1/t(G)]$ as a free module over $\mathbf{Z}[1/t(G)]$. Moreover, for any smooth scheme X of finite type over k with an action of G, $e_1, \ldots, e_m$ restrict to a basis for the motivic cohomology of $X//B$ as a free module over the motivic cohomology of $X//G$.*

A reductive group over a field k is *split* if it contains a maximal torus that is split over k. (So every reductive group over an algebraically closed field is split.) Chevalley made the remarkable discovery that the classification of split reductive groups is the same over all fields [18, 32]. In particular, split semisimple groups up to isogeny are classified by Dynkin diagrams.

We recall Grothendieck's definition of the torsion index [54, 139]. Let G be a split reductive group G over a field k. Let B be a Borel subgroup of G (that is, a maximal smooth connected solvable k-subgroup). Each homomorphism $B \to G_m$ determines a line bundle on the flag manifold G/B. (The group $\text{Hom}(B, G_m)$ is called the *weight lattice* of G.) Consider the subring S of $CH^*(G/B)$ generated by the first Chern classes of these line bundles. Since CH^*BB is the polynomial ring over $\mathbf{Z}$ generated by the weight lattice, we can also say that S is the image of the natural homomorphism $CH^*BB \to CH^*(G/B)$. Let N be the dimension of G/B; then $CH^N(G/B)$ is isomorphic to $\mathbf{Z}$, generated by the class of a point. The *torsion index* $t(G)$ is the least positive integer such that S contains $t(G)$ times the class of a point.

Proof Equivariant motivic cohomology is defined as the motivic cohomology of suitable G-spaces on which G acts freely with quotient a scheme. So it suffices to prove the theorem when G acts freely on X, with quotient $Y = X/G$ a smooth scheme over k.

The flag manifold G/B has an algebraic cell decomposition, the Bruhat decomposition [18, theorem 14.12, proposition 21.2.9]. As a result, the Chow ring $CH^*(G/B)$ is a finitely generated free abelian group. Demazure gave a combinatorial description of the Chow ring of G/B. In particular, he showed that the natural homomorphism $CH^*BB \to CH^*(G/B)$ becomes surjective after inverting the torsion index $t(G)$ [31, proposition 5]. Let $R = \mathbf{Z}[1/t(G)]$. It follows that there are elements $e_1, \ldots, e_m$ of $CH^*(BB) \otimes_{\mathbf{Z}} R$ that restrict to a basis for $CH^*(G/B) \otimes_{\mathbf{Z}} R$.

We now return to the situation where G acts freely on a smooth scheme X over k with quotient scheme $Y = X/G$. More generally, let $X \to Y$ be a principal G-bundle over a scheme Y of finite type over k; we have to consider singular schemes for our induction. We have a smooth proper morphism $X/B \to X/G$ with fiber G/B. Define a homomorphism of motivic homology groups (that is, higher Chow groups) $\varphi_Y\colon H_*(X/G, R(*))^{\oplus N} \to H_*(X/B, R(*))$ by $(x_1, \ldots, x_N) \mapsto \sum e_i \pi^*(x_i)$. For a trivial principal bundle $X = Y \times_k G$ over Y, φ is an isomorphism, since G/B has a cell decomposition.

Next, let $f\colon X \to Y$ be a principal G-bundle that is trivialized by a finite flat morphism $Z \to Y$ of degree d. Then φ_Z is an isomorphism. Let $X_Z = X \times_Y Z$; then we have finite flat morphisms $f\colon X_Z/B \to X/B$ and $f\colon X_Z/G = Z \to X/G = Y$ of degree d. For both maps, $f_* f^*$ equals multiplication by d on motivic homology. It follows that φ_Y becomes an isomorphism after inverting d.

Fix a prime number p not dividing $t(G)$. It suffices to show that φ is an isomorphism after localizing at p. By Grothendieck's interpretation of the torsion index, every principal G-bundle over a field is trivialized by a finite separable extension field of degree prime to p [54, theorem 2]. Therefore, for any principal G-bundle over a k-variety Y, there is a nonempty open subset $U \subset Y$ such that G is trivialized on some variety V with a finite etale morphism $V \to U$ of degree prime to p. So φ_U is an isomorphism p-locally.

Let S be a closed subscheme of a k-scheme Y, with $U = Y - S$. By the localization sequence for motivic homology, φ_Y is an isomorphism if φ_U and φ_S are isomorphisms. By induction on dimension, the previous paragraph implies that φ_Y is an isomorphism p-locally for every k-scheme Y of finite type. Since p was any prime not dividing $t(G)$, we have shown that $\varphi\colon H_*(X/G, R(*))^{\oplus N} \to H_*(X/B, R(*)) = H_*(X/T, R(*))$ is an isomorphism, where $R = \mathbf{Z}[1/t(G)]$. □

16.2 Leray spectral sequence for a divisor with normal crossings

In this section, we construct a spectral sequence converging to the motivic cohomology of the complement of a divisor with normal crossings in a smooth scheme. Deligne constructed the analogous spectral sequence for the ordinary cohomology of complex varieties [30, equation 3.2.4.1]. (The weight filtration of a smooth complex variety U is defined as the filtration of $H^*(U, \mathbf{Q})$ given by this spectral sequence, using any simple normal crossing compactification of U.) In that topological setting, the spectral sequence can be viewed as the Leray spectral sequence of the inclusion from U into its compactification. One could ask for a motivic Leray spectral sequence in much greater generality than Lemma 16.2.

All the spectral sequences we define have the standard cohomological numbering, meaning that the differential d_r on the E_r term has bidegree $(r, 1-r)$.

Lemma 16.2 *Let $\cup_{i=1}^n D_i$ be a divisor with simple normal crossings in a smooth scheme E over a field k. For each subset $I \subset \{1, \ldots, n\}$, let $D_I = \cap_{i \in I} D_i \subset E$. Then, for each integer j, there is a second-quadrant spectral sequence*

$$E_1^{pq} = \oplus_{|I|=-p} H_M^{2p+q}(D_I, \mathbf{Z}(j+p)) \Rightarrow H_M^{p+q}(E - \cup_{i=1}^n D_i, \mathbf{Z}(j)).$$

The E_1 term is concentrated in rows 0 to $2j$ and in columns $-n$ to 0.

Proof We write the proof in the language of higher Chow groups (section 6.2). Thus we want to define a spectral sequence, for every integer j:

$$E_1^{pq} = \oplus_{|I|=-p} CH^{j+p}(D_I, 2j-q) \Rightarrow CH^j(E - \cup_{i=1}^n D_i, 2j-p-q).$$

For each integer j and scheme X over k, let $z^j(X,*)$ be Bloch's chain complex of abelian groups, whose homology is $CH^j(X,*)$. For $D \subset X$ a divisor, we have a pushforward map $z^{j-1}(D,*) \to z^j(X,*)$. For $1 \leq i_1 < \cdots < i_a \leq n$ and $1 \leq b \leq a$, define a map

$$z^{j-a}(D_{i_1,\ldots,i_a}, *) \to z^{j-a+1}(D_{i_1,\ldots,\widehat{i_b},\ldots,i_a}, *)$$

as $(-1)^b$ times the pushforward map. Combining these gives a double complex

$$0 \to z^{j-n}(D_{1\cdots n}, *) \to \cdots \to \oplus_{i=1}^n z^{j-1}(D_i, *) \to z^j(E,*) \to 0,$$

where the summands are indexed by the subsets of $\{1, \ldots, n\}$. There is an obvious map from this double complex to $z^j(E - \cup D_i, *)$, given by the flat pullback map on $z^j(E,*)$. If we can show that this map is a quasi-isomorphism, then we have the desired spectral sequence converging to $CH^j(E - \cup D_i, *)$, as one of the standard spectral sequences associated to a double complex.

We show this by induction on n. It is trivial for $n = 0$, and for $n = 1$, $(z^{j-1}(D_1,*) \to z^j(E,*)) \to z^j(E - D_1, *)$ is a quasi-isomorphism as we want, by the localization theorem on higher Chow groups (Lemma 6.8). In general, consider the diagram

$$\begin{array}{ccccccc}
(z^{j-n}(D_{1\cdots n},*) & \longrightarrow & \cdots & \longrightarrow & z^{j-1}(D_1,*)) & \longrightarrow & z^{j-1}(D_1-\cup_{i=2}^n D_i,*) \; , \\
 & & & \downarrow & & & \downarrow \\
(z^{j-n+1}(D_{2\cdots n},*) & \longrightarrow & \cdots & \longrightarrow & z^{j}(E,*)) & \longrightarrow & z^{j}(E-\cup_{i=2}^n D_i,*) \\
 & & & \downarrow & & & \downarrow \\
(z^{j-n}(D_{1\cdots n},*) & \longrightarrow & \cdots & \longrightarrow & z^{j}(E,*)) & \longrightarrow & z^{j}(E-\cup_{i=1}^n D_i,*)
\end{array}$$

where the left complex on the first line is a sum over all subsets $I \subset \{1,\ldots,n\}$ containing 1, the left complex on the second line is a sum over all subsets not containing 1, and the third line runs over all subsets. We trivially have a distinguished triangle in the derived category of abelian groups on the left, and we have an distinguished triangle on the right by the localization theorem for the inclusion $D_1 - \cup_{i=2}^n D_i \to E - \cup_{i=2}^n D_i$ (Lemma 6.8). By induction, the first and second horizontal maps are quasi-isomorphisms. It follows that the third horizontal map is a quasi-isomorphism, by the five lemma. This completes the induction.

The E_1 term is clearly concentrated in columns $-n$ to 0. It is concentrated in rows 0 to j because $H^a_M(Y,\mathbf{Z}(b)) = 0$ for $a > 2b$ and Y any smooth scheme over k. □

16.3 Eilenberg-Moore spectral sequence in motivic cohomology

In this section, we prove a motivic Eilenberg-Moore spectral sequence for principal G-bundles, with G a split reductive group. As an application, the spectral sequence relates the motivic cohomology of BG with the motivic cohomology of $GL(n)/G$, for any affine group scheme G with a faithful representation $G \to GL(n)$.

The first step is to prove the Eilenberg-Moore spectral sequence for a torus action. This is a result of Krishna's [82, theorem 1.1].

Corollary 16.3 *Let X be a smooth scheme over a field k with a free action of the torus $T = (G_m)^n$ for some natural number n. Suppose that the quotient X/T exists as a scheme. For each integer j, there is a second-quadrant spectral sequence*

$$E_2^{pq} = \mathrm{Tor}^{CH^*BT}_{-p,q,j}(\mathbf{Z}, H^*_M(X/T,\mathbf{Z}(*))) \Rightarrow H^{p+q}_M(X,\mathbf{Z}(j)).$$

The E_2 term is concentrated in rows 0 to $2j$ and in columns $-n$ to 0.

Here we consider $CH^*BT = \oplus_j H^{2j}(BT, \mathbf{Z}(j)) = \oplus_j H^{2j,j}(BT)$ as a bigraded ring. For bigraded modules M and N over a bigraded ring R, $\mathrm{Tor}^R_{i,q,j}(M, N)$ denotes the (q, j)th bigraded piece of $\mathrm{Tor}^R_i(M, N)$. We write $M(a, b)$ for the bigraded module M with bidegrees moved down by (a, b).

Proof Consider the standard representation of $T = (G_m)^n$ on A^n. The A^n-bundle over X/T associated to this representation contains X as the complement of a divisor with normal crossings $D_1 \cup \cdots \cup D_n$, corresponding to the coordinate hyperplanes in A^n. Each intersection D_I of divisors is the total space of a vector bundle over X/T, and hence has the same motivic cohomology as X/T. So, for each integer j, the spectral sequence of Lemma 16.2 takes the form:

$$E_1^{pq} = H_M^{2p+q}(X/T, \mathbf{Z}(j+p))^{\oplus\binom{n}{-p}} \Rightarrow H_M^{p+q}(X, \mathbf{Z}(j)).$$

For $n = 2$, this E_1 term looks like:

$$\begin{array}{ccccc}
\vdots & \vdots & \vdots & \vdots & \\
0 & H^0(X/T, \mathbf{Z}(j-2)) \longrightarrow & H^2(X/T, \mathbf{Z}(j-1))^{\oplus 2} \longrightarrow & H^4(X/T, \mathbf{Z}(j)) & 0 \\
0 & 0 & H^1(X/T, \mathbf{Z}(j-1))^{\oplus 2} \longrightarrow & H^3(X/T, \mathbf{Z}(j)) & 0 \\
0 & 0 & H^0(X/T, \mathbf{Z}(j-1))^{\oplus 2} \longrightarrow & H^2(X/T, \mathbf{Z}(j)) & 0 \\
0 & 0 & 0 & H^1(X/T, \mathbf{Z}(j)) & 0 \\
0 & 0 & 0 & H^0(X/T, \mathbf{Z}(j)) & 0
\end{array}$$

Moreover, we have an explicit description of the d_1 differential in this spectral sequence in terms of pushforward maps for the inclusions $D_I \to D_{I-i_b}$. The Chow ring CH^*BT is a polynomial ring $\mathbf{Z}[u_1, \ldots, u_n]$ with $|u_i| = 1$, and the u_i map in $CH^*(X/T)$ to the first Chern classes of the n obvious line bundles on X/T. The pushforward map for the inclusion $D_I \to D_{I-\{m\}}$ is multiplication by u_m, when the motivic cohomology rings of both schemes are identified with the motivic cohomology of X/T. Therefore, the E_1 term with its differential is a complex that computes $\mathrm{Tor}^{CH^*BT}_{**}(\mathbf{Z}, H^*(X/T, \mathbf{Z}(*)))$, using the Koszul resolution of $\mathbf{Z}$ as a module over the bigraded polynomial ring $R = CH^*(BT)$

(Lemma 3.11):

$$0 \to R(-2n, -n) \to \cdots \to R^{\binom{n}{2}}(-4, -2) \to R^{\binom{n}{1}}(-2, -1) \to R \to \mathbf{Z} \to 0.$$

Thus the E_2 term of the spectral sequence is the Tor group we want. The E_2 term vanishes outside the range mentioned, by the corresponding vanishing in Lemma 16.2. □

We now deduce a motivic Eilenberg-Moore spectral sequence for actions of $GL(n)$. In fact, we get a weaker statement (inverting finitely many primes) for any split reductive group. For our applications, we only need the case of $GL(n)$ or more generally products of groups $GL(n_i)$, in which case the torsion index is 1. (The torsion index is defined in section 16.1.)

Theorem 16.4 *Let X be a smooth scheme over a field k with an action of a split reductive group G. Let $t(G)$ be the torsion index of G. For each integer j, there is a second-quadrant spectral sequence*

$$E_2^{pq} = \mathrm{Tor}_{-p,q,j}^{CH^*BG}(\mathbf{Z}, H_G^*(X, \mathbf{Z}(*)))[t(G)^{-1}] \Rightarrow H^{p+q}(X, \mathbf{Z}(j))[t(G)^{-1}].$$

The E_2 term is concentrated in rows 0 to $2j$ and in columns $-\mathrm{rank}(G)$ to 0.

Proof Equivariant motivic cohomology is defined as the motivic cohomology of suitable quotient schemes of free G-actions. So it suffices to prove the theorem when G acts freely on X, with quotient X/G a scheme over k.

Consider the map of fibrations

$$\begin{array}{ccccc} X & \longrightarrow & X/T & \longrightarrow & BT \\ \downarrow & & \downarrow & & \downarrow \\ X & \longrightarrow & X/G & \longrightarrow & BG. \end{array}$$

By Lemma 16.3, we have a spectral sequence

$$E_2^{pq} = \mathrm{Tor}_{-p,q,j}^{CH^*BT}(\mathbf{Z}, H^*(X/T, \mathbf{Z}(*))) \Rightarrow H^{p+q}(X, \mathbf{Z}(j)).$$

By Theorem 16.1, after inverting the torsion index $t(G)$, CH^*BT becomes a finitely generated free CH^*BG-module, with basis elements $e_1, \ldots, e_m$ in CH^*BT. Moreover, again with $t(G)$ inverted, the motivic cohomology of X/T is a free module over the motivic cohomology of X/G, with the same basis $e_1, \ldots, e_m$. By flat base change for Tor [149, proposition 3.2.9], we have

$$\mathrm{Tor}_*^{CH^*BT}(\mathbf{Z}, A \otimes_{CH^*BG} CH^*BT)[t(G)^{-1}] \cong \mathrm{Tor}_*^{CH^*(BG)}(\mathbf{Z}, A)[t(G)^{-1}]$$

for any CH^*BG-module A. Applying this with A the motivic cohomology of X/G, we deduce that the E_2 term of the spectral sequence above can be rewritten (after inverting $t(G)$) as

$$E_2^{pq} = \mathrm{Tor}_{-p,q,j}^{CH^*BG}(\mathbf{Z}, H_G^*(X, \mathbf{Z}(*)))[t(G)^{-1}].$$

Thus we have the spectral sequence of Eilenberg-Moore type for G, converging to the motivic cohomology of X. □

In particular, the Eilenberg-Moore spectral sequence of Theorem 16.4 allows us to relate the motivic cohomology of $GL(n)/G$ and BG for any group G, as follows.

Corollary 16.5 *Let G be an affine group scheme over a field k with a faithful representation $G \to GL(n)$. For each integer j, there is a spectral sequence*

$$E_2^{pq} = \mathrm{Tor}_{-p,q,j}^{CH^*BGL(n)}(\mathbf{Z}, H_M^*(BG, \mathbf{Z}(*))) \Rightarrow H_M^{p+q}(GL(n)/G, \mathbf{Z}(j)).$$

Proof This is exactly Theorem 16.4 for the action of $GL(n)$ on $GL(n)/G$, using that $GL(n)$ has torsion index 1. □

To emphasize that the motivic Eilenberg-Moore spectral sequence has good properties that do not hold in topology, note that Corollary 16.5 implies Theorem 5.1 as a very special case. We restate Theorem 5.1 as:

Theorem 16.6 *Let G be an affine group scheme over a field k with a faithful representation V of dimension n. Then*

$$CH^*GL(n)/G \cong CH^*BG/(c_1V, \ldots, c_nV).$$

As a result, CH^*BG is generated as a module over the Chern classes $\mathbf{Z}[c_1V, \ldots, c_nV]$ by elements of degree at most $n^2 - \dim(G)$. It follows that the ring CH^*BG is generated by elements of degree at most $\max(n, n^2 - \dim(G))$.

Proof We need to show that $CH^*(GL(n)/G) \cong \mathbf{Z} \otimes_{CH^*BGL(n)} CH^*BG$. Consider the spectral sequence of Theorem 16.5 converging to $H_M^*(GL(n)/G, \mathbf{Z}(j))$. The group $\mathbf{Z} \otimes_{CH^*BGL(n)} CH^*BG$ in degree j is at the upper right corner of the rectangle in which the E_2 term of the spectral sequence may be nonzero. So it is the only group contributing to $CH^j(GL(n)/G) = H_M^{2j}(GL(n)/G, \mathbf{Z}(j))$, and there are no differentials into or out of it. □

Remark 16.7 Let us check that the Eilenberg-Moore spectral sequence of Theorem 16.4 does not hold without inverting the torsion index. Suppose that the spectral sequence holds integrally, for a split reductive group G over a field k. By the proof of Theorem 16.6, the spectral sequence implies that the natural map $CH^*(X/G) \otimes_{CH^*BG} \mathbf{Z} \to CH^*X$ is an isomorphism, for any principal G-bundle $X \to X/G$ over k. Taking $X = G$, this would imply that the Chow ring of G as a variety is $\mathbf{Z}$ in degree 0. But Grothendieck showed that CH^*G is the quotient of $CH^*(G/B)$ by the ideal generated by $\mathrm{Hom}(B, G_m) = CH^1 BB$ [54]. It follows that CH^*G is nonzero p-locally in positive degrees for every prime number p that divides $t(G)$. (If $CH^{>0}(G/B)$ is generated by $CH^1 BB$ as an ideal, then $CH^*(G/B)$ would be generated by $CH^1 BB$ as an algebra, by induction on the grading of $CH^*(G/B)$.) For example, $CH^1 SO(3)_{\mathbf{C}}$ is $\mathbf{Z}/2$, not zero.

17

The Chow Künneth Conjecture

For smooth schemes X and Y over a field k, the product map $CH^*X \otimes_{\mathbf{Z}} CH^*Y \to CH^*(X \times Y)$ is rarely an isomorphism. The early calculations of Chow rings of classifying spaces raised the unexpected possibility that they might satisfy the Chow Künneth formula, meaning that $CH^*BG \otimes_{\mathbf{Z}} CH^*X \to CH^*(BG \times X)$ is an isomorphism for all smooth schemes X over a field k, provided that k contains enough roots of unity [138, section 6]. This would in particular imply that CH^*BG_K is the same for all fields K containing k. Thus the Chow Künneth property, when it holds, is a strong rigidity property for the Chow groups of classifying spaces.

Conjecture 17.1 *Let G be a finite group of exponent e. Let k be a field such that the order of G is invertible in k and k contains the eth roots of unity. Then the product map*

$$CH^*BG_k \otimes_{\mathbf{Z}} CH^*X \to CH^*(BG \times_k X)$$

is an isomorphism for every smooth scheme X of finite type over k.

Conjecture 17.1 is interesting even in the special case where k is algebraically closed. In that case, one can expect the conjecture to hold for all affine group schemes G of finite type over k.

Some assumption that k contains enough roots of unity is essential in Conjecture 17.1. For example, $B(\mathbf{Z}/n)_{\mathbf{Q}}$ does not satisfy the Chow Künneth formula for n odd. Indeed, $CH^1B(\mathbf{Z}/n)_{\mathbf{Q}} = \text{Hom}(\mathbf{Z}/n, \mathbf{Q}^*) = 0$, whereas $CH^1B(\mathbf{Z}/n)_K = \text{Hom}(\mathbf{Z}/n, K^*)$ is isomorphic to $\mathbf{Z}/n$ for a number field K containing the nth roots of unity. This disproves the Chow Künneth formula for the product

$$B(\mathbf{Z}/n)_{\mathbf{Q}} \times_{\text{Spec } \mathbf{Q}} \text{Spec } K = B(\mathbf{Z}/n)_K.$$

One can also check that $CH^*B(\mathbf{Z}/3)_{\mathbf{Q}} \otimes_{\mathbf{Z}} CH^*B(\mathbf{Z}/3)_{\mathbf{Q}} \to CH^*B(\mathbf{Z}/3 \times \mathbf{Z}/3)_{\mathbf{Q}}$ is not surjective.

The Suslin rigidity theorem says (in particular) that for any smooth scheme X over an algebraically closed field k, and any algebraically closed field K containing k, the motivic cohomology with finite coefficients of X maps isomorphically to that of X_K [126, corollary 2.3.3]. It follows that the Chow ring CH^*BG_k is unchanged under extensions of algebraically closed fields k, for a finite group G. But that does not explain why CH^*BG_k seems to remain unchanged for arbitrary field extensions.

One partial explanation for why classifying spaces over sufficiently large fields seem to satisfy the Chow Künneth property is that any *linear variety* X satisfies the Chow Künneth property: $CH_*X \otimes_{\mathbf{Z}} CH_*Y \to CH_*(X \times Y)$ is an isomorphism for every Y [140]. Moreover, Joshua showed that linear varieties satisfy a natural generalization of the Künneth property that applies to all motivic homology groups [74]. Linear varieties are defined inductively: a scheme over a field is linear if it can be stratified as a disjoint union of finitely many locally closed subsets, each of which is an affine space of some dimension minus a lower-dimensional linear variety. For abelian groups and wreath product groups as in Lemma 2.12, over a field k with enough roots of unity, the classifying space BG can be approximated by linear varieties, and that implies the Chow Künneth formula for such groups G. (By the comments above, it follows that $B(\mathbf{Z}/3)_{\mathbf{Q}}$ cannot be approximated by linear varieties over $\mathbf{Q}$.)

On the other hand, linear varieties cannot provide the full explanation, if we hope to prove the Chow Künneth formula for all finite groups. Indeed, Saltman and Bogomolov gave examples of p-groups G such that the quotient by G of a faithful complex representation is never a rational variety [17]. A fortiori, we cannot approximate BG by linear varieties for those groups G. Even for a group G such that quotient varieties are rational, we have to work hard (with no guarantee of success) to show that BG is approximated by linear varieties. It is natural to look for other ways to prove the Chow Künneth formula.

We get some information by relating Chow groups to representation theory. To state this, let e be the exponent of a finite group G. Brauer showed that the representation theory of a finite group G is "the same" over all fields K such that $|G|$ is invertible in K and K contains the eth roots of unity [124, theorem 24].

Lemma 17.2 *Let G be a finite group. The geometric filtration of the representation ring $K_0^G(K)$ is independent of the field K when K contains the algebraic closure $\overline{\mathbf{Q}}$ of $\mathbf{Q}$.*

Proof We know by Brauer that the inclusion $\overline{\mathbf{Q}} \subset K$ induces an isomorphism $K_0^G(\overline{\mathbf{Q}}) \cong K_0^G(K)$. Clearly $F^j_{\text{geom}} K_0^G(\overline{\mathbf{Q}})$ is contained in $F^j_{\text{geom}} K_0^G(K)$, and we want to show that equality holds. If not, then (looking at the smallest j for which the filtrations differ) there is a natural number i such that

$\mathrm{gr}^i_{\mathrm{geom}} K_0^G(\overline{\mathbf{Q}}) \to \mathrm{gr}^i_{\mathrm{geom}} K_0^G(K)$ is surjective but not injective. If we can show that $\mathrm{gr}^i_{\mathrm{geom}} K_0^G(\overline{\mathbf{Q}}) \to \mathrm{gr}^i_{\mathrm{geom}} K_0^G(\overline{K})$ is injective for all i, then we have the same injectivity for $\overline{\mathbf{Q}}$ mapping into K, and we are done.

Let $u \in \mathrm{gr}^i_{\mathrm{geom}} K_0^G(\overline{\mathbf{Q}})$ be an element that maps to zero in $\mathrm{gr}^i_{\mathrm{geom}} K_0^G(\overline{K})$. To state this concretely, let V be a representation of G over $\overline{\mathbf{Q}}$ such that G acts freely on V on an open subset U with $V - U$ of codimension at least $i + 1$. Then our assumption means that u can be represented by some coherent sheaf with support of codimension at least $i + 1$ on $(U/G)_{\overline{K}}$. It follows that u can be represented by some coherent sheaf of codimension at least $i + 1$ on $(U/G)_L$ for some finitely generated field L over $\overline{\mathbf{Q}}$. Think of L as the function field of a smooth variety Y over $\overline{\mathbf{Q}}$. Then, after possibly replacing Y by a dense open subset, u is represented by a coherent sheaf of codimension at least $i + 1$ on $(U/G) \times_{\overline{\mathbf{Q}}} Y$. By restricting to some $\overline{\mathbf{Q}}$-rational point of Y (these being Zariski dense in Y), it follows that u is represented by a coherent sheaf of codimension at least $i + 1$ on the $\overline{\mathbf{Q}}$-variety U/G. Thus $\mathrm{gr}^i_{\mathrm{geom}} K_0^G(\overline{\mathbf{Q}}) \to \mathrm{gr}^i_{\mathrm{geom}} K_0^G(\overline{K})$ is injective, as we want. □

Corollary 17.3 *Let G be a finite group. Then the p-local Chow ring $(CH^* BG_K)_{(p)}$ is independent of the field K containing $\overline{\mathbf{Q}}$ in degrees $\leq p$.*

Proof The surjection $CH^i BG_K \to \mathrm{gr}^i_{\mathrm{geom}} BG_K$ is an isomorphism p-locally for all $i \leq p$ by Theorem 2.25. Then the result follows from Lemma 17.2. □

It is tempting to try to extend this argument to prove that $(CH^i BG_K)_{(p)}$ is independent of the field K containing $\mathbf{Q}(\mu_e)$ for $i \leq p$, where e is the exponent of a p-Sylow subgroup of G.

Beyond these results in degree at most p, our methods explain the Chow Künneth formula in all degrees for a certain class of groups, in the version that the Chow ring is unchanged under field extensions.

Theorem 17.4 *Let G be a finite group, p a prime number, P a p-Sylow subgroup of G. Consider G as an algebraic group over $\overline{\mathbf{Q}}$. Suppose that P has a faithful representation of dimension n over $\overline{\mathbf{Q}}$ with c irreducible summands such that $n - c \leq p$. Then the homomorphism $(CH^* BG_{\overline{\mathbf{Q}}})_{(p)} \to CH^*(BG_k)_{(p)}$ is an isomorphism for all fields k containing $\overline{\mathbf{Q}}$.*

Proof By Suslin's rigidity theorem as discussed above, we know that this homomorphism is injective for all fields k containing $\overline{\mathbf{Q}}$. To prove this p-local surjectivity, it suffices to show that the homomorphism $\varphi\colon CH^*_{G_{\overline{\mathbf{Q}}}} \to CH^*_{G_k}$ of mod p Chow rings is surjective. By Theorem 11.1, $CH^* BG_k$ is generated by transferred Euler classes for every field k containing $\overline{\mathbf{Q}}$. These classes are all defined over $\overline{\mathbf{Q}}$. □

For many groups, our calculations show that the Chow ring is the same over all fields with enough roots of unity, not just those that contain $\overline{\mathbf{Q}}$. For example, we showed in Chapter 13 that for all 14 groups of order 16, the Chow ring is the same for all base fields of characteristic not 2 that contain the eth roots of unity, where e is the exponent of G. Likewise for the 5 groups of order p^3 with p odd (Section 13.2), the 15 groups of order 81, and 13 of the 15 groups of order p^4 with $p \geq 5$, those for which we could make the calculation in Chapter 14.

18

Open Problems

(1) Yagita conjectured that algebraic cobordism $\Omega^* BG$ maps isomorphically to $MU^* BG$ for every complex algebraic group G [154, conjecture 12.2]. Since $\Omega^i BG$ maps to the topological cobordism $MU^{2i} BG$, this would in particular imply that $MU^* BG$ is concentrated in even degrees. Another consequence would be that the Chow ring of BG maps isomorphically to $MU^* BG \otimes_{MU^*} \mathbf{Z}$. A natural test case is the finite group G of 4×4 strictly upper-triangular matrices over $\mathbf{F}_p$ for p odd, since the Morava K-theory $K(2)^{\text{odd}} BG$ is nonzero in that case by Kriz and Lee [83, 84].

(2) In this book, we have concentrated on the mod p Chow ring of classifying spaces BG. One can try to extend both the general methods and the explicit calculations to richer theories, such as the integral Chow ring, motivic cohomology, or algebraic cobordism.

It would be interesting to prove general bounds and make systematic calculations for the Bloch-Ogus spectral sequence $H^i_{\text{Zar}}(BG, H^j_{\mathbf{F}_p}) \Rightarrow H^{i+j}_{\text{et}}(BG, \mathbf{F}_p)$, for an affine group scheme G over a field k. Guillot proved several results in this direction [63]. The spectral sequence is closely related to the motivic cohomology of BG; in particular, $H^i_{\text{Zar}}(BG, H^i_{\mathbf{F}_p})$ is the mod p Chow group CH^i_G. The output of the spectral sequence is essentially the ordinary cohomology of BG. The group $H^0(BG, H^i_{\mathbf{F}_p})$ is the group of cohomological invariants for G-torsors over fields, as defined by Serre [47, part 1, appendix C].

(3) For an odd prime number p and a finite group G, viewed as a complex algebraic group, does the mod p Chow ring CH^*_G consist of transferred Euler classes? For $p = 2$, Guillot showed that the answer is no, using the extraspecial 2-group 2^{1+6}_+ [62].

(4) Let $V = V_1 \oplus \cdots \oplus V_c$ be a faithful complex representation of minimal dimension of a p-group G. Find an optimal bound in terms of p and the dimensions of the irreducible summands V_i for the degrees of generators of H^*_G and CH^*_G. Chapter 7 gives some bounds. One could also ask for bounds in terms of other invariants, such as the order and the p-rank of G.

(5) Give good bounds for Henn-Lannes-Schwartz's topological nilpotence degree $d_0(H_G^*)$ and $d_0(CH_G^*)$. Is $d_0(CH_G^*) \leq d_0(H_G^*)/2$? Can the bounds in Theorems 13.17 and 12.7 be improved? The examples in section 13.5 suggest that these bounds may be improvable for non-p-central p-groups with p odd.

(6) Let p be a prime number, and let k be a field of characteristic not p that contains the pth roots of unity. Let G be a finite group, viewed as an algebraic group over k. Is the topological nilpotence degree $d_0(CH_G^*)$ equal to the supremum of the natural numbers d such that CH_G^*, as a module over the Steenrod algebra $\mathcal{A}$, contains a nonzero submodule of the form $\Sigma^d M$ with M an unstable $\mathcal{A}$-module? The known partial results are listed after Conjecture 12.8. The analogous statement is true for H_G^* by Henn-Lannes-Schwartz (Theorem 12.2).

(7) Is CH^*BG a finitely generated $\mathbf{Z}$-algebra for every affine group scheme G over a field? It would suffice to show that CH^iBG is a finitely generated abelian group for each i, by Theorem 5.2. Is there an explicit class of generators that works for any finite group G? In view of Guillot's example, transferred Euler classes are not enough in general.

(8) Does the Chow Künneth formula hold for arbitrary finite groups (Conjecture 17.1)?

(9) Let G be a p-group of rank at most 2, say viewed as an algebraic group over $\mathbf{C}$. Is the cycle map $CH^*BG \to H^{\text{ev}}(BG, \mathbf{Z})$ an isomorphism? The assumption on the rank seems natural, since the map is not surjective when $G = (\mathbf{Z}/p)^3$. For a p-group G of rank 2, $CH^*BG \to H^{\text{ev}}(BG, \mathbf{Z})$ is surjective at least for $p \geq 5$, since Yagita showed that $H^{\text{ev}}(BG, \mathbf{Z})$ consists of transferred Chern classes in that case [152]. Theorem 14.3 shows that the map is an isomorphism mod p when, in addition, G has a faithful complex representation that is the sum of a p-dimensional irreducible representation and some 1-dimensional representations. Split metacyclic p-groups $\mathbf{Z}/p^n \ltimes \mathbf{Z}/p^m$ are a natural test case for this problem.

(10) For a prime number $p \geq 5$, compute the Chow ring $CH_{\widetilde{B}}^*$ for the 1-dimensional group $\widetilde{B}$ from section 14.5. The group $\widetilde{B}$ is a central extension of the extraspecial p-group p_+^{1+2} by the multiplicative group G_m. Since $CH_{\widetilde{B}}^*$ maps isomorphically to $H^{\text{ev}}(B\widetilde{B}, \mathbf{Z})/p$ (Lemma 14.5), it is more or less equivalent to compute the cohomology of $B\widetilde{B}$. As explained in section 14.5, computing $CH_{\widetilde{B}}^*$ would finish the computation of the Chow rings of all groups of order p^4.

(11) For a field k of characteristic $p > 0$, restricted Lie algebras over k are equivalent to group schemes of height at most 1 [32, section VIIA.8]. As a result, the definition of CH^*BG gives a definition of the Chow ring of a restricted Lie algebra. The problem is to understand the Chow rings of restricted Lie algebras, as systematically as possible. This is a model problem for trying to understand mod p Chow groups in characteristic p. One goal is to relate the mod p Chow rings of smooth varieties in characteristic p to some more computable theory related to de Rham cohomology.

Lemma 2.18 gives that the Chow ring of a unipotent restricted Lie algebra $\mathfrak{g}$ is trivial. One natural problem is to compute the Chow rings of the restricted simple Lie algebras, which were classified in characteristic $p \geq 11$ by Block and Wilson [16].

(12) Theorem 6.5 shows that for every finite group scheme G over a field k and every prime number p invertible in k, the mod p Chow ring CH_G^* has regularity at most zero. Does this hold for all affine group schemes of finite type over k and all prime numbers p? The proof of Theorem 6.5 seems not to work. (The group $S = (\mu_p)^n$ acts on the algebraic space $X = U\backslash GL(n)/G$, and Chow groups make sense for algebraic spaces by Edidin-Graham [38, section 6.1]. The difficulty in extending the proof of Theorem 6.5 to G of positive dimension is that the fixed point sets in X for subgroups of S need not be closed when G has positive dimension, because X need not be separated. For example, for the diagonal torus T in $GL(2)$, $U\backslash GL(2)/T$ is isomorphic to the line with two origins, a non-separated scheme [67, example II.2.3.6], and $U\backslash GL(2)/N(T)$ is an algebraic space that is not a scheme, illustrated by Artin [3] and named by Kollár a bug-eyed cover [78].)

More strongly, examples suggest that we may have $\text{reg}(CH_G^*) \leq -\dim(G_{\overline{k}}/B)$ for every affine group scheme G and every prime number p, where B is a Borel subgroup (a maximal smooth connected solvable subgroup) of $G_{\overline{k}}$.

Note added in proof: I found that questions (7) and (8) have negative answers in general.

Appendix

Tables

The following tables show several invariants of the Chow ring for the p-groups of order at most p^4. Each p-group G of exponent e is viewed as an algebraic group over any field of characteristic not p that contains the eth roots of unity. The format was suggested by Kuhn's tables of $d_0(H_G^*)$ and related invariants for the 2-groups of order at most 64 [85, appendix A]. Most of the information comes from Chapters 13 and 14, along with calculations in GAP or Macaulay2 [48]. One observation is that $d_0(CH_G^*)$ turns out to be small in these examples; compare Section 13.5.

The tables show the order of a p-group G, its number in GAP's Small Groups library [46, 52], the rank of G, the rank $c(G)$ of $C = Z(G)[p]$, the depth of the mod p Chow ring CH_G^*, the Chow type (defined in the proof of Theorem 12.7, to describe the image of $CH_G^* \to CH_C^*$), the topological nilpotence degree $d_0(CH_G^*)$ (defined in Section 12.2), and the ranks of all maximal elementary abelian subgroups of G up to conjugacy.

We say that a group is indecomposable if it cannot be written as the product of two nontrivial groups. In the list of ranks, "$p+1$ of 2" means $p+1$ conjugacy classes of maximal elementary abelian subgroups of rank 2. The Notes use several different names for familiar groups, including both $\mathbf{Z}/n$ and $C(n)$ for the cyclic group of order n. We write "?" in a few places, meaning that the calculation remains to be done.

Table 1. *Indecomposable nonabelian 2-groups of order* ≤ 16

Order	#	Rank	$c(G)$	Depth	Type	$d_0(CH_G^*)$	Ranks	Notes
8	3	2	1	2	[2]	0	2,2	$D(8)$
	4	1	1	1	[2]	1	1	$Q(8)$
16	3	3	2	3	[2,1]	0	3	
	4	2	2	2	[2,1]	1	2	$\mathbf{Z}/4 \ltimes \mathbf{Z}/4$
	6	2	1	1	[2]	1	2	Mod(16)
	7	2	1	2	[2]	0	2,2	$D(16)$
	8	2	1	1	[2]	1	2	$SD(16)$
	9	1	1	1	[2]	1	1	$Q(16)$
	13	2	1	2	[2]	0	2,2,2	$D(8) * C(4)$

Table 2. *Indecomposable nonabelian 3-groups of order* ≤ 81

Order	#	Rank	$c(G)$	Depth	Type	$d_0(CH_G^*)$	Ranks	Notes
27	3	2	1	2	[3]	0	2,2,2,2	$E(27)$
	4	2	1	1	[3]	1	2	Mod(27)
81	3	3	2	2	[3,1]	2	3	
	4	2	2	2	[3,1]	1	2	$\mathbf{Z}/9 \ltimes \mathbf{Z}/9$
	6	2	1	1	[3]	1	2	Mod(81)
	7	3	1	2	[3]	0	2,3	$\mathbf{Z}/3 \wr \mathbf{Z}/3$
	8	2	1	1	[3]	1	2,2	
	9	2	1	2	[3]	0	2,2,2,2	$\mathrm{Syl}_3(U_3(8))$
	10	2	1	1	[3]	2	2	
	14	2	1	1	[3]	1	2,2,2,2	$E(27) * C(9)$

Table 3. *Indecomposable nonabelian p-groups of order* $\leq p^4$ *for* $p \geq 5$

Order	#	Rank	$c(G)$	Depth	Type	$d_0(CH_G^*)$	Ranks	Notes
p^3	3	2	1	1	$[p]$	2	$p+1$ of 2	$E(p^3)$
	4	2	1	1	$[p]$	1	2	$\mathrm{Mod}(p^3)$
p^4	3	3	2	2	$[p, 1]$	2	3	
	4	2	2	2	$[p, 1]$	1	2	$\mathbf{Z}/p^2 \ltimes \mathbf{Z}/p^2$
	6	2	1	1	$[p]$	1	2	$\mathrm{Mod}(p^4)$
	7	3	1	?	$[p]$	?	p of 2; 3	
	8	3	1	?	$[p]$	?	3	
	9	2	1	1	$[p]$	1	2,2	
	10	2	1	1	$[p]$	1	2,2	
	14	2	1	1	$[p]$	1	$p+1$ of 2	$E(p^3) * C(p^2)$

References

[1] A. Adem and R. J. Milgram. *Cohomology of finite groups.* Springer (1994). (Cited on pages 1, 5, 28, 30, 123, and 141.)

[2] J. Aguadé and L. Smith. On the mod p torus theorem of John Hubbuck. *Math. Z.* **191** (1986), 325–326. (Cited on page 129.)

[3] M. Artin. Algebraic construction of Brieskorn's resolutions. *J. Alg.* **29** (1974), 330–348. (Cited on page 216.)

[4] M. Aschbacher. *Finite group theory.* Cambridge (2000). (Cited on pages 107 and 144.)

[5] A. Asok, B. Doran, and F. Kirwan. Yang-Mills theory and Tamagawa numbers: the fascination of unexpected links in mathematics. *Bull. London Math. Soc.* **40** (2008), 533–567. (Cited on page 202.)

[6] M. Atiyah. Characters and cohomology of finite groups. *Publ. Math. IHES* **9** (1961), 23–64. (Cited on pages xv, 177, and 185.)

[7] M. Atiyah and F. Hirzebruch. Vector bundles and homogeneous spaces. *Proc. Sympos. Pure Math.*, vol. 3, 7–38. American Mathematical Society (1961). (Cited on page 177.)

[8] M. Atiyah and F. Hirzebruch. Analytic cycles on complex manifolds. *Topology* **1** (1962), 25–45. (Cited on pages xii and 198.)

[9] M. Atiyah and G. Segal. Equivariant K-theory and completion. *J. Diff. Geom.* **3** (1969), 1–18. (Cited on pages 177, 179, and 183.)

[10] D. Benson. *Polynomial invariants of finite groups.* Cambridge (1993). (Cited on pages 60, 109, and 111.)

[11] D. Benson. The image of the transfer map. *Arch. Math. 61* (1993), 7–11. (Cited on page 100.)

[12] D. Benson. *Representations and cohomology*, 2 vols. Cambridge (1998). (Cited on pages 1, 4, 7, 30, 31, 49, 88, 91, 96, 105, 141, and 173.)

[13] D. Benson. Dickson invariants, regularity and computation in group cohomology. *Illinois J. Math.* **48** (2004), 171–197. (Cited on pages 35 and 49.)

[14] S. Bloch. Algebraic cycles and higher K-theory. *Adv. Math.* **61** (1986), 267–304. (Cited on pages 10, 70, and 74.)

[15] S. Bloch. The moving lemma for higher Chow groups. *J. Alg. Geom.* **3** (1994), 537–568. (Cited on pages 10 and 70.)

[16] R. Block and R. Wilson. Classification of the restricted simple Lie algebras. *J. Alg.* **114** (1998), 115–259. (Cited on page 216.)

[17] F. A. Bogomolov. The Brauer group of quotient spaces of linear representations. *Izv. Akad. Nauk SSSR* **51** (1987), 485–516, 688; translation in *Math. USSR-Izv.* **30** (1988), 455–485. (Cited on pages xii, 13, and 211.)

[18] A. Borel. *Linear algebraic groups.* Springer (1991). (Cited on pages 21, 26, 27, 28, and 202.)

[19] N. Bourbaki. *Algebra II. Chapters 4–7.* Springer (2003). (Cited on page 59.)

[20] T. Bröcker and T. tom Dieck. *Representations of compact Lie groups.* Springer (1995). (Cited on page 4.)

[21] P. Brosnan. Steenrod operations in Chow theory. *Trans. Amer. Math. Soc.* **355** (2003), 1869–1903. (Cited on pages 11 and 99.)

[22] C. Broto and H.-W. Henn. Some remarks on central elementary abelian p-subgroups and cohomology of classifying spaces. *Quart. J. Math.* **44** (1993), 155–163. (Cited on pages 46 and 129.)

[23] W. Browder and J. Pakianathan. Cohomology of uniformly powerful p-groups. *Trans. Amer. Math. Soc.* **352** (2000), 2659–2688. (Cited on page 172.)

[24] K. Brown. *Cohomology of groups.* Springer (1982). (Cited on pages 1, 7, 22, and 24.)

[25] H. Campbell, J. Hughes, R. Shank, and D. Wehlau. Bases for rings of coinvariants. *Transform. Groups* **1** (1996), 307–336. (Cited on page 101.)

[26] J. Carlson, L. Townsley, L. Valeri-Elizondo, and M. Zhang. *Cohomology rings of finite groups.* Kluwer (2003). (Cited on pages xv, 1, 46, and 99.)

[27] H. Cartan and S. Eilenberg. *Homological algebra.* Princeton (1956). (Cited on page 44.)

[28] S. Cernele, M. Kamgarpour, and Z. Reichstein. Maximal representation dimension for groups of order p^n. *J. Group Theory* **14** (2011), no. 4, 637–647. (Cited on page 123.)

[29] C. Curtis and I. Reiner. *Representation theory of finite groups and associative algebras.* AMS Chelsea (2006). (Cited on page 133.)

[30] P. Deligne. Théorie de Hodge. II. *Publ. Math. IHES* **40** (1971), 5–57. (Cited on page 203.)

[31] M. Demazure. Invariants symétriques entiers des groupes de Weyl et torsion. *Invent. Math.* **21** (1973), 287–301. (Cited on page 203.)

[32] M. Demazure and A. Grothendieck. *Schémas en groupes* (SGA 3), 3 vols. Lecture Notes in Mathematics 151–153, Springer (1970). Revised version of vol. 1 and vol. 3, edited by P. Gille and P. Polo, Soc. Math. de France (2011). (Cited on pages 27, 57, 67, 69, 202, and 215.)

[33] T. Diethelm. The mod p cohomology rings of the nonabelian split metacyclic p-groups. *Arch. Math.* **44** (1985), 29–38. (Cited on page 164.)

[34] J. Duflot. Depth and equivariant cohomology. *Comment. Math. Helv.* **56** (1981), 627–637. (Cited on page 46.)

[35] J. Duflot. Smooth toral actions. *Topology* **22** (1983), 253–265. (Cited on page 53.)

[36] M. Hochster and J. Eagon. Cohen-Macaulay rings, invariant theory, and the generic perfection of determinantal loci. *Amer. J. Math.* **93** (1971), 1020–1058. (Cited on page 109.)

[37] S. Eilenberg. Homological dimension and syzygies. *Ann. Math.* **64** (1956), 328–336. (Cited on page 41.)

[38] D. Edidin and W. Graham. Equivariant intersection theory. *Inv. Math.* **131** (1998), 595–634. (Cited on pages 8, 11, 16, 17, 21, 72, 202, and 216.)

[39] G. Ellingsrud and T. Skjelbred. Profondeur d'anneaux d'invariants en caractéristique p. *Compos. Math.* **41** (1980), 233-244. (Cited on page 81.)

[40] H. Esnault and E. Viehweg. *Lectures on vanishing theorems.* Birkhäuser (1992). (Cited on page 25.)

[41] L. Evens. On the Chern classes of representations of finite groups. *Trans. Amer. Math. Soc.* **115** (1965), 180–193. (Cited on pages 88, 124, 126, and 152.)

[42] P. Fleischmann. Relative trace ideals and Cohen-Macaulay quotients of modular invariant rings. *Computational methods for representations of groups and algebras* (Essen, 1997), 211–233. Birkhäuser (1999). (Cited on pages 104, 109, and 110.)

[43] W. Fulton. *Intersection theory.* Springer (1998). (Cited on pages 8, 9, 10, 11, 23, 25, 34, and 67.)

[44] W. Fulton and S. Lang. *Riemann-Roch algebra.* Springer (1985). (Cited on pages 178 and 179.)

[45] W. Fulton and R. MacPherson. Characteristic classes of direct image bundles for covering maps. *Ann. Math.* **125** (1987), 1–92. (Cited on pages 84, 88, 89, 98, 124, and 126.)

[46] The GAP Group. GAP – Groups, Algorithms, and Programming, Version 4.6.4 (2013). http://www.gap-system.org (Cited on pages 123, 144, and 217.)

[47] S. Garibaldi, A. Merkurjev, and J.-P. Serre. *Cohomological invariants in Galois cohomology.* Amer. Math. Soc. (2003). (Cited on pages xii and 214.)

[48] D. Grayson and M. Stillman. Macaulay2, a software system for research in algebraic geometry. http://www.math.uiuc.edu/Macaulay2/ (Cited on page 217.)

[49] D. Green. On Carlson's depth conjecture in group cohomology. *Math. Zeit.* **244** (2003), 711–723. (Cited on page 163.)

[50] D. Green. The essential ideal is a Cohen-Macaulay module. *Proc. Amer. Math. Soc.* **133** (2005), 3191–3197. (Cited on page 105.)

[51] D. Green and S. King. The computation of the cohomology rings of all groups of order 128. *J. Alg.* **325** (2011), 352–363. (Cited on pages xv and 7.)

[52] D. Green and S. King. The cohomology of finite p-groups. http://users.minet.uni-jena.de/cohomology/ (Cited on pages xv, 7, 144, 145, 146, 148, 149, 156, 162, 164, 174, 178, 180, 181, 182, and 217.)

[53] D. Green and I. Leary. The spectrum of the Chern subring. *Comment. Math. Helv.* **73** (1998), 406–426. (Cited on page 125.)

[54] A. Grothendieck. Torsion homologique et sections rationelles. Exposé 5 in *Anneaux de Chow et applications. Séminaire C. Chevalley 1958.* Paris (1958). (Cited on pages 202, 203, and 209.)

[55] A. Grothendieck. Éléments de géométrie algébrique. II. Étude globale élémentaire de quelques classes de morphismes. *Publ. Math. IHES* **8** (1961), 222 pp. (Cited on page 36.)

[56] A. Grothendieck. Éléments de géométrie algébrique. III. Étude cohomologique des faisceaux cohérents. I. *Publ. Math. IHES* **11** (1961), 167 pp. (Cited on page 36.)

[57] A. Grothendieck. Problèmes ouverts en théorie des intersections. *Théorie des intersection et théorème de Riemann-Roch,* SGA 6, Lecture Notes in Mathematics 225, Exposé XIV, 667–689. Springer (1971). (Cited on pages 33 and 34.)

[58] A. Grothendieck. Classes de Chern et représentations linéaires des groupes discrets. *Dix exposés sur la cohomologie des schémas*, 215–305. North-Holland (1968). (Cited on pages 90 and 94.)

[59] A. Grothendieck. *Cohomologie locale des faisceaux cohérents et théorèmes de Lefschetz locaux et globaux* (SGA 2). Soc. Math. de France (2005). (Cited on pages 36 and 38.)

[60] P. Guillot. Chow rings and cobordism of some Chevalley groups. *Math. Proc. Cambridge Philos. Soc.* **136** (2004), 625–642. (Cited on page 31.)

[61] P. Guillot. Steenrod operations on the Chow ring of a classifying space. *Adv. Math.* **196** (2005), 276–309. (Cited on page 163.)

[62] P. Guillot. Addendum to the paper: "The Chow rings of G_2 and Spin(7)" [*J. Reine Angew. Math.* **604** (2007), 137–158]. *J. Reine Angew. Math.* **619** (2008), 233–235. (Cited on pages xiv, 122, and 214.)

[63] P. Guillot. Geometric methods for cohomological invariants. *Doc. Math.* **12** (2007), 521–545. (Cited on pages 32 and 214.)

[64] P. Guillot. The computation of Stiefel-Whitney classes. *Ann. Inst. Fourier* **60** (2010), 565–606. (Cited on page 141.)

[65] P. Guillot. The cohomology of 2-groups. http://www-irma.u-strasbg.fr/~guillot/research/cohomology_of_groups/index.html (Cited on page 141.)

[66] J. Gunawardena, B. Kahn, and C. B. Thomas. Stiefel-Whitney classes of real representations of finite groups. *J. Alg.* **126** (1989), 327–347. (Cited on page 20.)

[67] R. Hartshorne. *Algebraic geometry.* Springer (1977). (Cited on pages 10, 14, 34, and 216.)

[68] A. Hatcher. *Algebraic topology.* Cambridge (2002). (Cited on page 5.)

[69] H.-W. Henn, J. Lannes, and L. Schwartz. Localizations of unstable A-modules and equivariant mod p cohomology. *Math. Ann.* **291** (1991), 191–203. (Cited on pages xiv, 127, 128, 129, and 162.)

[70] M. Hopkins, N. Kuhn, and D. Ravenel. Generalized group characters and complex oriented cohomology theories. *J. Amer. Math. Soc.* **13** (2000), 553–594. (Cited on page 124.)

[71] J. Huebschmann. Chern classes for metacyclic groups. *Arch. Math.* **61** (1993), 124–136. (Cited on page 179.)

[72] D. Husemoller. *Fibre bundles.* Springer (1966). (Cited on page 1.)

[73] S. Iyengar, G. Leuschke, A. Leykin, C. Miller, E. Miller, A. Singh, and U. Walther. *Twenty-four hours of local cohomology.* Amer. Math. Soc. (2007). (Cited on pages 36, 37, 40, 41, and 43.)

[74] R. Joshua. Algebraic K-theory and higher Chow groups of linear varieties. *Math. Proc. Cambridge Phil. Soc.* **130** (2001), 37–60. (Cited on page 211.)

[75] B. Kahn. Représentations orthogonales et symplectiques sur un corps de caractéristique différente de 2. *Comm. Alg.* **31** (2003), 133–196. (Cited on page 124.)

[76] B. Kahn and Nguyen Thi Kim Ngan. Modules de cycles et classes non ramifiées sur un espace classifiant. arXiv:1211.0304 (Cited on page xii.)

[77] M. Kervaire. Opérations d'Adams en théorie des représentations linéaires des groupes finis. Enseignement Math. **22** (1976), 1–28. (Cited on page 151.)

[78] J. Kollár. Cone theorems and bug-eyed covers. *J. Alg. Geom.* **1** (1992), 293–323. (Cited on page 216.)

[79] J. Kollár. *Resolution of singularities.* Princeton (2007). (Cited on page 26.)

[80] C. Kratzer. Opérations d'Adams et représentations de groupes. *Enseign. Math.* **26** (1980), 141–154. (Cited on page 158.)

[81] A. Kresch. Cycle groups for Artin stacks. *Invent. Math.* **138** (1999), 495–536. (Cited on page 17.)

[82] A. Krishna. Equivariant K-theory and higher Chow groups of smooth varieties. arXiv:0906.3109 (Cited on pages xvi, 201, and 205.)

[83] I. Kriz. Morava K-theory of classifying spaces: some calculations. *Topology* **36** (1997), 1247–1273. (Cited on pages xv and 214.)

[84] I. Kriz and K. Lee. Odd-degree elements in the Morava $K(n)$ cohomology of finite groups. *Topology Appl.* **103** (2000), 229–241. (Cited on pages xv and 214.)

[85] N. Kuhn. Primitives and central detection numbers in group cohomology. *Adv. Math.* **216** (2007), 387–442. (Cited on pages 134, 162, 164, 172, and 217.)

[86] N. Kuhn. Nilpotence in group cohomology. *Proc. Edinburgh Math. Soc.* **56** (2013), 151–175. (Cited on pages xiv, 127, 128, 129, 130, 133, and 163.)

[87] P. Landweber. Elements of infinite filtration in complex cobordism. *Math. Scand.* **30** (1972), 223–226. (Cited on page 22.)

[88] S. Lang. *Algebra.* Springer (2002). (Cited on page 175.)

[89] I. Leary. The cohomology of certain groups. PhD thesis (Cambridge, 1990) with one extra chapter added in 1992. arXiv:0711.5020 (Cited on pages 140, 182, and 183.)

[90] I. Leary. The integral cohomology of some p-groups. *Math. Proc. Camb. Phil. Soc.* **110** (1991), 25–32. (Cited on pages 140, 156, 161, 162, and 168.)

[91] I. Leary. The mod-p cohomology rings of some p-groups. *Math. Proc. Camb. Phil. Soc.* **112** (1992), 63–75. (Cited on page 168.)

[92] I. Leary. 3-groups are not determined by their integral cohomology rings. *J. Pure Appl. Algebra* **103** (1995), 61–79. (Cited on pages 180 and 181.)

[93] I. Leary and N. Yagita. Some examples in the integral and Brown-Peterson cohomology of p-groups. *Bull. London Math. Soc.* **24** (1992), 165–168. (Cited on pages xv, 179, and 186.)

[94] J.-M. Lemaire. Le transfert dans les espaces fibrés (d'après J. Becker et D. Gottlieb). *Séminaire Bourbaki, Vol. 1975/76.* Exp. No. 472, 23–37. Lecture Notes in Mathematics 567, Springer (1977). (Cited on page 24.)

[95] M. Levine. K-theory and motivic cohomology of schemes I. http://www.esage.uni-due.de.marc.levine/pub/ (Cited on pages 177, 197, and 198.)

[96] M. Levine and F. Morel. *Algebraic cobordism.* Springer (2007). (Cited on page xiii.)

[97] G. Lewis. The integral cohomology rings of groups of order p^3. *Trans. Amer. Math. Soc.* **132** (1968), 501–529. (Cited on pages 153, 157, 160, 161, 168, and 169.)

[98] J. McCleary. *A user's guide to spectral sequences.* Cambridge (2001). (Cited on page 201.)

[99] H. Matsumura. *Commutative ring theory.* Cambridge (1989). (Cited on page 13.)

[100] C. Mazza, V. Voevodsky, and C. Weibel. *Lecture notes on motivic cohomology.* Amer. Math. Soc. (2006). (Cited on pages 70 and 71.)

[101] A. Merkurjev. Comparison of equivariant and the standard K-theory of algebraic varieties. *Algebra i Analiz* **9** (1997), 175–214; translation in *St. Petersburg Math. J.* **9** (1998), 815–850. (Cited on pages 198 and 201.)

[102] A. Merkurjev. Weight two motivic cohomology of classifying spaces for semisimple groups. Preprint (2013). (Cited on page 201.)

[103] R. J. Milgram and M. Tezuka. The geometry and cohomology of M_{12}. II. *Bol. Soc. Math. Mexicana* **1** (1995), 91–108. (Cited on page 168.)

[104] J. Milne. *Etale cohomology.* Princeton (1980). (Cited on pages 25, 91, and 187.)

[105] J. Milnor and J. Moore. On the structure of Hopf algebras. *Ann. Math.* **81** (1965), 211–264. (Cited on pages 129 and 132.)

[106] J. Milnor and J. Stasheff. *Characteristic classes.* Princeton (1974). (Cited on pages 3 and 54.)

[107] F. Morel and V. Voevodsky. A^1-homotopy theory of schemes. *Publ. Math. IHES* **90** (1999), 45–143. (Cited on pages xi, 11, 13, 24, and 26.)
[108] D. Mumford, J. Fogarty, and F. Kirwan. *Geometric invariant theory.* Springer (1994). (Cited on page 15.)
[109] R. Palais. On the existence of slices for actions of non-compact Lie groups. *Ann. Math.* **73** (1961), 295–323. (Cited on page 1.)
[110] S. Priddy and C. Wilkerson. Hilbert's Theorem 90 and the Segal conjecture for elementary abelian p-groups. *Amer. J. Math.* **107** (1985), 775–785. (Cited on page 101.)
[111] D. Quillen. The Adams conjecture. *Topology* **10** (1971), 67–80. (Cited on page 28.)
[112] D. Quillen. The mod 2 cohomology rings of extra-special 2-groups and the spinor groups. *Math. Ann.* **194** (1971), 197–212. (Cited on page 105.)
[113] D. Quillen. The spectrum of an equivariant cohomology ring I. *Ann. Math.* **94** (1971), 549–572. (Cited on page 52.)
[114] D. Quillen. A cohomological criterion for p-nilpotence. *J. Pure Appl. Alg.* **1** (1971), 361–372. (Cited on pages 91 and 92.)
[115] D. Quillen. On the cohomology and K-theory of the general linear groups over a finite field. *Ann. Math.* **96** (1972), 552–586. (Cited on pages 30 and 106.)
[116] M. Rosenlicht. Questions of rationality for solvable algebraic groups over non-perfect fields. *Ann. Mat. Pure Appl.* (4) **61** (1963), 97–120. (Cited on page 27.)
[117] M. Rost. Chow groups with coefficients. *Doc. Math.* **1** (1996), 319–393. (Cited on page 32.)
[118] C. Schoen. Complex varieties for which the Chow group mod n is not finite. *J. Alg. Geom.* **11** (2002), 41–100. (Cited on page 31.)
[119] C. Schoen. On certain exterior product maps of Chow groups. *Math. Res. Lett.* **7** (2000), 177–194. (Cited on page 31.)
[120] B. Schuster. Morava K-theory of classifying spaces. Habilitationsschrift (2006). http://www.math.uni-wuppertal.de/.schuster/research/MKBG.pdf (Cited on page 123.)
[121] B. Schuster and N. Yagita. Transfers of Chern classes in BP-cohomology and Chow rings. *Trans. Amer. Math. Soc.* **353** (2001), 1039–1054. (Cited on page 122.)
[122] L. Schwartz. *Unstable modules over the Steenrod algebra and Sullivan's fixed point conjecture.* Chicago (1994). (Cited on page 128.)
[123] J.-P. Serre. Sur la dimension cohomologique des groupes profinis. *Topology* **3** (1965), 413–420. (Cited on page 93.)
[124] J.-P. Serre. *Linear representations of finite groups.* Springer (1977). (Cited on pages 31, 85, 194, and 211.)
[125] R. Shank and D. Wehlau. The transfer in modular invariant theory. *J. Pure Appl. Algebra* **142** (1999), 63–77. (Cited on page 101.)
[126] A. Suslin. Algebraic K-theory of fields. *Proceedings of the International Congress of Mathematicians* (Berkeley, 1986), vol. 1, 222–244. Amer. Math. Soc. (1987). (Cited on page 211.)
[127] A. Suslin. Higher Chow groups and motivic cohomology. *Cycles, transfers, and motivic homology theories,* 239–254. Princeton (2000). (Cited on page 76.)
[128] D. Sullivan. *Geometric topology: localization, periodicity, and Galois symmetry,* ed. A. Ranicki. Springer (2005). (Cited on page 3.)
[129] P. Symonds. A splitting principle for group representations. *Comment. Math. Helv.* **66** (1991), 169–184. (Cited on page 125.)

[130] P. Symonds. On the Castelnuovo-Mumford regularity of the cohomology ring of a group. *J. Amer. Math. Soc.* **23** (2010), 1159–1173. (Cited on pages xi and 49.)

[131] P. Symonds. On the Castelnuovo-Mumford regularity of rings of polynomial invariants. *Ann. Math.* **174** (2011), 499–517. (Cited on pages 35, 65, and 109.)

[132] P. Symonds. Group actions on rings and the Cech complex. *Adv. Math.*, **240** (2013), 291–301. (Cited on page 109.)

[133] D. Tambara. On multiplicative transfer. *Comm. Alg.* **21** (1993), 1393–1429. (Cited on page 89.)

[134] M. Tezuka and N. Yagita. Cohomology of finite groups and Brown-Peterson cohomology II. *Homotopy theory and related topics* (Kinosaki, 1988), 57–69. Lecture Notes in Mathematics 1418, Springer (1990). (Cited on page 179.)

[135] C. B. Thomas. *Characteristic classes and the cohomology of finite groups.* Cambridge (1986). (Cited on pages 152 and 179.)

[136] R. W. Thomason. Algebraic K-theory of group scheme actions. *Algebraic topology and algebraic K-theory* (Princeton, 1983), 539–563. Princeton (1987). (Cited on page 33.)

[137] B. Totaro. Torsion algebraic cycles and complex cobordism. *J. Amer. Math. Soc.* **10** (1997), 467–493. (Cited on pages xii, 32, 186, 189, and 200.)

[138] B. Totaro. The Chow ring of a classifying space. *Algebraic K-theory* (Seattle, 1997), 249–281. *Proc. Symp. Pure Math.* **67**, Amer. Math. Soc. (1999). (Cited on pages xi, xii, xiii, xiv, xv, 8, 11, 13, 15, 16, 19, 28, 29, 32, 56, 99, 122, 140, 174, 186, 189, 198, and 210.)

[139] B. Totaro. The torsion index of the spin groups. *Duke Math. J.* **129** (2005), 249–290. (Cited on pages 22 and 202.)

[140] B. Totaro. Chow groups, Chow cohomology, and linear varieties. Preprint. (Cited on pages 19 and 211.)

[141] A. Vavpetic and A. Viruel. On the mod p cohomology of $BPU(p)$. *Trans. Amer. Math. Soc.* **357** (2005), 4517–4532. (Cited on page 87.)

[142] G. Vezzosi. On the Chow ring of the classifying stack of $PGL_3(\mathbf{C})$. *J. Reine Angew. Math.* **523** (2000), 1–54. (Cited on page 24.)

[143] A. Vistoli. On the cohomology and the Chow ring of the classifying space of PGL_p. *J. Reine Angew. Math.* **610** (2007), 181–227. (Cited on pages xvi, 185, 189, 191, and 192.)

[144] V. Voevodsky. Reduced power operations in motivic cohomology. *Publ. Math. IHES* **98** (2003), 1–57. (Cited on pages 11, 72, and 95.)

[145] V. Voevodsky. Motivic cohomology with $\mathbf{Z}/l$-coefficients. *Ann. Math.* **174** (2011), 401–438. (Cited on pages xii and 71.)

[146] V. Voevodsky. Motivic Eilenberg-MacLane spaces. *Publ. Math. IHES* **112** (2010), 1–99. (Cited on pages 11 and 198.)

[147] W. Waterhouse. *Introduction to affine group schemes.* Springer (1979). (Cited on pages 12, 13, 16, 57, and 67.)

[148] P. Webb. A guide to Mackey functors. *Handbook of algebra*, vol. 2, 805–836. North-Holland (2000). (Cited on pages 22 and 24.)

[149] C. Weibel. *An introduction to homological algebra.* Cambridge (1994). (Cited on pages 1, 6, 42, and 207.)

[150] C. Weibel. 2007 Trieste lectures on the proof of the Bloch-Kato conjecture. *Some recent developments in algebraic K-theory*, 277–305. ICTP Lect. Notes 23, Trieste (2008). (Cited on page 95.)

[151] T. Weigel. p-central groups and Poincaré duality. *Trans. Amer. Math. Soc.* **352** (2000), 4143–4154. (Cited on page 172.)

[152] N. Yagita. Cohomology for groups of $\mathrm{rank}_p G = 2$ and Brown-Peterson cohomology. *J. Math. Soc. Japan* **45** (1993), 627–644. (Cited on pages 179, 182, and 215.)

[153] N. Yagita. Chow rings of classifying spaces of extraspecial p-groups. *Recent progress in homotopy theory* (Baltimore, 2000), 397–409. Amer. Math. Soc. (2002). (Cited on pages 90, 92, and 96.)

[154] N. Yagita. Coniveau filtration of cohomology of groups. *Proc. London Math. Soc.* **101** (2010), 179–206. (Cited on pages xv and 214.)

[155] N. Yagita. Chow rings of nonabelian p-groups of order p^3. *J. Math. Soc. Japan* **64** (2012), 507–531. (Cited on pages 141 and 150.)

[156] N. Yagita. Note on the filtrations of the K-theory. http://www.mathematik.uni-bielefeld.de/LAG/man/504.html (Cited on pages xv, 150, 185, and 196.)

[157] S. Yagunov. On some differentials in the motivic cohomology spectral sequence. MPIM preprint (2007). http://www.mpim-bonn.mpg.de/preblob/3521 (Cited on page 198.)

Index

For EU product safety concerns, contact us at Calle de José Abascal, 56–1°,
28003 Madrid, Spain or eugpsr@cambridge.org.

www.ingramcontent.com/pod-product-compliance
Lightning Source LLC
Chambersburg PA
CBHW021941150426
43210CB00001B/13
* 9 7 8 1 1 0 7 0 1 5 7 7 7 *